Advances in Intelligent Systems and Computing

Volume 347

Series editor

Janusz Kacprzyk, Polish Academy of Sciences, Warsaw, Poland
e-mail: kacprzyk@ibspan.waw.pl

About this Series

The series "Advances in Intelligent Systems and Computing" contains publications on theory, applications, and design methods of Intelligent Systems and Intelligent Computing. Virtually all disciplines such as engineering, natural sciences, computer and information science, ICT, economics, business, e-commerce, environment, healthcare, life science are covered. The list of topics spans all the areas of modern intelligent systems and computing.

The publications within "Advances in Intelligent Systems and Computing" are primarily textbooks and proceedings of important conferences, symposia and congresses. They cover significant recent developments in the field, both of a foundational and applicable character. An important characteristic feature of the series is the short publication time and world-wide distribution. This permits a rapid and broad dissemination of research results.

More information about this series at http://www.springer.com/series/11156

Radek Silhavy · Roman Senkerik
Zuzana Kominkova Oplatkova
Zdenka Prokopova · Petr Silhavy
Editors

Artificial Intelligence Perspectives and Applications

Proceedings of the 4th Computer Science On-line Conference 2015 (CSOC2015), Vol 1: Artificial Intelligence Perspectives and Applications

Springer

Radek Silhavy
Faculty of Applied Informatics
Tomas Bata University in Zlín
Zlín
Czech Republic

Roman Senkerik
Faculty of Applied Informatics
Tomas Bata University in Zlín
Zlín
Czech Republic

Zuzana Kominkova Oplatkova
Faculty of Applied Informatics
Tomas Bata University in Zlín
Zlín
Czech Republic

Zdenka Prokopova
Faculty of Applied Informatics
Tomas Bata University in Zlín
Zlín
Czech Republic

Petr Silhavy
Faculty of Applied Informatics
Tomas Bata University in Zlín
Zlín
Czech Republic

ISSN 2194-5357 ISSN 2194-5365 (electronic)
Advances in Intelligent Systems and Computing
ISBN 978-3-319-18475-3 ISBN 978-3-319-18476-0 (eBook)
DOI 10.1007/978-3-319-18476-0

Library of Congress Control Number: 2015938581

Springer Cham Heidelberg New York Dordrecht London

Printed on acid-free paper

Springer International Publishing AG Switzerland is part of Springer Science+Business Media (www.springer.com)

Preface

This book constitutes the refereed proceedings of the Artificial Intelligence Perspectives and Applications Section of the 4th Computer Science On-line Conference 2015 (CSOC 2015), held in April 2015.

The volume Artificial Intelligence Perspectives and Applications brings 36 of the accepted papers. Each of them presents new approaches and methods to real-world problems and exploratory research that describes novel approaches in the field of artificial intelligence.

Particular emphasis is laid on modern trends in selected fields of interest. New algorithms or methods in a variety of fields are also presented.

CSOC 2015 has received (all sections) 230 submissions, 102 of them were accepted for publication. More than 53% of all accepted submissions were received from Europe, 27% from Asia, 10% from America and 10% from Africa. Researches from 26 countries participated in CSOC2015 conference.

CSOC 2015 conference intends to provide an international forum for the discussion of the latest high-quality research results in all areas related to Computer Science. The addressed topics are the theoretical aspects and applications of Computer Science, Artificial Intelligences, Cybernetics, Automation Control Theory and Software Engineering.

Computer Science On-line Conference is held on-line and broad usage of modern communication technology improves the traditional concept of scientific conferences. It brings equal opportunity to participate to all researchers around the world.

The editors believe that readers will find the proceedings interesting and useful for their own research work.

March 2015

Radek Silhavy
Roman Senkerik
Zuzana Kominkova Oplatkova
Zdenka Prokopova
Petr Silhavy
(Editors)

Organization

Program Committee

Program Committee Chairs

Zdenka Prokopova, Ph.D., Associate Professor, Tomas Bata University in Zlin, Faculty of Applied Informatics, email: prokopova@fai.utb.cz

Zuzana Kominkova Oplatkova, Ph.D., Associate Professor, Tomas Bata University in Zlin, Faculty of Applied Informatics, email: kominkovaoplatkova@fai.utb.cz

Roman Senkerik, Ph.D., Associate Professor, Tomas Bata University in Zlin, Faculty of Applied Informatics, email: senkerik@fai.utb.cz

Petr Silhavy, Ph.D., Senior Lecturer, Tomas Bata University in Zlin, Faculty of Applied Informatics, email: psilhavy@fai.utb.cz

Radek Silhavy, Ph.D., Senior Lecturer, Tomas Bata University in Zlin, Faculty of Applied Informatics, email: rsilhavy@fai.utb.cz

Roman Prokop, Ph.D., Professor, Tomas Bata University in Zlin, Faculty of Applied Informatics, email: prokop@fai.utb.cz

Program Committee Members

Boguslaw Cyganek, Ph.D., DSc, Department of Computer Science, University of Science and Technology, Krakow, Poland.

Krzysztof Okarma, Ph.D., DSc, Faculty of Electrical Engineering, West Pomeranian University of Technology, Szczecin, Poland.

Monika Bakosova, Ph.D., Associate Professor, Institute of Information Engineering, Automation and Mathematics, Slovak University of Technology, Bratislava, Slovak Republic.

Pavel Vaclavek, Ph.D., Associate Professor, Faculty of Electrical Engineering and Communication, Brno University of Technology, Brno, Czech Republic.

Miroslaw Ochodek, Ph.D., Faculty of Computing, Poznan University of Technology, Poznan, Poland.

Olga Brovkina, Ph.D., Global Change Research Centre Academy of Science of the Czech Republic, Brno, Czech Republic & Mendel University of Brno, Czech Republic.

Elarbi Badidi, Ph.D., College of Information Technology, United Arab Emirates University, Al Ain, United Arab Emirates.

Luis Alberto Morales Rosales, Head of the Master Program in Computer Science, Superior Technological Institute of Misantla, Mexico.

Mariana Lobato Baes,M.Sc., Research-Professor, Superior Technological of Libres, Mexico.

Abdessattar Chaâri, Professor, Laboratory of Sciences and Techniques of Automatic control & Computer engineering, University of Sfax, Tunisian Republic.

Gopal Sakarkar, Shri. Ramdeobaba College of Engineering and Management, Republic of India.

V. V. Krishna Maddinala, Assistant Professor, GD Rungta College of Engineering & Technology, Republic of India.

Anand N Khobragade, Scientist, Maharashtra Remote Sensing Applications Centre, Republic of India.

Abdallah Handoura, Assistant Prof, Computer and Communication Laboratory, Telecom Bretagne - France

Technical Program Committee Members

Ivo Bukovsky
Miroslaw Ochodek
Bronislav Chramcov
Eric Afful Dazie
Michal Bliznak
Donald Davendra
Radim Farana
Zuzana Kominkova Oplatkova
Martin Kotyrba
Erik Kral
David Malanik
Michal Pluhacek
Zdenka Prokopova
Martin Sysel
Roman Senkerik
Petr Silhavy
Radek Silhavy
Jiri Vojtesek
Eva Volna
Janez Brest
Ales Zamuda

Roman Prokop
Boguslaw Cyganek
Krzysztof Okarma
Monika Bakosova
Pavel Vaclavek
Olga Brovkina
Elarbi Badidi

Organizing Committee Chair

Radek Silhavy, Ph.D., Tomas Bata University in Zlin, Faculty of Applied Informatics, email: rsilhavy@fai.utb.cz

Conference Organizer (Production)

OpenPublish.eu s.r.o.
Web: http://www.openpublish.eu
Email: csoc@openpublish.eu

Conference Website, Call for Papers

http://www.openpublish.eu

Contents

A Multiagent-Based Approach to Scheduling of Multi-component Applications in Distributed Systems

Absalom E. Ezugwu[1], Marc E. Frincu[2], and Sahalu B. Junaidu[3]

[1] Department of Computer Science, Federal University Lafia, Nasarawa, Nigeria
ezugwu.absalom@fulafia.edu.ng
[2] Department of Electrical Engineering, University of Southern California, Los Angeles, USA
frincu@usc.edu
[3] Department of Mathematics, Ahmadu Bello University, Zaria, Nigeria
sahalu@abu.edu.ng

Abstract. In this paper, we present a multiagent-based scheduling framework for several classes of multi-component applications. We consider this scheduling problem in today's heterogeneous distributed systems. The heterogeneous nature of most parallel applications and distributed computing resource environments, makes this a challenging problem. However, the current off-the-shelf scheduling software can hardly cope with the demands for high performance and scalable computing power required by these applications. This paper proposes a scheduling mechanism that integrates routing indices with multi-agent system, to perform global scheduling in a collaborative and coordinated manner. Our intent is to apply agent-based distributed problem solving technique to address the problem of multi-component system scheduling.

Keywords: Multi-component systems, routing indices, multiagents.

1 Introduction

The rapid innovation in distributed multi-component computing application frameworks, calls for an urgent need to build an equivalent multi-component distributed system infrastructure or meta-computing infrastructure. However, a number of research groups [1-6] have proposed and implemented multi-component infrastructure, targeted at achieving high throughput for large number of diverse compute intensive multi-component applications. Most of these scheduling solutions assume that, either communication between components can be ignored, or the application will be confined to run in a single execution site, or the number of execution sites and components are small enough to make a brute-force scheduling algorithm feasible [10]. The recent shift in paradigm from parallel applications, requesting for resources from single execution sites[1], to multi-component applications, requesting for available

[1] The term "Grid resource" and the term "site" are used in this article to refer to either a set of machines (single, dual or quad CPU) in the form of a cluster.

R. Silhavy et al. (eds.), *Artificial Intelligence Perspectives and Applications*,
Advances in Intelligent Systems and Computing 347, DOI: 10.1007/978-3-319-18476-0_1

resources from heterogeneous multi-component sites, can be attributed to the single goal of achieving high performance.

In this paper, we grouped multi-component applications into two classes based on their resource requirement needs. First are the single-component applications; these are classes of applications that their resource requirements can be handled by resources from single execution sites. The second class of applications is the multi-component applications; these are classes of applications that require resources from different execution sites. These resources include remote database servers, remote laboratory instruments, remote super computers, remote network servers, and humans-in-the-loop [10]. The challenges inherent in distributed heterogeneous computing environments are well known [8].

Exploiting the performance potential that comes with the heterogeneous computing environments, requires effective application scheduling. This, in essence, would require an appropriate and efficient selection, and allocation of candidate resources to user applications. This problem is particularly challenging, due to the heterogeneous and unpredictable nature of both the resources and the application itself. The problem of scheduling heterogeneous applications and resources can be made more effective, by applying some scheduling heuristics that best understand the complete structures of both the application and resource information. The scheduling heuristics should be able to automatically extract this information and forward it to the global scheduler for adequate scheduling decision. It is in this light, that we propose a more flexible scheduling structure that incorporates the agility of multiagent systems (MAS) problem solving capability, that are beyond the individual capabilities or knowledge of each entity [9]. MAS consist of multiple agents that are considered as computing entities which have definite purposes and can run in the distributed environment independently and persistently, and they generally have the following main characteristics: autonomy, reaction, interaction and initiative [17, 18]. Agents adapt very quickly to most dynamic, unpredictable, and highly unreliable distributed environments.

Our intent in this paper is to present a MAS-Based scheduling framework that is adaptive to the characteristics of distributed systems, more specifically to heterogeneous multi-component systems, considering the problem from the perspective of the distributed multi-component resources as well as the users who consume its resources. A decentralized scheduling model, which is based on the collaborative coordination of MAS is also proposed. The MAS-Based model is capable of dynamically scheduling single and multi-component heterogeneous applications, across diverse multi-platforms of heterogeneous resources, with the single aim of achieving high performance. The primary contributions of this paper are:

i. Development of a resource selection strategy, suitable for scheduling of multi-component applications. These types of applications would usually seek for either schedulable, fixed or both available multi-component heterogeneous resources.
ii. Proposal of a scheduling solution that is adaptive to the characteristics of heterogeneous distributed systems, considering the problem from the perspective of the multi-component resources and the application user, who consumes it.

iii. Integration of Routing Indices (RI) with multiagents systems, by leveraging the distributed problem solving potentials of the MAS and the dynamism of the RI query forwarding capability.

The rest of the paper is organized as follows: Section 2 provides an overview of related research works. Sections 3 and 4, respectively, provide detailed discussions of the system architecture and Integration of Routing Indices in MAS. Section 5 presents the framework simulation using agents grid Repast Simulator. Conclusions and future work are presented in Section 6.

2 Related Work

A lot of work has been done in the area of decentralized MAS problem solving techniques. However, the control of large-scale dynamic systems of cooperative agents is a hard problem in the general case. MAS in distributed environments perform joint activities which may involve temporarily interdependent subsidiary tasks, to achieve their common goal. For instance, the scheduling of heterogeneous applications in multi-component resource environments, would require that MAS perform joint activities that involve at least (a) searching: finding agents that have the required capabilities and resources, (b) task allocation: allocating tasks to appropriate agents, and (c) scheduling: constructing a commonly agreed schedule for these agents. These three issues are difficult problems to deal with.

Routing Indices (RI) were initially developed for document discovery in peer-to-peer systems, and have also been used to implement a grid information service in [2]. The intent of RI basically, is to assist system users discover documents with content of interest across potential peer-to-peer source efficiently [13]. This approach is based on a push-update technique, where each peer sends to its neighboring sites, information about its resources and constantly updates them whenever its resources changes. Similar approaches are exploited in the work done in [5, 16]. A Grid resource discovering system which is based on the concept of routing indexes was proposed and discussed in [16]. In this system, nodes are organized in a tree-structured overlay network, where each node maintains information about the set of resources it manages directly, and a condensed description of the resources present in the sub-trees rooted in each of its neighboring nodes.

To initiate communication (or interactions) among agent peers in a grid system spanning multiple execution sites, the RI can be used to make query forwarding decisions between agents attached to each of the site, and to avert the need for flooding the entire network. The RI represents the availability of data of a specific type in the neighbors' information base. In [13, 16], the hop-count routing index (HRI) scheme was used. This RI scheme takes into account the number of hops (job forwarding) required to find the next node, and in our case, to find the next agent with information on the best execution site. In [11], related version of *HRI* that is of interest to us was used to determine the aggregate quality of a neighboring domain, based on the number of machines, power, current load, and so on.

There are several search techniques applicable when discovering certain peer nodes with content of interest in a peer-to-peer system, each with its advantages and disadvantages, see [13, 11]. The search techniques can be classified into three main categories. The first is the non index or flooding search mechanism (e.g. Gnutella [19]). One major disadvantage of this is that, it encourages network flooding. The second mechanism is one with specialized index nodes known as the centralized index search (e.g. Napster [2]). Also, one major limitation of these techniques is the single point of failure. Finally, we have the mechanism with indices at each node (distributed search); an example of this type of search is the *RIs*.

3 System Architecture

Multi-component applications consist of a set of distinct application components, that may communicate and interact over the course of the application execution period [10]. By component in this case, we refer to schedulable resources that may include; remote application servers, high performance computing servers, specialized remote instruments, remote databases and so on.

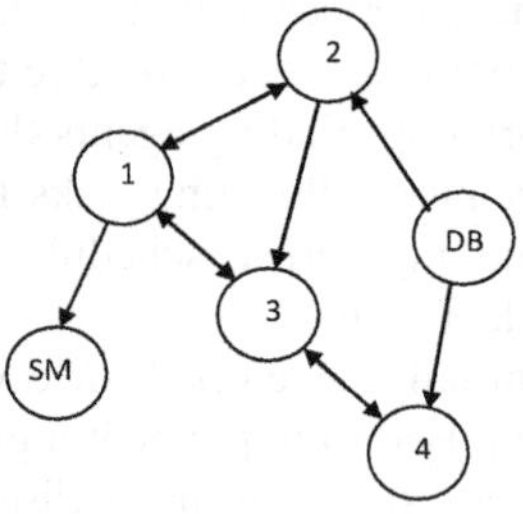

Fig. 1. Multi-component Application Model

The computational components referred to in this paper, may be sequential or parallel in nature. Some components are fixed and do not require scheduling. An instance of this type of component is the remote database server. However, the non-schedulable components may have the tendency of impacting the scheduling of other components. For example, the placement of components 2 and 4 in Fig. 1 may be influenced by the amount of data transmitted by a database to 2 and 4. If a great deal of data is moved, then a high-speed link between 2, 4 and the database may be required. It is also possible that other components are fixed due to scheduling constraints, such as a given program component must be restricted to run in a particular site.

3.1 Multi-component Resource Model

The underlying multi-component resource model contains a collection of heterogeneous sites connected by network bandwidth links (see Fig. 2). A site S is defined as a remote execution location, where an application can be run based on the site resource capability. Therefore, a site resource offer may consist of multiple resource components (with each performance evaluation based on the following: CPU speed, memory size, associated bandwidth, storage size, certain number of processors, and so on) and have a point-to-point bandwidth link to each site resources ($bw_{i,j}$) that may differ in their connectivity strength. We have been able to determine resource capability empirically, using the mathematical model presented later in Section 4.

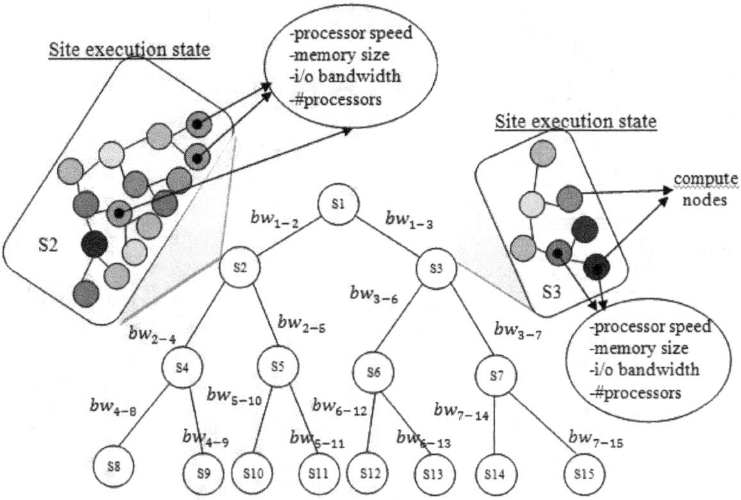

Fig. 2. Multi-component Resource Model

3.2 Multi-component Scheduling Framework

The multi-component scheduling system, shown in Fig. 3, incorporates a number of components and features:

i. A heterogeneous resource environment that incorporates agent daemons for each execution site. The agent daemon runs continuously and exists for the purpose of handling periodic job query service requests that an execution site expects to receive. The daemon forwards the requests to other agent daemons in other sites as appropriate.
ii. A resource database that provides a fine-grained description of the available machines, associated network bandwidths and storage devices, including capacity and performance parameters.
iii. A library of machine performance parameters that automatically configures for other subsequent detected machines.
iv. Distributed global schedulers that push scheduling logic to intelligent agents at each execution site to perform global scheduling in a collaborative and coordinated manner. The multi-component schedulers perform three basic functions of, resource discovery, resource mapping and resource selection. The multi-component schedulers, process descriptions of multi-component application task sets, and allocate them to the selected resources for execution based on the available fine-grained component descriptions.

v. A database of multi-component application performance models, which is used by the multi-component scheduler to make decision on which resources is more appropriate for the user submitted application.

The overall system architecture described above is as shown in Fig. 3 below.

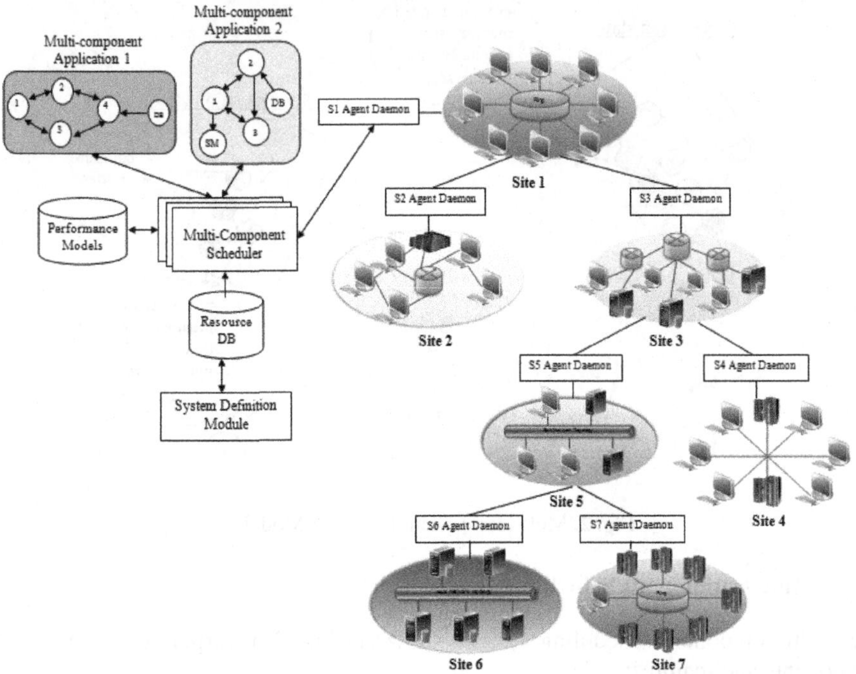

Fig. 3. A Multi-Component Scheduling Architecture

At the user level, we have the multi-component applications consisting of five components in application 2; three computations (1, 2, 3), a database server and a streaming media server (SM), while application 1 consists of five components, in which four are computations (1, 2, 3, 4) and one database server. The presence of an arc indicates data flow.

4 Integration of Routing Indices in MAS

In this paper, we consider sets of agents with specific capabilities that are attached to each execution site, to perform global scheduling and tasks that require the use of a certain heterogeneous resources. We assume that each agent can compute the capability of the site is attached to, and is able to pass the computed information to its neighbors, based on the user query given to it. In [12], it is stated that, MAS comprises network of intelligent agents, whose, neighborhood network can be modeled as a graph $AN = (N, E)$, where N is the set of agents and E is a set of bidirectional edges denoted as non-ordered pairs(A_i, A_j). The immediate neighborhood of an agent A_i

includes all the one-hop away agents. What this implies is that, $\forall A_j \exists (A_i, A_j) \in E$. The set of agent's A_i neighbors is denoted by $N(A_i)$.

Given a network of agents $G = (N, E)$, and the set of neighbors $(N(A))$ of agent A in N, the routing index (RI) of A denoted by $RI(A)$ is a collection of $|N(A) \cup \{A\}|$, vector of resources and capabilities availability, each corresponding to a neighbor of A or to A itself [12]. Given an agent A_i in $N(A)$, then the vector in $RI(A)$ that corresponds to A_i is an aggregation of the resources that are available to A via A_i. The key idea is that, given a request, A will forward this request to A_i if the resource capability available via A_i can meet the demand of the request.

Let k be the number of resource variables V for machine $N \in S$. The variables that can be considered for any resource include, processing speed, memory size, and intra-node communication bandwidth. To estimate the capability of each site S, a cost function (res_cap) is defined for machine $j \in S$:

$$res\_cap_j(V_1, V_2, \dots, V_k) \qquad \textit{(Eq. 1)}$$

Where, $V_i, 1 \leq i \leq k$, is a variable of the cost function.

The capability of a neighboring site can thus, be computed by an agent using a cost model presented in *Eq.1*. For example, a neighboring site S_i may be preferable over neighboring site S_j for a job execution request requiring compute intensive machines, if the aggregate quality of S_i in this regard is higher than that of S_j, based on the number of available processors, processing speed, memory size, and the associated intra-node communication bandwidth. The aggregate capability of the targeted neighboring site S_j is computed by the host agent A_i as shown in *Eq.2* and *3*:

$$I_{S_j}^{S_i} = \left(\sum_{k=1}^{N_S} res\_cap_k \right) \times eff\_bw\left(S_i, S_j\right) \qquad \textit{(Eq. 2)}$$

$$res\_cap = \left(k_1 \times ps + k_2 \times bw + k_3 \times ms\right) \times N_S \qquad \textit{(Eq. 3)}$$

$$k_1 = \frac{w_1}{\sum_{\forall m_i \in S_j} ps_i}, k_2 = \frac{w_2}{\sum_{\forall m_i \in S_j} bw_i}, k_3 = \frac{w_3}{\sum_{\forall m_i \in S_j} ms_i}$$

where,

- $I_{S_j}^{S_i}$ is the information that the host agent A_i keeps about the neighboring site S_j, which is provided by the neighboring agent $A_j \in S_j$.
- N_s is the number of machines site S_j has.
- ps is the CPU clock speed.
- bw is the associated machine i/o bandwidth,
- ms is the memory size,
- $eff\_bw(S_i, S_j)$ is the effective maximum bandwidth (Mbps) between the local site i and the neighboring site j. The value of this parameter is acquired from the Net Weather Service (NWS) tool in Grid Information Service (GIS).

- w_1, w_2, and w_3 are three weight constants that show the importance of each normalized parameters respectively, (processor speed, i/o bandwidth, and memory size). The three weight parameters sum up to one. The values of the three parameters are decided by experiments on different combinations of the three machine parameter values. The combination with the best performance is adopted for actual use. The values of the parameter ps and ms are acquired by the Meta Directory Service (MDS) tool, while bw is acquired by NWS which is part of Globus GIS tool.

From the expression given above, $w_1 + w_2 + w_3$ represents the maximum number of processing capability a particular machine should have. That is, if we consider the machine with the fastest processing speed, largest memory size and highest i/o communication bandwidth, $w_1 + w_2 + w_3$ should represent the maximum number of processing capability of that machine. These maximum values of processing speed, memory size and i/o bandwidth must be propagated between peer machines of neighboring sites, so that all the targeted machines share the same values for them. This is to allow for more objective comparison between resources held by different neighboring machines.

A local agent in site S_1 with lower aggregate resource quality performs the action of forwarding a job to another agent in higher quality site S_2, by computing the aggregate resource quality of S_2. However, there are other sites within the neighbourhood that are also worth considering before forwarding the job by S_1. For instance, if peer S_1 wants to forward a job to one of its neighbours, it will have to decide between S_2 and S_3, which has a better resource quality and effective network bandwidth. Before we can apply *RI* into use, there is need to first consider one major key component that determines the goodness of path of a neighboring site, known as the goodness function [13].

The notion of goodness here simply reflects the quality of the path that leads to a neighboring site. To calculate the goodness function (*Eq.4*) of a neighboring site S_i with respect to a job query Q, we represent the expression as follows.

$$goodness\left(S_i, Q\right) = \sum_{j=1\ldots H} \frac{\varphi_{i[j],Q}}{F^{j-1}} \qquad (Eq.\ 4)$$

where, S_i is the targeted execution site, $\varphi_{i,[j]}$is the RI entry for j hop through site S_i, and are j hops away from the local site, H is the horizon for $HRIs$ (that is the maximum number of aggregated RIs for each hop or job forwarding), and F is the fanout of the topology (the maximum number of neighbors a site has).

The value of sites that can be reached through the site φ_i and are j hops away from the local site S_1 (in Fig. 4), can be calculated as shown in *Eq.5* and is as described in [11]. A typical illustration of this is $S_{2.3}$ which represents the quality of sites which can be reached through S_2, whose distance from the local site is 3 hops:

$$\varphi_{k,j} = \begin{cases} I_{S_k}^{S_l}, & \text{when } j = 1 \\ \sum_i I_{S_i}^{S_l}, \forall S_i,\ d(S_l, S_i) = j \wedge d(S_l, S_t) = j - 1 \wedge d(S_t, S_i) = 1, & \text{otherwise} \end{cases}$$

(*Eq. 5*)

Where, $d(S_l, S_i)$ is the distance (in number of hops) between S_l and S_i. $\varphi_{k,j}$ is calculated based on the distance from some local site. When the distance is 1, then $\varphi_{k,j} = I_{S_k}^{S_l}$, because the only peer that can be reached from local S_l through S_k within one hop is S_k. Otherwise, for those peers S_l whose distance from the local site is j, the information that each peer S_t (which is the neighbour of S_i) keeps about them has to be added.

4.1 Searching Mechanisms

In distributed search mechanism, more effective searches can be performed by agents based on distributed indices. In these configurations, each agent attached to a node holds part of the index. The indexes help agents optimize the probability of finding quickly the requested information, by keeping track of the availability of required resources at each neighboring site. In this paper, HRI is integrated with MAS as a search mechanism for discovering quality resources for every job forwarded by an agent from one local site that does not meet the current job resource requirements, to another neighboring site which meets such needs.

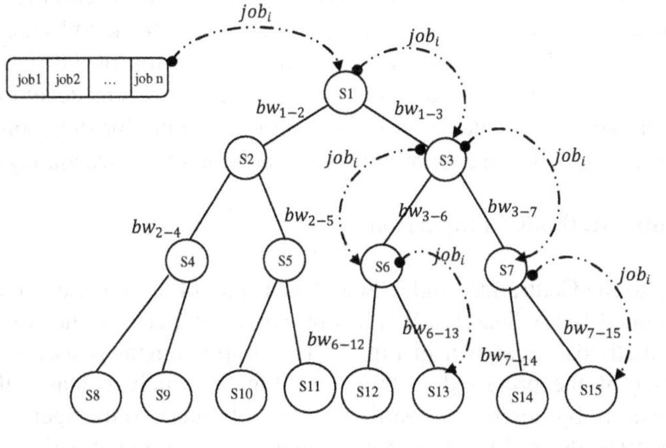

Fig. 4. Job forwarding from one peer to another peer using the RIs query forwarding mechanism

Fig.4 represents the processes involved when a local agent forwards a job request to an agent in neighboring site. Each targeted agent of the neighboring site calculates its goodness function and also considers the network link quality between the sender site and the recipient sites. As an instance, a job is forwarded from S_1 by the host agent, to the best agent neighbors attached to S_3, S_6, S_7, and S_{15}. These agents compute the capability of each of these sites, until the best site is discovered and job allocation made to the best discovered execution site.

The HRI agent-based resource searching mechanism is formulated purposely for discovering resource with high capability of executing user job. The searching logic is being shifted to agents attached to each of the execution sites, following the tree model presented in Fig.4. The distributed scheduling agents make use of this concept in order to first, discover, and assign user jobs to the best suited computing resources. It is important to note that in this kind of system, a neighboring site is only contacted when the local site does not have enough resources to process the job. In which case, the nearest agent neighbor is contacted for information about its local resources, otherwise, the process continues until a suitable site is discovered and allocated to the requested job.

5 Simulation Using Agent Grid Repast Simulator

The experiments were conducted using an open source, generic agent-based Grid simulator (Agent Grid Repast Simulator [14, 15]), specifically built for developing agent coordination mechanism in Grid systems. The simulator focuses on the development of agents for coordination of Grid activities such as scheduling, load balancing, virtual organization management and others. It is built on top of the repast agent simulation engine. Agent Grid Repast Simulator provides two types of models; Grid Model and Agent Framework Model.

The Grid Model simulates a Grid with sites hosting resources and a physical network of links between the sites. Grid activities such as job scheduling to the machines, job processing in the resources and data transfers over the links are simulated. The Agent Framework Model simulates an agent layer on top of the Grid Model. It completely abstracts the Grid Model, shifting the focus to agent development. The framework facilitates the development of autonomous agents for the coordination of the Grid activities, as well as the learning/communication processes amongst agents.

5.1 Searching Methods Comparison

Comparatively to the Centralized Index Search (CIS) and Flooding Search (FS) methods, the Distributed Index Search (DIS) has proven to outperform the two aforementioned search methods, as shown in Fig. 5. This improvement is as a result of the branching factor of the proposed method (see Fig.4), which increases the parallel power of the search, by iteratively distributing the job query to the agent's neighbors in parallel when the desired result is not obtained. As mentioned earlier, in the distributed index search strategy, each agent holds a copy of the index which often optimizes the probability of finding much faster, the requested information about a specific site or resource capability, by keeping track of the availability of such computed data by each neighboring site's agents.

The search comparison is based on the consideration of the elapse time from the point of job submission to the point of successful reception acknowledgement, by the sender or requester. Different job submissions were made between the range of 15 and 20. As shown in the Fig. 5, the DIS search method is the fastest to reach an almost complete (90%) semantic overlay network, while the FS and the CIS search methods have a very similar behavior. However, it was observed that the CIS seems to lose some performance to the other two methods, towards the total completeness of the network, as a result of the waiting factor induced by the centralized index search method (see Fig. 5 right).

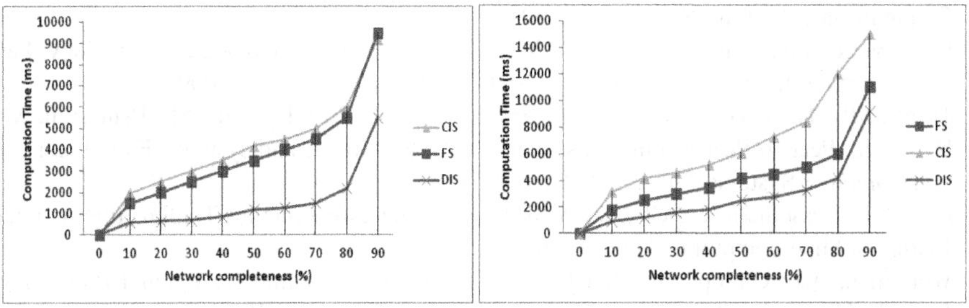

Fig. 5. Network completeness comparison between FS, CIS, and DIS when considering 15 (left) and 20 (right) job submissions by a user

6 Conclusion and Future Work

This paper presented a scheduling solution to the distributed resource problem, which involves the use of resources in different sites connected with each other via active network links. We proposed a multi-component scheduling solution, through the design and implementation of an intelligent agent-based system that collaborates with each other, to find a global scheduling solution to each problem.

More precisely, the proposed scheduling strategy is based on the integration of Routing Indices (RI) with intelligent agents. By incorporating agents into the system, global scheduling logic is pushed to cooperative agents that carry out coordination of job query forwarding, to neighbors that are more likely to have the desired computing resources. Usually, each execution site is assigned an agent coordinator, whose responsibility is to look for the suitable resource that matches the user job queries, by computing that site's resource capability. If an agent cannot find a suitable computing resource for a user job within its local site, it forwards the job query to a subset of its neighbors based on its local *HRI*, rather than by selecting neighbors at random, or flooding the network by forwarding the query to all neighbors.

The results obtained from the simulation experiment that was carried out, show that effective scheduling of multi-component applications is possible, if sufficient application and resource cost information is provided to the scheduler apriori. Our future work would focus on validating the proposed scheduling model in a real-time distributed resource environment.

Acknowledgments. The authors wish to thank Nneoma Okoroafor for her kind collaboration, and Bridget Pwajok for having read and commented carefully this paper.

References

1. Ghodsi, A., Zaharia, M., Hindman, B., Konwinski, A., Shenker, S., Stoica, I.: Dominant resource fairness: fair allocation of multiple resource types. In: NSDI (2011)
2. Jones, K.C.: International dragnet targets illegal music file-sharing. Information week, `http://www.informationweek.com/showArticle.jhtml?articleID=184428675` (November 4, 2014)
3. Grimshaw, S.A., Wulf, A.W.: The Legion Vision of a Worldwide Virtual Computer. Communications of the ACM 40(1) (1997)
4. Isard, M., Prabhakaran, V., Currey, J., Wieder, U., Talwar, K., Goldberg, A.: Quincy: Fair scheduling for distributed computing clusters. In: SOSP (November 2009)
5. Trunfio, P., Talia, D., Papadakis, H., Fragopoulou, P., Mordacchini, M., Pennanen, M., Haridi, S.: Peer-to-Peer resource discovery in Grids: Models and systems. Future Generation Computer Systems 23(7), 864–878 (2007)
6. Foster, I., Kesselman, C.: Globus: A Metacomputing Infrastructure Toolkit. International Journal of Supercomputing Applications 11(2) (1997)
7. Weissman, B.J.: Gallop: The Benefits of Wide-Area Computing for Parallel Processing. Journal of Parallel and Distributed Computing 54(2) (November 1998)
8. Freund, F., Siegel, J.H.: Heterogeneous Processing. IEEE Computer (1993)
9. Peng, Y., Finin, T., Labrou, Y., Chu, B., Long, J., Tolone, J.W., Boughannam, A.: A multi-agent system for enterprise integration. In: Proceedings of the Third International Conference on the practical Applications of Agents and Multi-Agent Systems (PAAM 1998), pp. 1–14 (1998)
10. Weissman, B.J.: Scheduling multi-component applications in heterogeneous wide-area networks. In: Proceedings of the 9th Heterogeneous Computing Workshop (HCW 2000). IEEE (2000)
11. Caminero, A., Omer, R., Blanca, C., Carmen, C.: Network-aware heuristics for inter-domain meta-scheduling in Grids. Journal of Computer and System Sciences 77(2), 262–281 (2011)
12. Theocharopoulou, C., Partsakoulakis, I., Vouros, G.A., Stergiou, K.: Overlay networks for task allocation and coordination in dynamic large-scale networks of cooperative agents. In: Proceedings of the 6th International Joint Conference on Autonomous Agents and Multiagent Systems, p. 55. ACM (2007)
13. Crespo, A., Garcia-Molina, H.: Routing indices for peer-to-peer systems. In: Proceedings of the 22nd International Conference on Distributed Computing Systems, pp. 23–32. IEEE (2002)
14. `https://sourceforge.net/projects/agentGridrepast` (accessed October 24, 2014)
15. Chao, I., Ardaiz, O., Sangüesa, R.: A group selection pattern optimizing job scheduling in decentralized grid markets. In: Meersman, R., Tari, Z., Herrero, P. (eds.) OTM 2007 Ws, Part I. LNCS, vol. 4805, pp. 37–39. Springer, Heidelberg (2007)
16. Marzolla, M., Mordacchini, M., Orlando, S.: Resource discovery in a dynamic Grid environment. In: Proc. DEXA Workshop 2005, pp. 356–360 (2005)
17. Chen, S.R., Tu, R.M.: Development of an agent-based system for manufacturing control and coordination with ontology and RFID technology. Expert Systems with Applications 36(4), 7581–7593 (2009)
18. Fazel Zarandi, H.M., Ahmadpour, P.: Fuzzy agent-based expert system for steel making process. Expert Systems with Applications 36(5), 9539–9547 (2009)
19. Ripeanu, M.: Peer-to-peer architecture case study: Gnutella network. In: Proceedings of the First International Conference on IEEE Peer-to-Peer Computing, pp. 99–100 (2001)

A Cellular Automaton Based Approach for Real Time Embedded Systems Scheduling Problem Resolution

Fateh Boutekkouk

IResearch Laboratory on Computer Science's Complex Systems ReLa(CS)^2,
Oum El Bouaghi University, BP 358, Algeria
fateh_boutekkouk@yahoo.fr

Abstract. Real Time Embedded Systems are becoming ubiquitous. Since these systems have autonomous batteries, their design must minimize power consumption in order to extend batteries life time. On the other hand, Cellular Automaton (CA) appears a good choice to simulate the future behavior of complex dynamic and parallel systems. Due to some intrinsic characteristics such as neighborhood and local transitions, CA can exhibit some complex behaviors. In this work, we apply CA to model the well known problem of Real time scheduling and eventually to optimize the power consumption of a Real time multi-cores embedded system with periodic tasks. CA algorithm is focused on the so-called technique: Dynamic Voltage Scaling (DVS). The proposed CA is a 2D grid.

Keywords: Real time embedded systems, Real time scheduling, Cellular automata, DVS, Power consumption.

1 Introduction

Real Time Embedded Systems (RTES) are reactive systems. They interact with the external world via sensors and actuators and must provide correct results but without missing a specified deadline. RTES may be hard, soft or firm. RTES are generally subjected to a variety of constraints beyond temporal ones such as surface, weight, power consumption and cost. RTES include a logical part (or application) which is composed of a set of dependent and/or independent tasks, periodic and/or aperiodic tasks. Each task is characterized by a set of parameters such as period, arrival time, execution time which may be expressed as WCET (Worst Case Execution Time), ACET (Average Case Execution Time) or BCET (Best Case Execution Time), priority, etc. The hardware part is composed of a set of embedded processors which are connected by a communication support such as buses or an embedded network. Each processor is also characterized by a set of parameters like clock frequency, local memory size, power consumption, programmability level, etc. Since RTES have autonomous batteries, their design must minimize power consumption in order to extend batteries life time. Thus power consumption management and reduction become a challenge. To deal with this problem, researchers have proposed a set of power estimation and reduction techniques at many levels of abstraction and at both software and hardware stages.

R. Silhavy et al. (eds.), *Artificial Intelligence Perspectives and Applications*,
Advances in Intelligent Systems and Computing 347, DOI: 10.1007/978-3-319-18476-0_2

Among these techniques, Dynamic Voltage Scaling or DVS is becoming more and more interesting for power reduction [2, 3, 6, 11]. The idea behind DVS is to enable a processor to change its voltage level dynamically (during execution) thus the processor operates on a set of modes and the transition from a level to another level consumes a time and a power (we call this the overhead due to transition between modes). A high voltage implies a high clock frequency. Voltage reduction leads to a considerable reduction in power consumption but on the prize of clock frequency reduction that means a longer execution time. In a real time context, a longer execution time can lead to a deadline missing. Finding the right voltage level is a serious problem. To solve this dilemma, we will assume that each embedded processor have three functioning modes: The high frequency mode with the highest power consumption level, the middle frequency mode with an acceptable power consumption level and finally the low frequency mode with the lowest power consumption level. Using some temporal information during tasks execution, we can choose the right modes that lead to optimal results. For instance by measuring the distance between the task deadline and its end time or measuring the processor usage ratio, we can find a good distribution of tasks on processors with appropriate frequencies levels. In this work, we investigate the idea of using cellular automaton (CA) to model and simulate real time tasks scheduling and allocation problems in multi-cores embedded systems.

CA [8] is a computation model used to model and simulate behaviors of complex dynamic and parallel systems. We can see a CA as a grid of cells together form a neighborhood. Each cell has a state and can change its state according to its neighborhood state. This is called a local transition rule. By applying local rules on cells synchronously or asynchronously, the CA changes its global state and some complex emerging behaviors can be produced. This is a primary inherent characteristic of CA. Informally, our proposed CA is a two dimensional grid where each cell represents the information of the task allocation on a processor with a certain frequency mode. Thus a cell state is a triplet (task, processor, frequency mode). Of course when more that one task is allocated to the same processor, a scheduling policy (example Rate Monotonic, Deadline Monotonic, etc.) must be defined. In order to simulate the CA, we have to define some local transition rules. Such transition rules can for instance, change the allocation, the priority or the frequency mode of a task. By applying these rules continually, we can observe some emerging behaviors or some good configurations with minimal power consumption. The rest of paper is organized as follows: section two is devoted to some pertinent works. The formal definition of our proposed CA is presented in section three. In section four, we discuss our implementation and some results before the conclusion.

2 Related Work

Literature on employing CA to solve multiprocessors scheduling problem is rich. However, we can state that:

1. Most of these works cope with classical multiprocessors systems but not Real time embedded systems [4, 7, 10, 12].

2. Most of existing works do not take into consideration power consumption.
3. Some works target only bi-processors systems. Thus the corresponding CA is supposed binary [9].
4. Some work, hybrid CA with some optimization evolutionary heuristics such as genetic algorithms or ant colony to discover the best local rules [1, 5]. We think that this technique may consume a huge amount of time to find good solutions.

In this work, we try to formulate the Real time scheduling problem in embedded multi-cores systems including DVS strategy using a simple CA characterized by graphical capabilities and defined with some local rules to reduce the power consumption under temporal constraints (deadlines) and balancing between the usage ratios of processors.

3 Formal Definition

Formally a cellular automaton is defined as CA = (D, S, N, £, F) where:

D is the grid dimension = 2. Each column of the grid represents a task and each line represents a processor number to which the task is allocated.

S is the CA cells states. Each state is defined as a triplet (task name, processor number, frequency mode). N is the neighborhood of a cell, it is defined as follows:

In dynamic priority based scheduling, the neighborhood N of a task T is equal to the set of tasks allocated to the same processor as T and it includes T itself. Since tasks priorities change over time, all tasks allocated to the same processor have an impact on their states. For this reason, N includes all tasks allocated to the same processor.

In static priority based scheduling, the neighborhood N of a task T is equal to the set of tasks allocated to the same processor as T having higher priorities than T and it includes T itself. Since tasks priorities are pre-known, only higher priorities tasks have impact on lower priorities tasks allocated to the same processor. For this reason, N includes higher priorities tasks. **£** is the local transition function. **F** is the global transition function of CA.

£ : NOM x PROC x MODE --> NOM x PROC x MODE
NOM designate tasks domain (string).
PROC designate processors domain (natural).
MODE designate frequency mode (enumerate = (low, middle, high)).

We defined three transition functions as follows:

The first function is applied on tasks missing their deadlines in case of dynamic priority based scheduling EDF (Earliest Deadline first). The role of this function is to change the processor and/or frequency mode of a task. Normally, when applying this function, we expect the task will respect its deadline in next periods.

The second function is similar to the first one but it is applied in case of static priority based scheduling DM (Deadline Monotonic). The third function is applied on tasks respecting their deadlines in the case of dynamic and static priority based scheduling.

The role of this function is to change the processor and/or frequency mode of a task. Normally, when applying this function, we expect the power consumption will be reduced. The section bellow details the algorithms of the three transition functions.

```
Algorithm1
begin
 choose randomly a task T missing its deadline allocated to processor Pi
  if (mode (T) = high) then
    if (exists a processor  Pj)  and  (its rate_usage  < 0.33)) then
              migrate T on Pj
              set mode(T) to 'low'
       else
          if (exists a processor Pj) and (its rate_usage < 0.66)) then
               migrate T on Pj
               set  mode(T)  to 'middle'
          else
            if (exists a processor Pj) and  (its rate_usage < 1 )) then
               keep T on Pi
               set mode(T) to 'high'
             else
               if (mode(T) = middle) then
                   keep T on Pi
                   set mode(T) to 'high'
                 else
                   if (mode(T) = low) then
                      keep T on Pi
                      set mode (T) to 'middle'
                     end if
end
```

The idea behind algorithm 1 is to migrate a task missing its deadline on a new processor which is not very busy and according to its ratio usage, we fix the right frequency mode so we keep energy consumption at lower levels if it is possible. If there is no chance to find such a processor, we keep the task on the same processor and set the frequency mode to lower levels.

The idea of algorithm 2 is similar to the first one with a bit difference in neighborhood.

Algorithm 3 tries to reduce the power consumption by measuring the distance between the task deadline and its end execution time (Tend_exe) and according to the task frequency mode and processor ratio usage, we decide whether we migrate or keep the task on the same processor.

```
Algorithm2
begin
choose randomly a task allocated to processor Pi whose neighborhood includes at least
a task T missing its deadline
  if (mode(T)= high) then
    if (exists a processor Pj) and (its rate_usage < 0.33)) then
          migrate T on Pj
          set mode(T) to 'low'
      else
        if (exists a processor Pj) and (its rate_usage < 0.66)) then
            migrate T on Pj
            set mode(T) to 'middle'
          else
            if (exists a processor Pj) and (rate_ usage <1)) then
              keep T on Pi
              set mode(T) to 'high'
            else
              if (mode(T) = middle) then
                keep T on Pi
                set mode (T) to 'high'
               else
                 if (mode (T) = low) then
                   keep T on Pi
                   set mode (T) to 'middle'
                 end if
  end
```

Algorithm3

begin

choose randomly a task T respecting its deadline allocated to processor Pi

if (mode(T) = high and $\frac{\text{deadline}-\text{Tend\_exe}}{\text{deadline}} > 0.5$) then

  keep T on Pi

  set mode(T) to 'middle'

  else

   if (mode (T)= high and $\frac{\text{deadline}-\text{Tend\_exe}}{\text{deadline}} < 0.5$) then

     if (exists a processor Pj) and (its rate_usage < 0.33) then

        migrate T on Pj

        set mode (T) to 'middle'

       else

        if (mode(T)=middle and $\frac{\text{deadline}-\text{Tend\_exe}}{\text{deadline}} > 0.5$) then

          keep T on Pi

          set mode (T) to 'low'

         else

           if (mode (T) = low) then

             keep T on Pi

           end if

  end

4 Motivational Example

In order to test our approach, we will consider an example of a system including 20 periodic tasks with P (period) = 30 cycles and 5 processors. All tasks information are indicated in table 1. We note that arrival dates of tasks are generated randomly between [0, αi] where αi is given (i is the number of the period). Initially tasks allocation is done randomly. Each processor is characterized by a color and three frequency modes that are: High frequency mode or H (frequency = 1, powerPerCycle = 2 watt).

Middle frequency mode or M (frequency = 1.5, powerPerCycle = 1.5 watt).

Low frequency mode or L (frequency = 2, powerPerCycle = 1 watt).

The actual WCET of a task, noted AWCET = WCET * frequency.

The power consumed by a task = AWCET * powerPerCycle. The overhead due to transition between modes is given in table 2. All tasks allocated to the same processor will be colored by the processor color. Tasks missing their deadline will be colored by black. Simulation time = 15 periods.

Table 1. Tasks parameters

Task Id.	Processor	WCET	Deadline	Mode	Arrival
Task0	P0	3	5	H	0
Task1	P4	3	6	M	2
Task2	P2	3	4	M	2
Task3	P1	5	9	H	0
Task4	P4	5	8	M	3
Task5	P3	4	5	M	1
Task6	P3	3	3	H	0
Task7	P4	4	5	M	1
Task8	P3	2	5	M	2
Task9	P3	3	5	M	2
Task10	P1	4	6	H	1
Task11	P4	2	4	L	0
Task12	P0	2	3	M	2
Task13	P3	3	3	H	1
Task14	P3	2	4	L	1
Task15	P0	5	7	L	2
Task16	P0	2	2	L	0
Task17	P3	3	6	H	1
Task18	P4	4	8	H	0
Task19	P3	1	3	L	0

Table 2. Transitions overheads

Transition	Transition Time (cycles)	Transition Power (watt)
H to M	1	1
H to L	1	1
M to H	1	1
M to L	1	1
L to H	1	1

Figures from 1 to 5 show respectively the CA state in the first period, CA progression in an arbitrary period, processors usage ratios, processors power consumption, number of tasks missing their deadlines for processor P0 using EDF. For the sake of space, we do not show results of other processors.

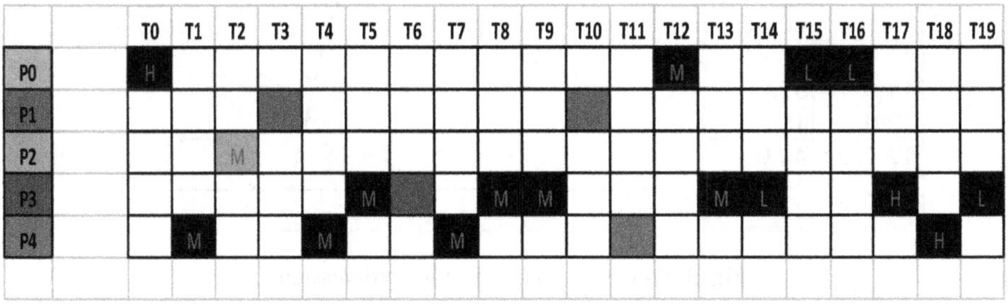

Fig. 1. CA state in the first period using EDF

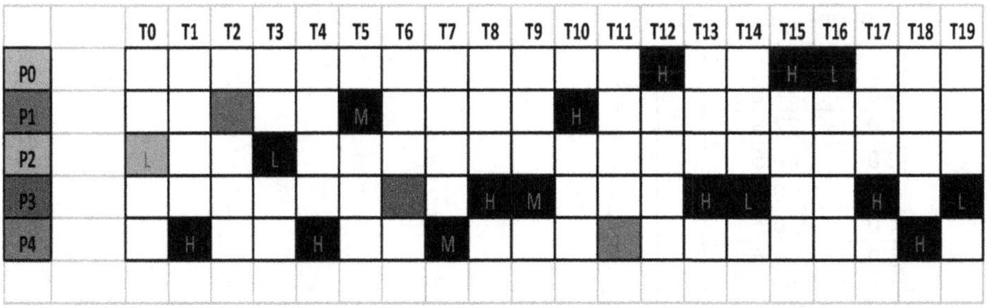

Fig. 2. CA progression

Processor P0 shows a decreasing in its usage ratio, power consumption, and the number of tasks beyond their deadlines over early periods, however, when the CA progresses, these parameters become more stable. Figure 6 shows the overall system power consumption evolution over simulation periods. We can remark that during first simulation periods, the consumed power was in its low level because frequencies modes were set to 'low' that means a big number of tasks missed their deadlines. When the CA progresses, the consumed power increases to its high level which corresponds to high frequencies modes. At this stage, all the tasks respect their deadlines but the consumed power is maximal. After that, we observe a certain alternation between low and high levels before power falling. This alternation is due to the fact that CA applies local rules to compromise between power consumption and tasks deadlines respect.

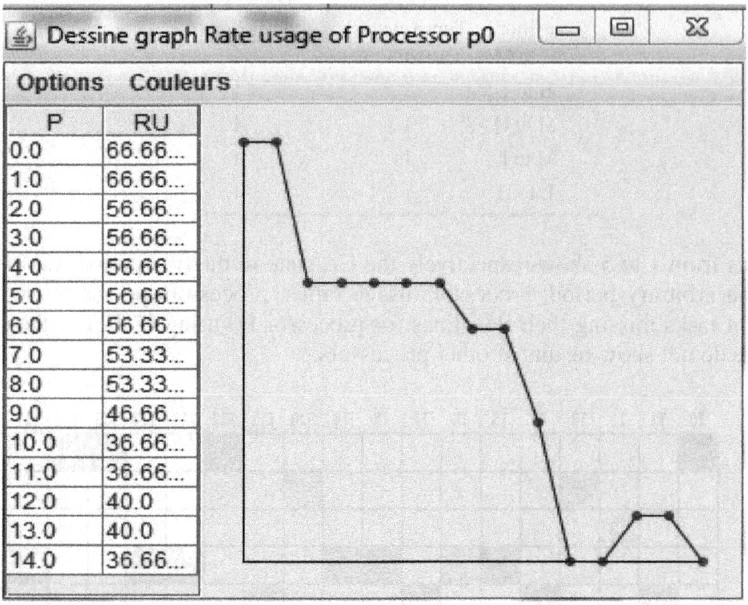
Dessine graph Rate usage of Processor p0
Options
Couleurs
P
RU
0.0
66.66...
1.0
66.66...
2.0
56.66...
3.0
56.66...
4.0
56.66...
5.0
56.66...
6.0
56.66...
7.0
53.33...
8.0
53.33...
9.0
46.66...
10.0
36.66...
11.0
36.66...
12.0
40.0
13.0
40.0
14.0
36.66...

Fig. 3. Processor P0 usage ratios progression

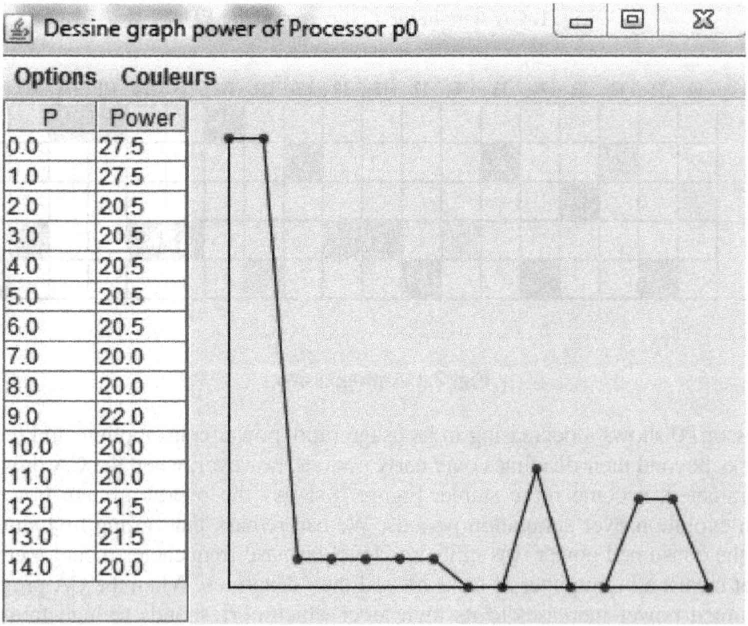
Dessine graph power of Processor p0
Options
Couleurs
P
Power
0.0
27.5
1.0
27.5
2.0
20.5
3.0
20.5
4.0
20.5
5.0
20.5
6.0
20.5
7.0
20.0
8.0
20.0
9.0
22.0
10.0
20.0
11.0
20.0
12.0
21.5
13.0
21.5
14.0
20.0

Fig. 4. Processor P0 power consumption progression

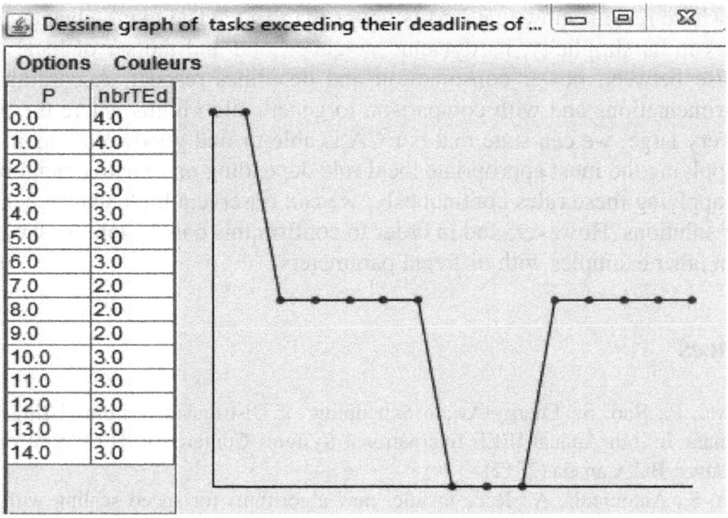

P	nbrTEd
0.0	4.0
1.0	4.0
2.0	3.0
3.0	3.0
4.0	3.0
5.0	3.0
6.0	3.0
7.0	2.0
8.0	2.0
9.0	2.0
10.0	3.0
11.0	3.0
12.0	3.0
13.0	3.0
14.0	3.0

Fig. 5. Number of tasks missing their deadlines progression for P0 processor

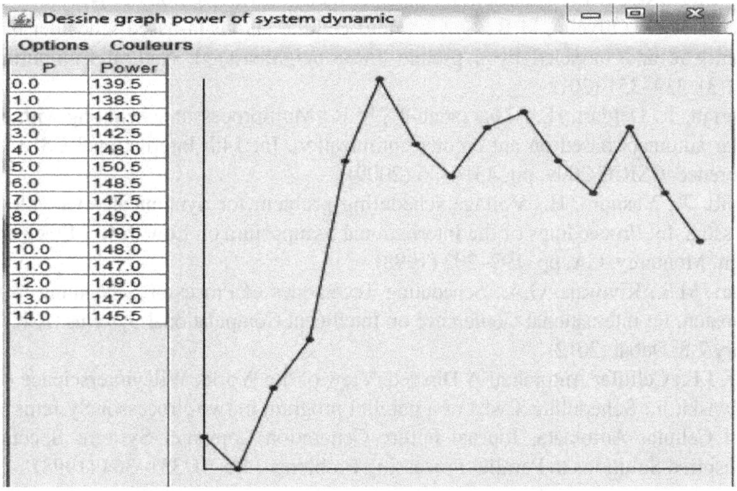

P	Power
0.0	139.5
1.0	138.5
2.0	141.0
3.0	142.5
4.0	148.0
5.0	150.5
6.0	148.5
7.0	147.5
8.0	149.0
9.0	149.5
10.0	148.0
11.0	147.0
12.0	149.0
13.0	147.0
14.0	145.5

Fig. 6. System power consumption progression

5 Conclusion

In this paper, we presented a new approach to resolve the scheduling and allocation problems in real time embedded systems with periodic tasks using cellular automata. Our CA is 2D grid. We defined three local rules to change the state of cells. These rules try to optimize power consumption, balancing usage ratios of processors and minimize the number of tasks missing their deadlines. We adopted two well known real time scheduling algorithms that are DM and EDF. We tested our CA on a simple example whose results show that DM is better than EDF. The main advantage of our CA resides in its simplicity, its dynamic neighborhood and the set of local rules to compromise between power consumption and deadlines respect. According to our first experimentations and with comparison to genetic algorithms where the research space is very large, we can state that our CA is able to find good solutions in a short time by applying the most appropriate local rule depending on dynamic neighborhood state. By applying these rules continuously, we can observe a big enhancement in the quality of solutions. However, and in order to confirm this conclusion, we have to test our CA on other examples with different parameters.

References

1. Agrawal, P., Rao, S.: Energy-Aware Scheduling of Distributed Systems Using Cellular Automata. In: 6th Annual IEEE International Systems Conference (IEEE SysCon 2012), Vancouver, BC, Canada (2012)
2. Albers, S., Antoniadis, A.: Race to idle: new algorithms for speed scaling with a sleep state. In: Proceedings of the Twenty-Third Annual ACM-SIAM Symposium on Discrete Algorithms, SODA 2012, pp. 1266–1285. SIAM (2012)
3. Aydin, H., Melhem, R., Mosse, D., Mejıa-Alvarez, P.: Power-aware scheduling for periodic real-time tasks. IEEE Transactions on Computers 53(5), 584–600 (2004)
4. Carneiro, M.G., Oliveira, M.B.: Synchronous cellular automata-based scheduler initialized by heuristic and modeled by a pseudo-linear neighborhood. Natural Computing Journal 12(3), 339–351 (2013)
5. Ghafarian, T., Deldari, H., Akbarzadeh-T., M.R.: Multiprocessor scheduling with evolving cellular automata based on ant colony optimization. In: 14th International CSI Computer Conference, CSICC 2009, pp. 431–436 (2009)
6. Ishihara, T., Yasuura, H.: Voltage scheduling problem for dynamically variable voltage processors. In: Proceedings of the International Symposium on Low Power Electronics and Design, Monterey, CA, pp. 197–202 (1998)
7. Laghari, M.S., Khuwaja, G.A.: Scheduling Techniques of Processor Scheduling in Cellular Automaton. In: International Conference on Intelligent Computational Systems (ICICS 2012), January 7-8, Dubai (2012)
8. Schiff, J.L.: Cellular Automata: A Discrete View of the World. Wileyinterscience (2007)
9. Seredynski, F.: Scheduling Tasks of a parallel program in two-Processor Systems with the use of Cellular Automata. Journal Future Generation Computer Systems Special Issue: Bio-inspired Solutions to Parallel Processing Problems 14(5-6), 351–364 (1998)
10. Seredynski, F., Zomaya, A.: Sequential and parallel cellular automata-based scheduling algorithms. IEEE Transactions on Parallel and Distributed Systems 13(10), 1009–1023 (2002)
11. Shin, Y., Choi, K., Sakurai, T.: Power optimization of real-time embedded systems on variable speed processors. In: Proceedings of the International Conference on Computer Aided Design, pp. 365–368 (November 2000)
12. Swiecicka, A., Seredynski, F., Zomaya, A.: Multiprocessor scheduling and rescheduling with use of cellular automata and artificial immune system support. IEEE Transactions on Parallel and Distributed Systems 17(3), 253–262 (2006)

Ways of Increasing of the Effectiveness of the Making Decisions by Intelligent Systems Using Fuzzy Inference

Olga Dolinina and Aleksandr Shvarts

Yury Gagarin State Technical University of Saratov, Saratov, Russia
{odolinina09,shvartsaleksandr}@gmail.com

Abstract. There are considered rule-based expert systems using fuzzy inference. Comparative analysis of different approaches and algorithms of making decisions on the base of fuzzy logic is given. Building of graph of the dependence of the rules and the graph of dependence of the linguistic variables are suggested. On the base of the developed groups of rules and defuzzification of the linguistic variables it is planned to increase the effectiveness of the decision making with using of parallel calculations for each group what is considered to be an actual problem for the large dimension knowledge bases.

Keywords: expert systems, fuzzy inference, increasing of the effectiveness of the making decisions, algorithms of the building of the rules' dependence.

1 Introduction

Intelligent systems are one of the most rapidly developing branches of science and technologies. This is caused not only by the growth of knowledge accumulated by humanity, but also by the need for using this knowledge and data for automatic reasoning, complex systems control and modern tools for human-computer interaction.

The most popular sub-branch of intelligent systems is expert systems (ES), which allow storing, accumulating and using expert knowledge on order to automate reasoning in different areas. ES core is knowledge base with reasoning engine. It should be noted, that since the appearance of the first ES in 1970's and up to today there has been long evolutionary way from simple determined rule-based knowledge bases towards different combinations of knowledge representation models considering aspects of the subject area.

In the majority of cases expert is not able to determinately describe some facts or express the causation between them. The most popular method to design systems with such uncertainties was introduced by Zadeh [1,2].

2 Fuzzy Intelligent Systems

Fuzzy rule base is presented in the form of

$$(V,I,G,P,R), \tag{1}$$

R. Silhavy et al. (eds.), *Artificial Intelligence Perspectives and Applications*,
Advances in Intelligent Systems and Computing 347, DOI: 10.1007/978-3-319-18476-0_3

with V- set of linguistic variables of the system,
I - set of input linguistic variables, $I \subset V$,
G - set of output linguistic variables, $G \subset V$,
P – set of intermediate linguistic variables, $P \subset V$,
$I \cup G \cup P = V, I \cap G = \emptyset,\ I \cap P = \emptyset, P \cap G = \emptyset$.
R - set of fuzzy rules

$$< r_j, u_j > IF\ v_{j,1}\ is\ t_{j,1}\ AND\ ...AND\ v_{j,n_j-1}\ is\ t_{j,n-1} \\ THEN\ v_{j,n_j}\ is\ t_{j,n_j} \quad (2)$$

with r_j – unique rule name,
u_j – linguistic certainty degree of the r_j rule;
n_j – number of linguistic variables in the r_j rule;
$v_{j,1},\dots v_{j,n_j-1}$ – premises (input variables) for the r_j rule;
$t_{j,1},\dots t_{j,n_j-1}$ – values of input linguistic variables (terms) for the r_j rule,
v_{j,n_j} – output linguistic variable (goal) for the r_j rule,
t_{j,n_j} – value of output linguistic variable for the r_j rule.

Fuzzy set is A is presented in the form (3)

$$A = \{(x, \mu_A(x)) | x \in X\} \quad (3)$$

with X – universal set;

x – element of X set;

$\mu_A(x)$ – membership function.

Membership function $\mu_A(x)$ describes the degree of membership of each $x \in X$ to the fuzzy set A, with μ: X → [0,1]. Thereby, $\mu_A(x)$ is 1, if element x is completely included in A, and 0, if element x is not included in A. Membership function values on interval (0,1) describe fuzzy included elements.

Diagram of reasoning engine in fuzzy ES in presented on Fig. 1.

A number of reasoning methods and algorithms have been introduced and discussed since the appearance of fuzzy sets theory. Each of algorithms [3-10] differs from the point of view of sequence and type of operators applied in fuzzyfication, fuzzy inference and defuzzyfication stages. They also use fuzzy rules of different form.

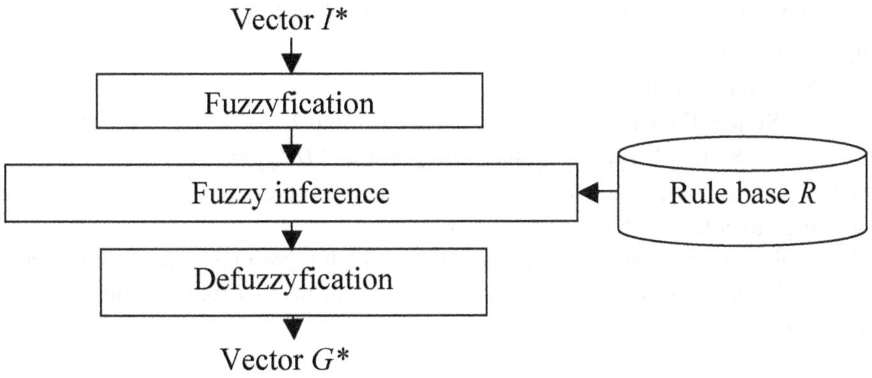

Fig. 1. Diagram of reasoning engine in fuzzy ES

Mamdani [3], Tsukamoto [4], product-sum-gravity [5] and Larsen [6] methods use fuzzy rules, which form is very close to the way of human thinking. Therefore, it would be easier for experts and knowledge engineers to design the rule base.

Tsukamoto inference method also requires only monotonous membership functions, that is unacceptable in many cases.

Sugeno [7], simplified fuzzy inference [8], singleton [9] and SIRMs [10] methods requires crisp number values or coefficients, which calculation is not a trivial task for some areas. However, those methods involve only simple algebraic operators, that leads to lower computational cost in comparison with the first group of methods.

SIRMs fuzzy systems contains separate groups of rules (modules) and each of the rules must have only one premise.

Most of the modern intelligent systems using fuzzy inference has rule bases of high dimension (thousands of rules). This causes the need to use special methods of inference management in order to improve efficiency and minimize calculation time. Existing methods of fuzzy inference allows parallel execution on a number of computational nodes, that can reduce time spent on reasoning process. This paper introduces algorithms for improving efficiency of fuzzy inference in rule-based intelligent systems.

In fuzzy rule bases there is a wide range of cases, when it is impossible to process one rule without processing some other rules before. This problem appears, if the rule's premises contains variables, which are not received from the input of the whole system, but calculated after different rules are processed.

In order to solve this problem a method, which allows to define the order of the rules processing, is introduced.

3 Exploring Dependencies in the Rule Base

3.1 Rules Dependencies

The first stage is to generate the matrix of rules dependencies. Following algorithm is used:

Step 1. Initialize zero-filled matrix DR with dimension of $|R| \times |R|$

($|R|$ – cardinal number of R set).

Step 2. For each $r_j \in R$ execute Step 3.

Step 3. For each $r_q \in R$, $r_q \neq r_j$ execute Step 4.

Step 4. If $v_{j,n_j} \in G_q$, then $DR_{j,q} = 1$ and $DR_{q,j} = -1$.

DR is an adjacency matrix of the directed graph of rules dependencies. This graph can be used to define each rule predecessor rules.

Each rule $r_j \in R$ is mapped to a value $\rho(r_j)$, which represents sequence number of a group of processed rules. This sequence number means that rule r_{j_1} cannot be processed before r_{j_2}, if $\rho(r_{j_2}) < \rho(r_{j_1})$.

Calculating the Sequence Number of a Group of Processed Rules

Step 1. Initialize zero-filled set RO, $|RO| = |R|$.

Step 2. $j = 1$.

Step 3. If $j \leq |R|$, then go to Step 4, else go to Step 10.

Step 4. $i = 1$.

Step 5. If $i \leq |R|$, then go to Step 6, Step 8.

Step 6. If $DR_{j,i} = -1$, then go to Step 9.

Step 7. $i = i + 1$.

Step 8. $ro_j = 1$.

Step 9. $j = j + 1$. Go to Step 3.

Step 10. $j = 1, c = 0$.

Step 11. If $j \leq |R|$, then go to Step 12, else go to Step 20.

Step 12. $i = 1, m = 1$.

Step 13. If $i \leq |R|$, then go to Step 14, else go to Step 17.

Step 14. If $DR_{j,i} = -1$, then go to Step 15, else go to Step 16.

Step 15. If $ro_i > m$, то $m = ro_i$.

Step 16. $i = i + 1$. Go to Step 13.

Step 17. If $ro_j < m + 1$, то $c = 1$.

Step 18. $ro_j = m + 1$.

Step 19. $j = j + 1$. Go to Step 11.

Step 20. If $c = 0$, then go to Step 21, else go to Step 10.

Step 21. For each $r_j : \rho(r_j) = ro_j$.

Using calculated values of $\rho(r_j)$ following sets are formed

$$RG_m : r_j \in RG_m \Leftrightarrow \rho(r_j) = m, m = \overline{1, \ldots L_R} \quad (4)$$

with L_R – total number of groups of processed rules.

3.2 Linguistic Variables Dependencies

Fuzzy inference involves not only calculation of membership functions for each rule r_j, but also defuzzyfication of rules' goals. Sequence of goal variables defuzzyfication is not simply obtained from rules dependencies graph, because, each variable can be used in the right part of many rules with different value of $\rho(r_j)$.

So there is important to discover dependencies between linguistic variables. Following algorithm is used to generate the matrix of dependencies:

Step 1. Initialize zero-filled matrix D with dimension of $|V| \times |V|$.
Step 2. For each $v_k \in V$ execute Step 3.
  Step 3. For each $v_h \in V$, $v_h \neq v_k$ execute Step 4.
    Step 4. For each $r_j \in R$ execute Step 5.
      Step 5. If $v_k \in G_j$ and $v_h = v_{j,n_j}$, then $DV_{k,h} = 1$ and $DV_{h,k} = -1$.

DV is an adjacency matrix of directed graph of linguistic variables dependencies. This graph can be used to find each variable predecessors.

Each rule $r_j \in R$ is mapped to a value $\rho(r_j)$, which represents sequence number of a group of processed rules. This sequence number means that rule r_{j_1} cannot be processed before r_{j_2}, if $\rho(r_{j_2}) < \rho(r_{j_1})$.

Each linguistic variable $v_k \in V$ is mapped to a value $\rho(v_k)$, which represents sequence number of a group of defuzzyfied variables. This sequence number means that variable v_{k_1} cannot be defuzzyfied before v_{k_2}, if $\rho(v_{k_2}) < \rho(v_{k_1})$.

Calculating the Sequence Number of a Group of Defuzzyfied Variables

Step 1. Initialize zero-filled set VO, $|VO| = |V|$.
Step 2. $k = 1$.
Step 3. If $k \leq |V|$, then go to Step 4, else then go to Step 10.
  Step 4. $h = 1$.
  Step 5. If $h \leq |V|$, then go to Step 6, else then go to Step 8.
    Step 6. If $DV_{k,h} = -1$, then go to Step 9.
    Step 7. $h = h + 1$.
  Step 8. $vo_k = 1$.
  Step 9. $k = k + 1$. Go to Step 3.
Step 10. $k = 1, c = 0$.
Step 11. If $k \leq |V|$, then go to Step 12, else then go to Step 20.
  Step 12. $h = 1, m = 1$.
  Step 13. If $h \leq |V|$, then go to Step 14, else then go to Step 17.
    Step 14. If $DV_{k,h} = -1$, then go to Step 15, else then go to Step 16.
      Step 15. If $vo_h > m$, то $m = vo_h$.
    Step 16. $h = h + 1$. Go to Step 13.
  Step 17. If $vo_k < m + 1$, то $c = 1$.
  Step 18. $vo_k = m + 1$.

Step 19. $k = k + 1$. Go to Step 11.

Step 20. If $c = 0$, then go to Step 21, else then go to Step 10.

Step 21. For each $v_k : \rho(v_k) = vo_k$.

Using calculated values of $\rho(v_k)$ following sets are formed

$$VG_m : v_k \in VG_m \Leftrightarrow \rho(v_k) = m, m = \overline{1, \dots L_V} \tag{5}$$

with L_V – total number of groups of defuzzyfied variables.

Any variable $v_k: \rho(v_k) = 1$ is input variable ($v_k \in I$) and there is no need to defuzzyfy it.

Formed groups of processing and defuzzyfication can be used to improve effectiveness of fuzzy inference by applying parallel calculations for each group. Proper parallel processing is impossible without grouping the rules in advance, because transitive dependencies often allow only sequential processing. Introduced methods were successfully implemented in expert system GAZDETECT for fault detection in gas compressor plants [12].

References

1. Zadeh, L.A.: Fuzzy Sets. Information and Control 8, 338–353 (1965)
2. Zadeh, L.A.: The Concept of a Linguistic Variable and its Application to Approximate Reasoning. Information Sciences 8, 199–249, 301–357; 9, 43–80 (1975)
3. Mamdani, E.H., Assilian, S.: An Experiment In Linguistic Synthesis With Fuzzy Logic Controller. International Journal of Man-Machine Studies 7(1), 1–13 (1975)
4. Tsukamoto, Y.: An Approach To Fuzzy Reasoning Method. In: Gupta, M.M., Ragade, R.K., Yager, R.R. (eds.) Advances In Fuzzy Set Theory and Applications, pp. 137–149 (1979)
5. Seki, H., Mizumoto, M.: On the Equivalence Conditions of Fuzzy Inference Methods – Part 1: Basic Concept and Definition. IEEE Transactions on Fuzzy Systems 9(6), 1097–1106 (2011)
6. Larsen, P.M.: Industrial Applications of Fuzzy Logic Control. International Journal of Man-Machine Studies 12, 3–10 (1980)
7. Takagi, T., Sugeno, M.: Fuzzy Identification of Systems and its Applications to Modeling and Control. IEEE Transactions on Systems, Man, and Cybernetics 15(1), 116–132 (1985)
8. Mizumoto, M.: Fuzzy controls under various fuzzy reasoning methods. Information Science 45, 129–151 (1988)
9. Sugeno, M.: On stability of fuzzy systems expressed by fuzzy rules with singleton consequents. IEEE Transactions on Fuzzy Systems 7(2), 201–224 (1999)
10. Yubazaki, N., Yi, J., Hirota, K.: SIRMs (Single Input Rule Modules) connected fuzzy inference model. Journal of Advanced Computational Intelligence and Intelligent Informatics 1(1), 23–30 (1997)
11. Jang, J.-S.R.: ANFIS: Adaptive-Network-Based Fuzzy Inference System. IEEE Transactions on Systems & Cybernetics 23, 665–685 (1993)
12. Dolinina, O.N., Antropov, P.G., Kuzmin, A.K., Shvarts, A.Y.: Using Intelligent Systems for fault detection in gas compressor plants. Modern Problems of Science and Education (6) (2013) (in Russian), http://www.science-education.ru/113-11252

A Hybrid Model Based on Mutual Information and Support Vector Machine for Automatic Image Annotation

Cong Jin, Jinan Liu, and Jinglei Guo

School of Computer Science, Central China Normal University,
Wuhan, Hubei, 430079, China
jincong@mail.ccnu.edu.cn

Abstract. Automatic image annotation (AIA) has been widely studied during recent years and a considerable number of approaches have been proposed. However, the performance of these approaches is still not satisfactory. The main purpose of this paper is to use the feature selection (FS) approach for improving the performance of AIA. The mutual information (MI) is used to measure contribution of each image feature. We introduce a nonlinear factor to evaluation function of the feature selection approach to measure its performance. The experiment results show that proposed AIA approach achieves higher performance than the existing most AIA approaches, its performance is satisfactory.

Keywords: Automatic image annotation, Feature selection, Mutual information, Performance.

1 Introduction

With exponential growth of digital images on online image sites, automatic image annotation (AIA) now become an important issue of image management and image analysis. However, AIA is not an easy task, and is a challenging problem dealing with the textual description of images, i.e. associating tags or even better descriptive text to images [1]. The difficulty of AIA has mainly as follows. The first difficulty of AIA is that regions that are visually different can be associated to the same concept. The second difficulty of AIA is that some regions that are visually similar may denote different concepts. In addition, content-based image retrieval (CBIR) computes relevance only based on the visual similarity of low-level image features such as colors, textures and shapes [1]. In fact, people prefer retrieving images according to high-level semantic content. The problem is that visual similarity does not equal semantic similarity. There is a gap between low-level image features and high-level semantic contents. Therefore, the performances of many existing image annotation algorithms are not so satisfactory. In fact, some image features are redundant or irrelevant for describing the image semantic content, their presence not only increases the time cost of AIA, but also decreases the performance of AIA. Therefore, AIA algorithm must

R. Silhavy et al. (eds.), *Artificial Intelligence Perspectives and Applications*,
Advances in Intelligent Systems and Computing 347, DOI: 10.1007/978-3-319-18476-0_4

solve a basic problem: which features are more appropriate than the others in order to express the concept of the current query and improve the performance of AIA, whose essence is FS. FS is expected to improve performance of AIA, particularly in situations of huge collection images. FS has been applied to various areas, e.g., natural language processing [2], and intrusion detection [3] and so on. The main contributions of this work are the following. Firstly, we propose a FS approach based on MI to improve the annotation accuracy of AIA and to simplify the representation of the image. Secondly, we use an ensemble classifier based on majority voting scheme for implement image annotation, i.e., the classification results of several two-class support vector machine (SVM) classifiers are fused into an ensemble classifier, which classification result is the final annotation result.

2 Feature Selection

When we annotate images, the features of image are not all effective. In other words, some features are redundant or irrelevant. These extra features seriously affect the effectiveness of image annotation performance. FS is used for searching the optimal subset of image features for AIA task. In the existing FS algorithms, MI is applied to estimate the correlation and redundancy between candidate features and selected features [4]. Generally, we put an image feature into the selected feature subset if it maximizes MI between inputs and output. This procedure is executed until predefined feature number T is met. In this FS process, an evaluation function is used to decide how to choose the best combinations from all possible image features. Some evaluation functions based on MI have been proposed for FS [4].

In this paper, we let W be a set of word labels, *i.e.*, $W = \{W_1, W_2, ..., W_p\}$, F the candidate image features subset, *i.e.*, $F = \{F_1, F_2, ..., F_m\}$, S the subset of selected image features, *i.e.*, $S = \{S_1, S_2, ..., S_T\}$. f and s is elements of F and S respectively, and $J(f)$ is the evaluation function of feature f. Generally, one let MI $I(W; f) = J(f)$. However, its disadvantage is obvious, because it does consider the redundancy among the selected features. In [5], an improved method was proposed as follows

$$J(f) = I(W; f) - \beta \sum_{s \in S} I(s; f) \tag{1}$$

Where, β is a parameter, and its value should be (0.5, 1). However, this evaluation function does not consider the effect of the size $|S|$ of selected feature subset. We suggest an evaluation function with a nonlinear factor showed in eq.(2) as follows

$$J(f) = I(W; f) - a \cdot |S|^d \sum_{s \in S} I(s; f), \quad a \in R, d \in Z \tag{2}$$

Experiments results [6] will show that adding nonlinear factor is appropriate for evaluation functions. Now, we can acquire FS approach.

Algorithm IMIFS: Improved MI-based FS

Step 1. Initialization: set F (F is a set includes all image features), $S = \Phi$.

Step 2. For each candidate feature f in F, compute MI of f with the word labels C, $I(W; f)$.

Step 3. Find the feature f with the maximize $I(W;f)$, i.e., $f = arg\ max\{I(W;f)\}$, then we have following operates

$$F = F - \{f\},\ S = S + \{f\} \tag{3}$$

Step 4. While $|S| < T$ do
Step 5. Output selected image feature subset S.

3 Image Annotation

3.1 Annotated Image Data Set

In order to implement AIA with SVM, the images in training set need be preprocessed to get annotated image database. Preprocessing step is as follows.

Algorithm PA: Preprocessing Algorithm
Step 1. For each image in training set, it is captioned using a few keywords that describe image semantic content. Each image is segmented into some target regions. So, we can obtain the image region set for each image. After gathering all regions of each image, we may get the region set of all training images, denoted by $B=\{B_1, B_2, \ldots B_P\}$. Where, P is the number of all regions.
Step 2. First of all, we cluster regions set B using k-means algorithm, the class set is generated by the cluster centers, which is denoted by $C=\{C_1, C_2, \ldots C_p\}$. Where, p is the number of all classes.
Step 3. For an image with caption, after extracting all nouns denoted by K from its caption, we let K be candidate keywords set of this image, which is denoted by $K=\{K_1, K_2, \ldots K_L\}$. Where, L is the number of all keywords.
Step 4. We calculated the correlation between the keywords and class as follows

$$C(K_i, C_j) = \frac{\alpha}{L+P}(l_1 \times l_2) + \frac{1-\alpha}{L+p}(l_3 \times l_4) \tag{4}$$

Where, l_1 is the number of K_i appearing in B_j, l_3 is the number of K_i appearing in C_j, l_4 is the average number of K_i appearing in C, and α is an adjustment parameter. l_2 is a value to describe B_j's content, and it is calculated by the following formula

$$l_2 = \begin{cases} 1, & \textit{if } K_i \textit{ is a noun to describe } B_j\textit{'s background} \\ 2, & \textit{if } K_i \textit{ is a noun to describe } B_j\textit{'s foreground} \end{cases} \tag{5}$$

Step 5. Propagate keyword K_i to class C_j when

$$C(K_i, C_j) = \max_{j \in \{1,2,\ldots,p\}} \{C(K_i, C_j)\} \tag{6}$$

So, for each keyword K_i, it belongs to itself class. Therefore, we can get the annotated image dataset for training SVM.

3.2 Image Low-Level Features Vector

In this paper, we extract image low-level features with MPEG-7 standard. In MPEG-7 standard, image features such as color, texture and shape are used for the description of image low-level feature content. These MPEG-7 descriptors are Scalable Color (64 features), Edge Histogram (80 features), Homogeneous Texture (62) features, Color Layout (12 features) and Color Structure (32 features). The total number of image feature is 250. These image features constitute a feature vector. So, for each image region, there is always such an image feature vector corresponding it.

In these features, different feature has different value range and different dimensional number, if some features have a fairly large original value, it will bring a considerable importance, while those features which have small original values are ignored. Obviously, it is inconsistent with the actual. Thus, each feature was normalized in order to eliminate error caused by inconsistent magnitude and large gap between original values. Normalization equation is as follows:

$$f' = \frac{f - \min(f)}{\max(f) - \min(f)} \tag{7}$$

3.3 Image Annotation Approach

AIA correlates label that describes semantic concepts to image. In this paper, AIA is viewed as the classification process. Usually, annotation is realized by exploiting the correspondence between image features and semantic concepts of the images. SVM is used as the classifier for AIA task. The main steps of proposed model are described as follows.

Algorithm AIAFS: AIA based on features selection

Step 1. For given an image I without caption, it is segmented into N regions, i.e., $I = \{ R_1, R_2, \ldots, R_N\}$. Feature vector of each region i can be extracted using MPEG-7 standard and denoted by $R_i = \{ R_{i1}, R_{i2}, \ldots, R_{i250}\}$. Normalize image feature vector according to eq.(7).

Step 2. Select feature subset using MI.

Step 3. Stopping condition of selecting feature is to meet a certain predefined number of features to be selected.

Step 4. Because SVM is a two-class classifier, we use the annotated data set to generate a SVM classifier for every two classes. So, $p(p-1)/2$ SVM classifiers need to be generated.

Step 5. Get an ensemble classifier using all $p(p-1)/2$ SVM classifiers to annotate images.

Step 5. We used the ensemble classifier to annotate I. Specifically, if a region of the image I is classified into the class C_i, then all keywords of class C_i are propagated to this region.

Step 6. After keywords of all regions of the image I are obtained, we let the set of these keywords as the semantic annotation words. So, the unlabeled image is automatically annotated.

3.4 Construct Ensemble Classifier

In algorithm AIAFS, the ensemble classifier is constructed based on majority voting scheme by the following method. We let

$$G = \left|g_{ij}\right| = \begin{vmatrix} - & g_{12} & \cdots & g_{1p} \\ g_{21} & - & \cdots & g_{2p} \\ \cdots & \cdots & - & \cdots \\ g_{p1} & g_{p2} & \cdots & - \end{vmatrix}$$

where $i, j \in Y = \{1, 2, \ldots, p\}$, $i \neq j$, g_{ij} *is the SVM classifier of classes* C_i *and* C_j, $g_{ij} = g_{ji}$. *Each SVM classifier* g_{ij} *has the output '1' for positive instance and '0' for negative instance. For selected image feature vector X of a certain region, we let ith row of G be* $g_i = \{g_{i1}, g_{i2}, \ldots, g_{ip}\}$, *where* $g_i(X) = \sum_{g_{ij}(X)=i, i \neq j} g_{ij}(X)$, *then* the ensemble classification result of this *region* is

$$EC(X) = \arg\max_{i \in Y} g_i(X)$$

4 Experimental Results

4.1 Image Datasets and Evaluate Annotation Accuracy

In the experiments, in order to test the effectiveness and accuracy of the proposed AIA, we conduct experiments on three image datasets, i.e., Corel5k, Corel30k and IAPRTC12.

AIA performance is evaluated by comparing the captions automatically generated for the test set with the human-produced ground truth. We compute the recall and precision of every class C_i in the test set respectively as follows:

$$recall(C_i) = \frac{|a_{correct}(C_i)|}{|a_{total}(C_i)|}, \quad precision(C_i) = \frac{|a_{correct}(C_i)|}{|a_{system}(C_i)|}$$

Where $|a_{correct}(C_i)|$ represents the number of images correctly annotated with the class C_i, the $|a_{total}(C_i)|$ represents the number of images having that class C_i in ground truth annotation, and $|a_{system}(C_i)|$ represents the number of image given class C_i. The average recall (*AR*), average precision (*AP*) and *F*-measure are as follows:

$$AR = \frac{\sum_{i=1}^{L} recall(C_i)}{L}, \quad AP = \frac{\sum_{i=1}^{L} precision(C_i)}{L}, \quad F-measure = \frac{2 \cdot AR \cdot AP}{AR + AP}$$

4.2 Parameters Initialization

During the experiments, the training set is used to build the classifiers and the testing set is used to evaluate the annotation accuracy of AIA. For comparing annotation accuracy of proposed AIA, the ensemble classifier is practiced in the two given image data sets, respectively. (1) SVM: The regularization parameter was set to 1 and the bandwidth of the kernel function was set to 0.5. (2) Algorithm IMIFS: For evaluation function, after a large number of experiments, we let a=1.012, d=−2.892. (3) In the experiments, the annotation accuracy is defined by F-measure.

4.3 Experimental Results and Comparison

4.3.1 Results for Corel5k

We compare the annotation accuracy of proposed AIA and AIA non-FS approach (denoted by AIANO) with other approaches such as TM [7], JEC [8], GM-PLSA [9], TGLM [10], MSC [11], TagProp [12], HGDM [13], SSLM [14], CMRM [15], CRM [16], MBRM [17], MRS-MIL [18], BMHM [19], MRFA [20], HSVM-DT [21], and PLSA-WORDS [22]. The annotation accuracy of proposed AIA is evaluated by comparing the captions automatically generated with the original manual annotations. Similar to [18], we computer the precision, recall, and F-measure of every keyword in the testing set and use the F-measure values to summarize the performance of AIA. AIAs' comparison results are shown in Table 1.

Table 1. Comparison results of different approaches on Corel5k

Approaches	Average recall	Average precision	F-measure
TM [7]	0.040	0.060	0.048
JEC [8]	0.320	0.270	0.293
GM-PLSA [9]	0.250	0.260	0.255
TGLM [10]	0.290	0.250	0.269
MSC [11]	0.320	0.250	0.281
TagProp [12]	0.423	0.327	0.369
HGDM [13]	0.320	0.280	0.299
SSLM [14]	0.290	0.270	0.280
CMRM [15]	0.100	0.090	0.095
CRM [16]	0.194	0.163	0.177
MBRM [17]	0.250	0.240	0.245
MRS-MIL [18]	0.190	0.160	0.174
BMHM [19]	0.290	0.230	0.257
MRFA [20]	0.230	0.270	0.248
HSVM-DT [21]	0.359	0.360	0.359
PLSA-WORDS [22]	0.200	0.140	0.165
AIANO	0.316	0.308	0.312
AIAFS	0.369	0.361	0.365

Table 1 shows the comparison results of proposed AIAFS and AIANO with other approaches on Corel5k. It can be seen in Table 1 that AIAFS is one of the AIA approaches achieving the best performance, however not the best one on Corel5k. From Table 1, the average recall of AIAFS is 0.369, which is higher than the average recall 0.359 of HSVM-DT, and the improvement is at least 0.01. However, compared with TagProp, AIAFS loses a little in the average recall. It is due to the average recall declines with the average precision raising. Furthermore, AIAFS achieves the best result in average precision for all compared approaches. This proves the feasibility of proposed AIAFS. For *F*-measure, TagProp has not been exceeded (0.369), however there is not much difference between TagProp and AIAFS (0.365). Their difference is only 0.004, which shows that the performance of AIAFS is satisfactory.

Compare AIAFS and AIANO, we notice that the average recall, average precision, and *F*-measure of AIAFS are higher than AIANO, which shows that FS can remove redundant features and it is possible to greatly improve performance of AIA approach. Therefore, FS plays a very important role for image annotation. Moreover, Figure 1 shows some examples of annotation by a particular keyword.

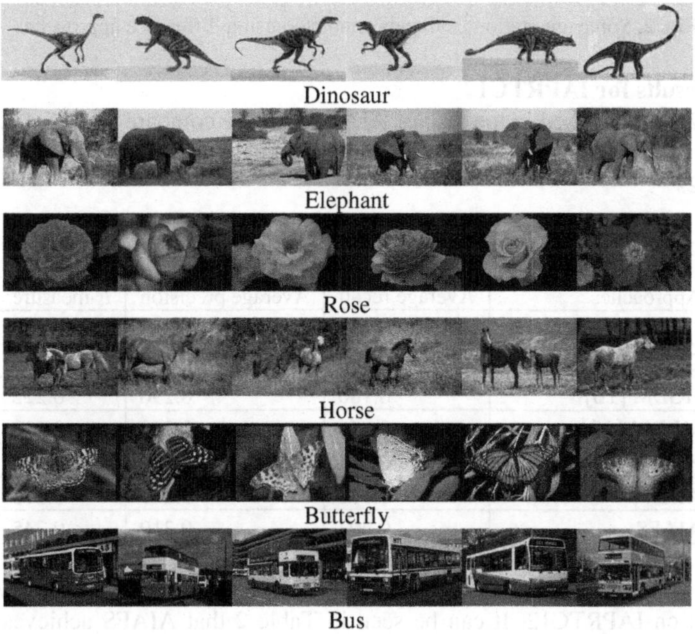

Fig. 1. Some examples of annotation results using AIAFS in Corel5k. Each row shows annotated images with first 6 *F*-measure using a particular keyword.

Figure 2 shows some annotation examples annotated using difference approaches. We can see that annotations generated by AIAFS are more reasonable than other compared approaches.

Image	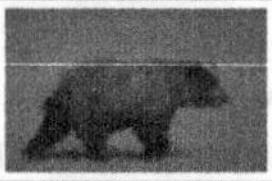	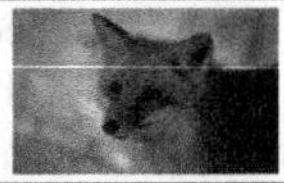		
Ground truth	grizzly, meadow, grass, bear	head, fox, snow, close-up	landscape, trees, garden, flowers	blue-footed, rock, booby, bird
PLSA-WORDS[23]	meadow, grizzly, bear, horse, sand	sculpture, clouds, rabbit, stone, sky	flowers, garden, farm, trees, bench	booby, bird, tree, sky, rock
AIAFS	grass, meadow, grizzly, bear	dog, snow, head, fox	flowers, grass, trees, garden	booby, bird, rock, meadow
Image				
Ground truth	trees, ice, sky, frost	courtyard, trees, sky, building	mountain, snow, sky, peak	people, water, sky, sand
PLSA-WORDS [23]	sculpture, desert, path, ice, grass	building, temple, wall, sand, sky	mountain, peak, sky, landscape, snow	iceburg, snow, ice, beach, water
AIAFS	trees, sky, frost, ice	building, trees, sky, wall	mountain, clouds, sky, snow	people, water, sky, sailboat

Fig. 2. Some annotation examples annotated using difference approaches

4.3.2 Results for IAPRTC12

IAPRTC12 is firstly used in the task of AIA by [5]. We compare AIAFS and AIANO with other AIAs, and Table 2 gives the experimental results for 1980 keywords in IAPRTC12.

Table 2. Comparison results of different approaches on IAPRTC12

Approaches	Average recall	Average precision	F-measure
JEC [5]	0.160	0.250	0.195
GM-PLSA [6]	0.160	0.230	0.189
HGDM [10]	0.180	0.290	0.222
MBRM [14]	0.140	0.210	0.168
PLSA-WORDS [19]	0.120	0.180	0.144
AIANO	0.178	0.280	0.218
AIAFS	0.203	0.310	0.245

Table 2 shows the comparison results of proposed AIAFS and AIANO with other approaches on IAPRTC12. It can be seen in Table 2 that AIAFS achieves the best result in average recall, average precision and F-measure for all compared approaches, which proves that the performance of proposed AIAFS is reliable.

Compare AIAFS and AIANO, we notice that the average recall, average precision and F-measure of AIAFS are higher than AIANO, which again shows that FS is possible to greatly improve performance of AIA.

Figure 3 shows some examples of annotation using by PLSA-WORDS, HGDM and AIAFS respectively. Comparing the annotations made by PLSA-WORDS,

HGDM and the ground truth, we can see that AIAFS is often more reasonable than other approaches for annotated image.

Image				
Ground truth	people, meadow, fence, tree, sky	girl, skirt, house, waistcoat, hill	waterfall, forest, tourist, river, ravine	lawn, path, tree, cloud, sky
PLSA-WORDS [22]	grass, tree, sky, stone, building	sky, sand, wall, desert, landscape	water, forest, tree, grass, cloud	tree, cloud, house, sky, river
HGDM	tree, sky, meadow, people, house	girl, sky, people, cloud, hill	waterfall, forest, snow, river, ravine	lawn, garden, tree, cloud, sky
AIAFS	people, meadow, lawn, tree, sky	girl, skirt, house, land, hill	waterfall, forest, mountain, river, ravine	lawn, tree, cloud, grassland, sky

Fig. 3. Some annotation examples annotated using difference approaches

5 Conclusions

In this paper, we propose a hybrid model based on MI and SVM for AIA, named as AIAFS. The main advantages of AIAFS are as follows. (1). In the proposed FS, MI is used to measure contribution of each image feature, which ensures the proposed FS can achieve better performance. (2). The proposed FS can indeed remove the redundant features and improve the performance of AIA. Experiment results confirm that proposed FS is helpful for improving the performance of AIA. (3). Majority voting is used for creating the ensemble classifier, which is used for annotating images. Experiment results confirm that performance of the ensemble classifier is more superior than compared most other classifiers.

Acknowledgment. This work was supported by Natural Social Science Foundation of China (Grant No.13BTQ050).

References

1. Zhang, D.S., Islam, M.M., Lu, G.J.: A review on automatic image annotation techniques. Pattern Recognition 45(1), 246–262 (2012)
2. Faheema, A.G., Rakshit, S.: Feature selection using bag-of-visual-words representation. In: IEEE 2nd International Advance Computing Conference (IACC), Patiala, February 19-20, pp. 151–156 (2010)
3. Amiri, F., Yousefi, M.M.R., Lucas, C., et al.: Mutual information-based feature selection for intrusion detection systems. Journal of Network and Computer Applications 34(4), 1184–1199 (2011)

4. Liu, H.W., Sun, J.G., Liu, L., Zhang, H.J.: Feature selection with dynamic mutual information. Pattern Recognition 42(7), 1330–1339 (2009)
5. Battiti, R.: Using mutual information for selecting features in supervised neural net learning. IEEE Transactions on Knowledge and Data Engineering 17(9), 1199–1207 (2005)
6. Wang, P., Jin, C., Jin, S.W.: Software defect prediction scheme based on feature selection. In: 2012 International Symposium on Information Science and Engineering, Shanghai, China, pp. 477–480 (2012)
7. Duygulu, P., Barnard, K., de Freitas, J.F.G., Forsyth, D.: Object recognition as machine translation: Learning a lexicon for a fixed image vocabulary. In: Heyden, A., Sparr, G., Nielsen, M., Johansen, P. (eds.) ECCV 2002, Part IV. LNCS, vol. 2353, pp. 97–112. Springer, Heidelberg (2002)
8. Makadia, A., Pavlovic, V., Kumar, S.: Baselines for image annotation. International Journal of Computer Vision 90(1), 88–105 (2010)
9. Li, Z., Shi, Z., Liu, X., Shi, Z.: Modeling continuous visual features for semantic image annotation and retrieval. Pattern Recognition Letters 32(3), 516–523 (2011)
10. Liu, J., Li, M., Liu, Q., Lu, H., Ma, S.: Image annotation via graph learning. Pattern Recognition 42(2), 218–228 (2009)
11. Wang, C., Yan, S., Zhang, L., Zhang, J.H.: Multi-label space coding for automatic image annotation. In: IEEE Computer Society Conference on Computer Vision and Pattern Recognition, pp. 1643–1650 (2009)
12. Guillaumin, M., Mensink, T., Verbeek, J., Schmid, C.: Tagprop: discriminative metric learning in nearest neighbor models for image auto-annotation. In: IEEE 12th Conference on Computer Vision, pp. 309–316 (2009)
13. Li, Z.X., Shi, Z.Z., Zhao, W.Z., Li, Z.Q., Tang, Z.J.: Learning semantic concepts from image database with hybrid generative/discriminative approach. Engineering Applications of Artificial Intelligence 26(9), 2143–2152 (2013)
14. Zhu, S., Liu, Y.: Semi-supervised learning model based efficient image annotation. IEEE Signal Processing Letters 16(11), 989–992 (2009)
15. Jeon, J., Lavrenko, V., Manmatha, R.: Automatic image annotation and retrieval using cross-media relevance models. In: 26th Annual International ACM SIGIR Conference on Research and Development in Information Retrieval, Toronto, Canada, pp. 119–126 (2003)
16. Lavrenko, V., Manmatha, R., Jeon, J.: A model for learning the semantics of pictures. In: 7th Annual Conference on Neural Information Processing Systems, Vancouver, Canada, pp. 446–453 (2003)
17. Feng, S.L., Manmatha, R., Lavrenko, V.: Multiple bernoulli relevance models for image and video annotation. In: IEEE Conference on Computer Vision and Pattern Recognition, Washington, DC, USA, pp. 1002–1009 (2004)
18. Zhao, Y., Zhao, Y., Zhu, Z., Pan, J.S.: MRS-MIL: minimum reference set based multiple instance learning for automatic image annotation. In: International Conference on Image Processing, San Diego, California, USA, pp. 2160–2163 (2008)
19. Stathopoulos, V., Jose, J.M.: Bayesian mixture hierarchies for automatic image annotation. In: Boughanem, M., Berrut, C., Mothe, J., Soule-Dupuy, C. (eds.) ECIR 2009. LNCS, vol. 5478, pp. 138–149. Springer, Heidelberg (2009)
20. Xiang, Y., Zhou, X., Chua, T., Ngo, C.: A revisit of generative model for automatic image annotation using Markov random fields. In: IEEE Conference on Computer Vision and Pattern Recognition, Miami, FL, pp. 1153–1160 (2009)
21. Chen, Z., Hou, J., Zhang, D.S., Qin, X.: An annotation rule extraction algorithm for image retrieval. Pattern Recognition Letters 33(10), 1257–1268 (2012)
22. Monay, F., Perez, D.G.: Modeling semantic aspects for cross-media image indexing. IEEE Transactions on Pattern Analysis and Machine Intelligence 29(10), 1802–1817 (2007)

Pose Estimation Using Local Binary Patterns for Face Recognition

Nhat-Quan Huynh Nguyen and Thai Hoang Le

Department of Computer Science, VNUHCM - University of Science
1012337@student.hcmus.edu.vn,
lhthai@fit.hcmus.edu.vn

Abstract. One of the most challenging factors in face recognition is pose variation. This paper proposes an appropriate framework to identify images across pose. The input profile image is classified to a pose range. After that, this image is matched with the same pose range images from a certain database to determine its identification. In this framework, descriptors which are based on Local Binary Patterns are used to extract features of all images. Experiments on the FERET database prove the robustness of the proposed framework based on comparison with other approaches.

Keywords: Face recognition, Local binary patterns, Pose estimation.

1 Introduction

Face recognition is one of the most important biometric techniques which attracts many researchers to solve its different issues. With the advantages of natural and passive sample collecting, this area seems to be more appreciated than others such as fingerprint and iris recognition. However, this benefit also brings the challenge of uncontrolled images. *Pose* is one of the key factors that can significantly affect face recognition systems' performances [1] due to the fact that the differences between images of one person from different poses are often larger than those of different people at a similar yaw angle. In many methods proposed for analyzing face images, Local Binary Patterns (LBP) is widely used thanks to its potential for speed and accuracy. LBP can be applied to not only face recognition, but also many applications such as: texture classification, facial expression recognition and gender recognition. In 2011, Moore et al. [18] reported the robustness of LBP for multi-view facial expression recognition, which promises the applications of LBP for profile face recognition.

Techniques of face recognition across poses can be classified into three categories: general algorithms, 2D techniques and 3D approaches. General algorithms do not contain specific tactics of handling pose variations, which is proposed for general purposes of face recognition through all image condition like illumination, expression or pose [2,10,16]. The other techniques such as 2D transformation [4,5,13] and 3D reconstruction [6,11,14] tend to eliminate difficulties caused by image variations according to its own properties.

R. Silhavy et al. (eds.), *Artificial Intelligence Perspectives and Applications*,
Advances in Intelligent Systems and Computing 347, DOI: 10.1007/978-3-319-18476-0_5

After being analyzed, three categories above can be re-grouped into two classes. The general algorithms tend to tackling all condition of images while the other methods (2D transformation and 3D reconstruction) require a large number of training images to construct a general model. These characteristics are also drawbacks of these two trends. The general methods is inefficient in case matching frontal faces with profile ones. While approaches based on sample models reported many very impressive results, the cost of preparing data for training is so large.

This paper focuses on constructing a novel framework using Local Binary Patterns to recognize profile faces. This framework includes a step of estimating pose range of face in order to choose a suitable image gallery for recognition in the step later. It does not require any training data. Moreover, it pays attention to pose of facial images in procedure of recognition. This approach is different from traditional methods [17] that the predicted pose does not need to be exactly the same as the pose of gallery images, which avoids error propagation to a recognition step.

Our research is organized as follows: the basic theories of Local Binary Patterns and pose range estimation are introduced in section 2; the proposed framework is then illustrated in section 3; the following section presents our experiments on the FERET database; finally, conclusion and references are in the rest of this paper.

2 Base Theory

2.1 Local Binary Patterns

Being introduced by Ojala et al. [20] in 1996, LBP operator is a powerful tool for texture description. This operator labels the pixels of an image and thresholds each neighbourhood of 3 × 3 pixels by using the central pixel value. The gray value of each pixel g_p in the neighborhood is compared to the gray value g_c of the central pixel. If g_p is greater than g_c, then it is assigned 1, and 0 if not. The LBP label for the central pixel of image is obtained as [2]:

$$LBP = \sum_{p=0}^{7} S(g_p - g_c)2^p, \quad \text{where } S(x) = \begin{cases} 1, & x \geq 0 \\ 0, & x < 0 \end{cases} \tag{1}$$

The basic operator was then extended to use neighbourhood of different sizes to capture dominant features at different scales. Notation $LBP(P, R)$ denotes a neighborhood of P equally spaced sampling points on a circle of radius of R. For example, the original LBP operator can be denoted $LBP(8, 1)$.

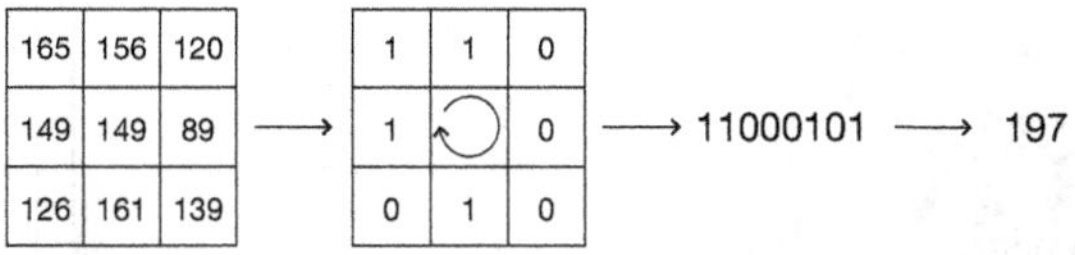

Fig. 1. $LBP(8,1)$ operator

According to [19], there is a small subset of 2^P patterns, called uniform patterns, accounted for the majority of the texture of images, over 90% of all patterns for $LBP(8,1)$ and about 70% for $LBP(16,2)$. These patterns contain at most two bitwise transitions from 0 to 1 or vice versa for a circular binary pattern. For instance, '00000000' (0 transition), '01111100' (2 transitions), and '10001111' (2 transitions) patterns are uniform while '11001110' pattern is not uniform due to its four bit transitions. The uniform patterns can be used to find the pixels that belong some texture primitives such as spot, flat area, edge, and corner. It can be denoted as $LBP^{u2}_{(P,R)}$, which was mentioned in [3].

LBP Feature and Distance Measurement. In the face recognition procedure of Ahonen et al. [3], a facial image is divided into m small same sized non-overlapping regions called patches from which LBP histograms are extracted and concatenated into a single vector, which enhanced the facial representative ability. A patch histogram is defined as [2]:

$$H_i = \sum_{x,y} I\{LBP(x,y) = i\}, \quad i = 0, 1, ..., n-1 \tag{2}$$

in which n is the number of different labels produced by the LBP operator and

$$I(A) = \begin{cases} 1, & A \text{ is true} \\ 0, & A \text{ is false.} \end{cases} \tag{3}$$

In using classifier like k-Nearest Neighbour to label an input image, distance computing formula must be used in order to calculate distance between images which is the difference between feature vectors of them. In this research, we use Chi square statistics to build a dissimilarity measure of face.

$$\chi^2(S,M) = \sum_i \frac{(S_i - M_i)^2}{S_i + M_i} \tag{4}$$

where S and M are two LBP histogram.

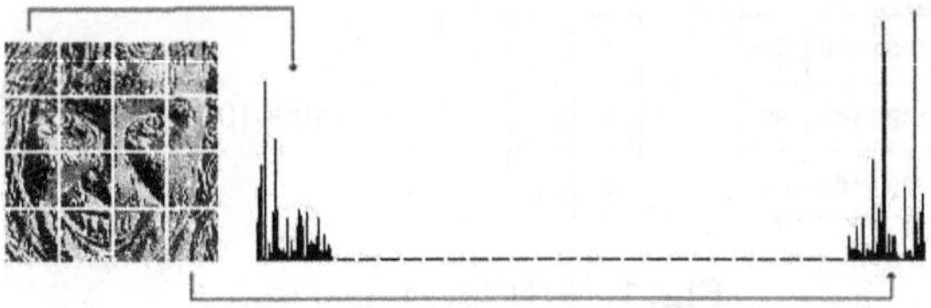

Fig. 2. LBP feature extracting procedure

2.2 Applying Local Binary Patterns for Profile Face Recognition

Traditional approaches did not pay enough attention to yaw angle of images. They recognize the profile faces using frontal view or other random poses' images as gallery set [23]. This leads to a weak performance when processing images with large yaw angle. For example, Eigenfaces [15] with 26.3%, Fisherfaces [15] with 25.7% and KPDA [9] with 44.32%. Our work contains a process of predicting pose of image before identifying it.

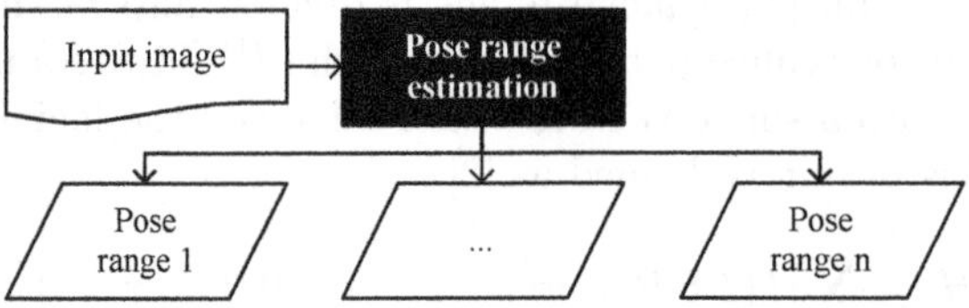

Fig. 3. Determining pose range of a face image

Pose Range Estimation. In many previous researches, predicting yaw angle of a face image is a work of returning an exact value of pose. However, the final target of a face recognition system is finding a suitable label for an input image, not a value of an angle. In our work, this issue is defined as follows: given a face image and n pose ranges which are parts of total range of images in a database, return a right pose range of that image using a procedure of *pose range estimation* (Figure 3). The value of n and the total range are given by experts and based on a certain database.

Figure 4 illustrates the process of estimating pose range of an original face image. Firstly, a feature vector of the input image is generated using a descriptor. This is also done for all sample images coming from ranges of pose. After that, a distance measurement is applied to calculate distances of input image's feature vector with sample features. Finally, input image's pose range is determined as the same range with the nearest sample image based on their distance. In this process, *Argmin* is a stage of finding a sample image that has the minimum distance to the input image.

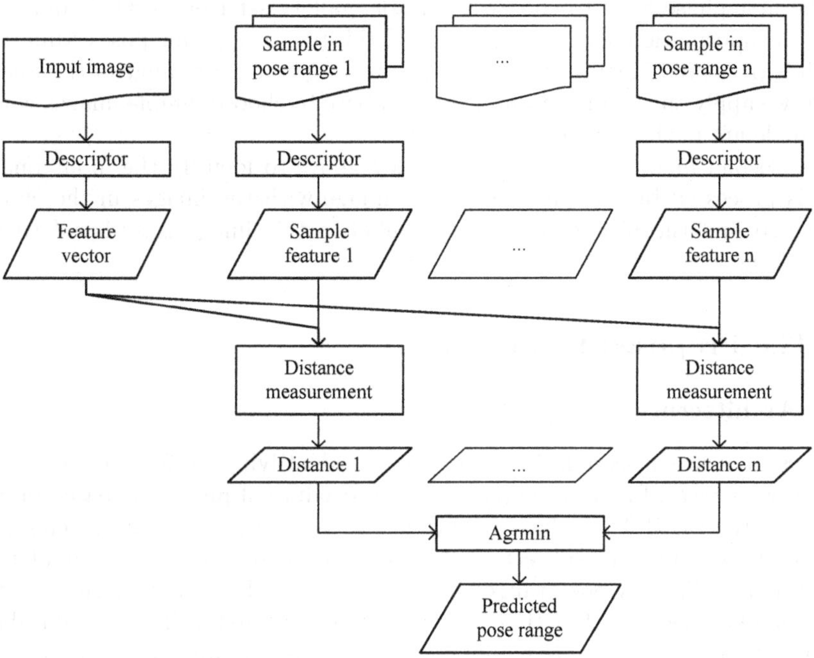

Fig. 4. Pose estimation in details

Applying Predicted Pose Range for Profile Face Recognition. The procedure of identifying a profile face image is illustrated in Figure 5, which begins with an input image and ends with a predicted label of that image. It can be divided into two phases: choosing gallery set and identifying.

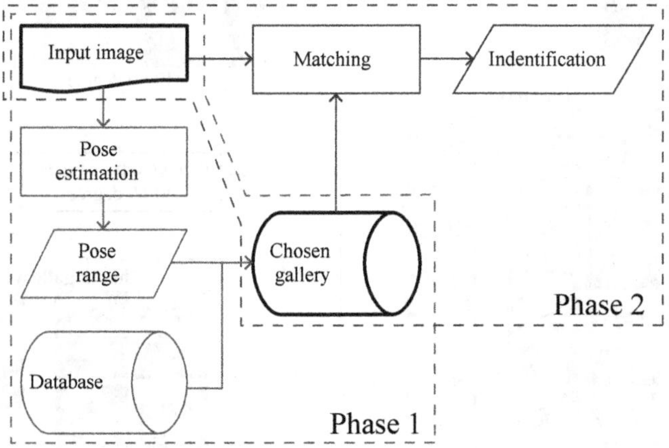

Fig. 5. Applying pose range information in recognition process of profile faces

The first phase aims to choose a suitable gallery set from a whole database using the information of the given image. In this phase, applying pose estimation, which is presented in subsection 2.2, helps us to gain a pose range of the image. Then, we apply this information in the database to choose sample images whose yaw angle are in that range.

The second phase uses the chosen galley above to identify the input image. This is processed by matching the input image with the images in the chosen gallery to find the nearest image. The label of this image is assigned to the original one.

3 The Proposed Framework

3.1 Architecture

As presented in subsection 2.2, procedure of identifying profile face image includes two phases. Figure 6 represents the procedure of phase one based on the images of the FERET database (mentioned in subsection 4.1). After being preprocessed (presented in subsection 3.2), the image becomes an input of pose estimation finding its pose range. In this example, the given image is in the range of 40-60 degree. After that, we use the pose range for filtering a database to get suitable sample images whose yaw angle are in the same range of the input image. This image set is called *chosen gallery.*

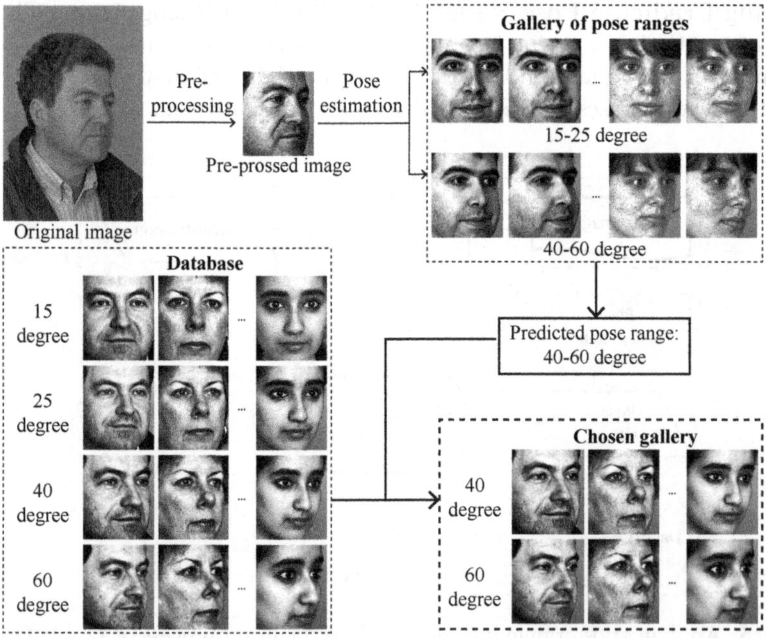

Fig. 6. Choosing gallery for a profile face image

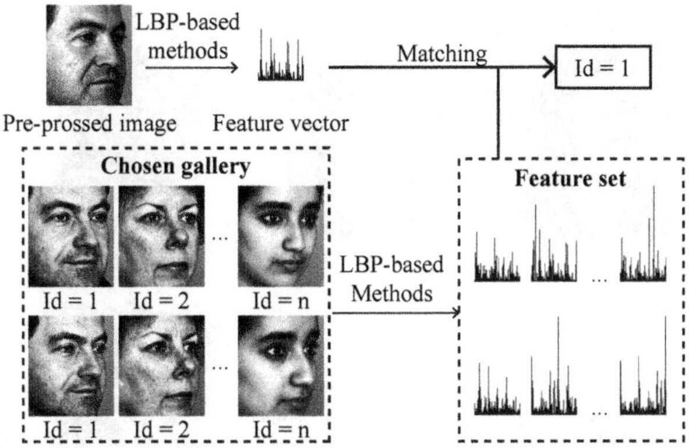

Fig. 7. Identifying an image based on the chosen gallery

The identifying process (phase two) of the example image is presented in Figure 7. Feature vectors of all images, original image and all images in *chosen gallery*, are extracted using a LBP-based method. Feature vector of the input image is then matched with other features using the distance measurement of Chi-square (mentioned in 2.1) to find the nearest image and get its label.

3.2 Pre-processing

Based on the concept of CSU [7] when pre-processing frontal images in FERET database, we constructed a similar procedure to process profile images in our system (Figure 8). This is a simple approach which can be easily found on the Internet [8, 22]. For each image, first the coordinates of eyes are detected. Then, we align all images so that the eyes positions are fixed by using affine transformations like rotation and scaling. Next, these images are cropped to extract regions of face. Finally, histogram equalization is applied to each image in order to cope with illumination changes.

4 Experiments

4.1 Database

In our experiments, we use frontal and profile images from FERET database [21]. This database contains 14051 face images which are divided into many subsets based on their condition. According to FRVT 2000 Tests May 2000, some profile image subsets were classified as below based on their yaw angle in comparison with frontal view.

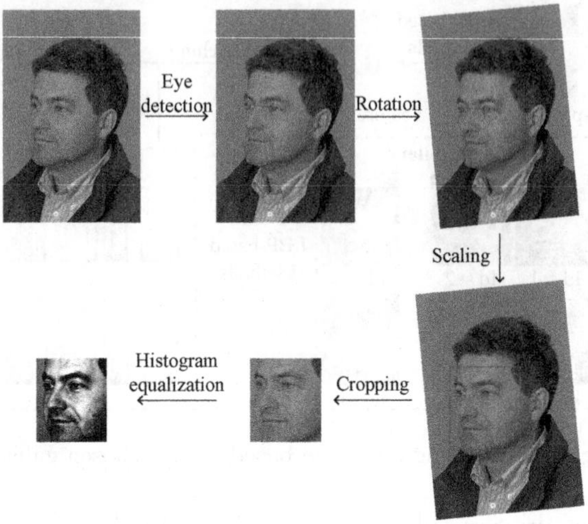

Fig. 8. Pre-processing a profile face image

Table 1. FERET profile image subsets

Subset	Quantity	Description
Gallery	200	Single frontal images of all subjects
P1	400	Pose at 15 degree
P2	400	Pose at 25 degree
P3	400	Pose at 40 degree
P4	400	Pose at 60 degree

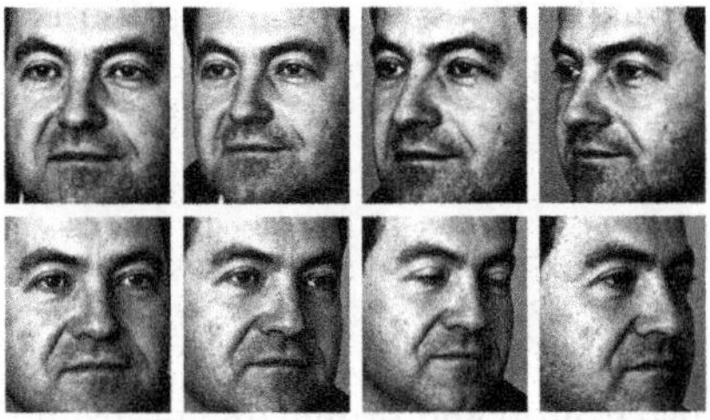

Fig. 9. Pre-processed profile images of one subject in the FERET database

From these image sets, we choose images from 20 subjects (160 images) to make a gallery set for pose estimation. The remaining images are divided into the gallery and testing set for recognition work, which can be swapped to get a general result by calculating an average value.

4.2 Result

Pose Estimation. The mistakes when estimating poses of profile images using LBP are shown in Table 2. It is easy to realize that the mistakes of predicted poses are nearby the right values. If we use the pose range as the result of pose estimation, the issue of pose false are generally solved.

Table 2. Pose mistakes when predicting the exact value of yaw angle

		Predicted angle			
		15	**25**	**40**	**60**
True angle	**15**	1.0	0	0	0
	25	0.2	0.8	0	0
	40	0	0.25	0.75	0
	60	0	0	0.15	0.85

Table 3. Performance when applying pose estimation in profile face recognition

Method	Number of subjects	Poses	Performance
Eigenfaces [12]	100	9 poses within ±40 degree in yaw	39.40
KPDA [9]	200	7 poses: 0, ±15, ±25 and ±45 degree in yaw	44.32
Proposed framework	180	2 pose ranges: 15 – 25 and 40 – 60 degree	72.22

Profile Face Recognition. Table 3 presents our framework's performance in comparison with other approaches. As can be seen from this table, the method of combining pose range estimation to choose gallery images and recognition based on these images gains a promised performance for profile face recognition.

In comparison with modern approaches like 3DMM [6] with a result of 95.8%, our framework cannot get a similar result. However, 3DMM is a method that needs a period of building shape models for faces, which is so complex and takes a lot of time for the data to be ready. The proposed framework does not need a training time like that. This is an advantage of our method when applying to real-time applications.

5 Conclusion

This study introduces an appropriate framework for identifying a profile face image, which predicts the pose range of image to choose the suitable gallery for recognition. A suitable pre-processing method for this kind of image is also presented. Estimating face image's pose range helps us to avoid error propagation when pose predicting returns false results. Recognizing image with gallery set of surrounding pose range increases chance of input image to match its subject's sample image, which means the performance of system is improved. The experiments on the FERET database and comparison with other approaches shows a robustness of our framework.

Acknowledgement. This research is funded by Vietnam National University of Ho Chi Minh City (VNU-HCMC) under the project "Features descriptor under variation condition for real-time face recognition application", 2014.

References

1. Abate, A.F., Nappi, M., Riccio, D., Sabatino, G.: 2D and 3D face recognition: A survey. Pattern Recognition Letters 28(14), 1885–1906 (2007)
2. Ahonen, T., Hadid, A., Pietikäinen, M.: Face recognition with local binary patterns. In: Pajdla, T., Matas, J. (eds.) ECCV 2004. LNCS, vol. 3021, pp. 469–481. Springer, Heidelberg (2004)
3. Ahonen, T., Hadid, A., Pietikainen, M.: Face description with local binary patterns: Application to face recognition. IEEE Transactions on Pattern Analysis and Machine Intelligence 28(12), 2037–2041 (2006)
4. Beymer, D., Poggio, T.: Face recognition from one example view. In: Proceedings of the Fifth International Conference on Computer Vision, pp. 500–507. IEEE (1995)
5. Beymer, D.J.: Face recognition under varying pose. In: Proceedings of the 1994 IEEE Computer Society Conference on Computer Vision and Pattern Recognition, CVPR 1994, pp. 756–761. IEEE (1994)
6. Blanz, V., Vetter, T.: Face recognition based on fitting a 3D morphable model. IEEE Transactions on Pattern Analysis and Machine Intelligence 25(9), 1063–1074 (2003)
7. Bolme, D.S., Beveridge, J.R., Teixeira, M., Draper, B.A.: The CSU face identification evaluation system: Its purpose, features, and structure. In: Crowley, J.L., Piater, J.H., Vincze, M., Paletta, L. (eds.) ICVS 2003. LNCS, vol. 2626, pp. 304–313. Springer, Heidelberg (2003)
8. Cao, D., Chen, C.: Hot or Not? (2011), `http://www.csee.wvu.edu/~gidoretto/courses/2011-fall-cp/assignments/final_project/results/cao_chen/index.html` (Online; accessed July 20, 2014)
9. Shin, H.C., Park, J.H., Kim, S.D.: Combination of warping robust elastic graph matching and kernel-based projection discriminant analysis for face recognition. IEEE Transaction on Multimedia 9(6), 1125–1136 (2007)
10. Etemad, K., Chellappa, R.: Discriminant analysis for recognition of human face images. In: Bigün, J., Borgefors, G., Chollet, G. (eds.) AVBPA 1997. LNCS, vol. 1206, pp. 125–142. Springer, Heidelberg (1997), `http://dx.doi.org/10.1007/BFb0015988`
11. Gao, Y., Leung, M., Wang, W., Hui, S.: Fast face identification under varying pose from a single 2-D model view. IEE Proceedings-Vision, Image and Signal Processing 148(4), 248–253 (2001)
12. Gross, R., Matthews, I., Baker, S.: Appearance-based face recognition and light-fields. IEEE Transactions on Pattern Analysis and Machine Intelligence 26(4), 449–465 (2004)

13. Huang, J., Yuen, P.C., Chen, W.S., Lai, J.H.: Choosing parameters of kernel subspace lda for recognition of face images under pose and illumination variations. IEEE Transactions on Systems, Man, and Cybernetics, Part B: Cybernetics 37(4), 847–862 (2007)
14. Ishiyama, R., Hamanaka, M., Sakamoto, S.: An appearance model constructed on 3-D surface for robust face recognition against pose and illumination variations. IEEE Transactions on Systems, Man, and Cybernetics, Part C: Applications and Reviews 35(3), 326–334 (2005)
15. Jiang, D., Hu, Y., Yan, S., Zhang, L., Zhang, H., Gao, W.: Efficient 3D reconstruction for face recognition. Pattern Recognition 38(6), 787–798 (2005)
16. Lawrence, S., Giles, C., Tsoi, A.C., Back, A.: Face recognition: a convolutional neural-network approach. IEEE Transactions on Neural Networks 8(1), 98–113 (1997)
17. Li, S.Z., Jain, A.K.: Handbook of face recognition (2011)
18. Moore, S., Bowden, R.: Local binary patterns for multi-view facial expression recognition. Computer Vision and Image Understanding 115(4), 541–558 (2011), `http://www.sciencedirect.com/science/article/pii/S1077314210002511`
19. Ojala, T., Pietikainen, M., Maenpaa, T.: Multiresolution gray-scale and rotation invariant texture classification with local binary patterns. IEEE Transactions on Pattern Analysis and Machine Intelligence 24(7), 971–987 (2002)
20. Ojala, T., Pietikäinen, M., Harwood, D.: A comparative study of texture measures with classification based on featured distributions. Pattern Recognition 29(1), 51–59 (1996), `http://www.sciencedirect.com/science/article/pii/0031320395000674`
21. Phillips, P.J., Moon, H., Rizvi, S.A., Rauss, P.J.: The feret evaluation methodology for face-recognition algorithms. IEEE Transactions on Pattern Analysis and Machine Intelligence 22(10), 1090–1104 (2000)
22. Wagner, P.: Aligning face images (2012), `http://www.bytefish.de/blog/aligning_face_images/` (Online; accessed July 12, 2014)
23. Zhang, X., Gao, Y.: Face recognition across pose: A review. Pattern Recognition 42(11), 2876–2896 (2009)

The Bioinspired Algorithm of Electronic Computing Equipment Schemes Elements Placement

V.V. Kureichik and D.V. Zaruba

Southern Federal University, Rostov-on-Don, Russia
vkur@sfedu.ru, daria.zaruba@gmail.com

Abstract. One of the important design problems - the problem of ECE schemes elements placement is considered in this article. It belongs to the class of NP-hard problems. Statement of a placement problem is made; the complex criterion considering boundary conditions and restrictions is entered. The modified hybrid architecture of the bioinspired search using multilevel evolution and migration mechanism is offered. The genetic and evolutionary algorithms allowing receiving sets of quasi-optimum decisions, for polynomial time are developed. The program environment is created and computing experiment is made. The series of tests and experiments have allowed specifying theoretical estimations of placement algorithms running time and their behavior for schemes of various structures. The best running time of algorithms is O (n log n), the worst one is $O(n^3)$.

Keywords: ECE, Design, Elements placement, Optimization, Bioinspired search, Genetic algorithm, Evolutionary algorithm.

1 Introduction

The intellectual CAD systems construction is connected with software development. It includes mathematical models, methods and intellectual design algorithms. CAD systems are considered to be the basic part of intellectual CAD systems. The basic stages here are the following: blocks configuration, elements placement, connections routing and schemes verification [1]. The placement stage defines efficiency and quality of the design being created.

The problem of ECE designs typical elements (DTE) placement can be considered as a purposefulness of actions of acceptance of design decisions resulting in realization of the projected object scheme set of the given degree of detailed elaboration.

The placement problem is NP-hard and full, therefore there is a necessity for working out of the heuristic algorithms inspired by natural systems, allowing receiving sets of quasi-optimal decisions for polynomial time. The bioinspired search of optimum decisions of a problem of ECE schemes elements placement is offered in this paper.

R. Silhavy et al. (eds.), *Artificial Intelligence Perspectives and Applications*,
Advances in Intelligent Systems and Computing 347, DOI: 10.1007/978-3-319-18476-0_6

2 Problem Definition

In general a kind of a placement problem is possible to be presented informally as follows. The set of the elements which are in the relation of connectivity, according to the essential electric scheme of object being created is given. It is required to place elements in switching space so that the set criterion function reached local or optimum value.

The main objective of placement algorithms is to minimize a total area of switching space where elements are placed, to minimize the general total length of all circuits and to minimize the length of critical communications. Often also the problem (or restriction) is put to minimize the general number of in-circuit crossings of connecting wires [1, 2, 3].

The model of a plane for a placement problem is shown on fig.1. The Cartesian system of co-ordinates with axes s and t is overlapped on an initial plane. The element of the switching scheme can be placed in each cell of the plane. The distance between scheme elements is calculated on the basis of one of known formulas [3, 4].

$$d_{ij} = |s_i - s_j| + |t_i - t_j|, \tag{1}$$

$$d_{ij} = \sqrt{|s_i - s_j|^2 + |t_i - t_j|^2}, \tag{2}$$

$$d_{ij} = (s_i - s_j)^a + (t_i - t_j)^a, \tag{3}$$

Here (s_i, t_i); (s_j, t_j) are co-ordinates of elements x_i, x_j, d_{ij} is a distance between elements x_i, x_j on the set plane. The expression (1) allows defining Manhattan distance, i.e. distance between two points, defined on vertical and horizontal directions. The horizontal and vertical step (distance) between two nearby laying elements is considered to be equal to 1. Expression (2) allows calculating rectilinear distance between two points. In case of an iterative circuit its length is counted up, as semi perimeter of the rectangle covering its trailer points a =1-4, depending on complexity of the switching scheme. Expression (3) allows considering communications of the maximum length indirectly. Then the total length of all connections (model edges) is defined by the known formula [5]

$$L(G) = \tfrac{1}{2}\textstyle\sum_{i=1}^{n}\sum_{j=1}^{n} d_{ij}c_{ij}/c_{ij} \tag{4}$$

Here L (G) is a general total length of edges of the graph; c_{ij} is a quantity of the edges connecting elements x_i and x_j. The number of edges crossings of the graph model is defined by the formula

$$\Pi(G) = \tfrac{1}{2}\textstyle\sum_{u_{ij}\in U}^{n} \Pi(u_{ij})/c_{ij} \tag{5}$$

Here Π (G) is a number of edge crossings of scheme graph model.

$$B(G) = \tfrac{1}{2}\textstyle\sum_{u_{ij}\in U}^{0} b(u_{ij})/o \tag{6}$$

where b ($u_{i,j}$) is a length of critical communication, B (G) is a total length of all critical communications; o is a quantity of critical communications, o - m, and m is a quantity of all communications of the switching scheme.

$$C(G) = \frac{1}{2}\sum_{u_{ij}}^{q} q(u_{ij})/c_{ij} \tag{7}$$

where q ($u_{i,j}$) is a number of bends of one connection, C(G) is a total number of all bends of connections.

At designing of ECE schemes the algorithms minimizing criteria (4-7) development is necessary. Let us notice, that at elements placement the minimization of wires total length usually results in minimization of number of in-circuit crossings, numbers of bends and total length of critical communications. Such simultaneous minimization of criteria (4-7) occurs to some limit starting from which the length reduction results in number of bends and crossings increasing. Therefore let us enter complex criterion A (G).

$$A(G) = \alpha L(G) + \beta \Pi(G) + \gamma B(G) + \delta C(G). \tag{8}$$

Here α, β, γ, δ are the factors considering degree of importance of this or that criterion ($\alpha+\beta+\gamma+\delta = 1$). Values α, β, γ and δ can be defined on the basis of experience of the designer or expert estimations.

Sets of designs typical elements (DTE) which are represented by tops of graph or hypergraph model $X = \{x_1, x_2..., x_n\}$; a set of signal circuits represented by hypergraph edges $E = \{e_1, e_2..., e_m\}$; a set of positions $Q = \{q_1, q_2..., q_f\}$ will be initial data in a placement problem.

The problem of elements placement is considered to be an optimization problem, therefore it is necessary to define objective function (criterion) of restriction and boundary conditions.

Let us formulate a placement problem as appointment of each element in a unique position (cell) of a switching field so that to optimize objective function (OF). As OF a value of expression (8) is chosen. Positions of appointment of each DTE will be target data. As restrictions the following conditions get out. For every $x_i \in X$ one position $q_i \in Q$ is appointed only. Accordingly for each position $q_i \in Q$ one element $x_i \in H$ corresponds at least. Limit condition is expression $|X| \leq |Q|$ if elements placement on the set positions is made. Sometimes instead of DTE placement the signal circuits' one is made, in this case the following condition is boundary – the quantity of positions should be ≥ of quantities of placed circuits trailer points [1, 2, 3, 4, 5].

3 The Description of Search Architecture

The basic difficulty of the decision of elements placement problem with a considerable quantity of local optimum is preliminary convergence of algorithms [1, 2, 3, 4, 5]. In other words, decision hits in one local optimum which is not the best. For the effective decision of this problem the idea of multilevel evolution where the bioinspired search (BS) is used at two levels (consistently or in parallel) is offered. It will allow reducing quantity of computer resources, search time and will allow receiving optimum and quasi-optimum results for polynomial time. The modified hybrid architecture of the search is presented on fig. 1.

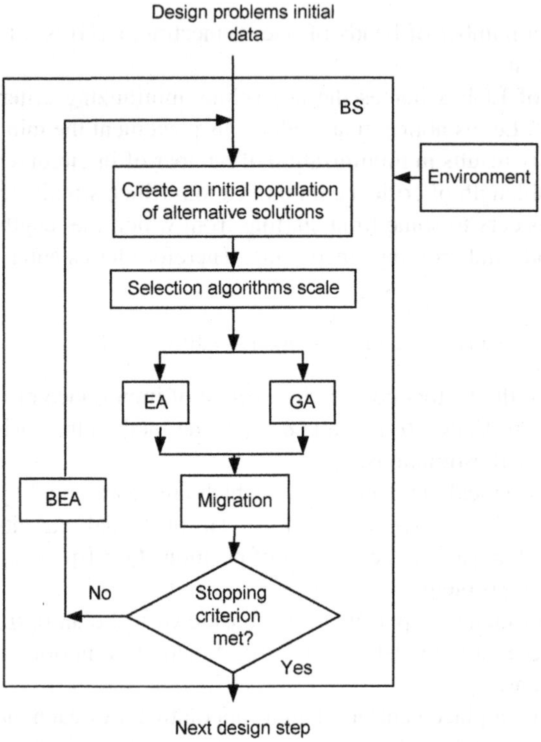

Fig. 1. Hybrid architecture of the bioinspired search

At the consecutive approach it is offered to use the evolutionary algorithm using only one operator of a mutation or its various modifications and allowing to receive quickly sets of quasi-optimum decisions. In case of reception of unsatisfactory placement results the transition through the evolutionary adaptation block to performance of genetic algorithm has to be made.

At the parallel approach population in genetic algorithm (GA) [5, 6] and in evolutionary algorithm (EA) evolve during several generations set independently from each other on certain "islands", and then the exchange of a genetic material in the migration block and independent evolution is made again. In the search scheme the adaptive filter cutting the decisions with low OF value is entered. And bottom OF border is adaptive in sense of dependence on decisions on each generation [7].

The following bioinspired search technology is offered. At first on the basis of design and technological restrictions the area of admissible decisions search is reduced.

A specification, the parameters describing the switching block or the scheme go in an architecture input. On the basis of operating signals of an environment the casual, directed or combined image creates initial population of the alternative decisions which size is the entrance constant parameter set by the user. Search of decisions in casually chosen direction often does not lead to the effective real decision. Thereupon here special stages of a search direction are entered. On the basis of a placement problem statement OF is defined. Further on the basis of OF values the analysis of population and selection of the best alternative decisions is made for the further search. In the considered scheme of the bioinspired search the genetic or evolutionary algorithm is chosen depending on expert system or the person making solution (scale). In genetic algorithm new decisions are formed by realization of various genetic operators (crossover, mutation, inversion, segregation, a translocation, a transposition, removals and inserts) [5, 6, 7]. Further in the block "migration" an exchange of potentially useful genetic material occurs which defines search quality and applied algorithm efficiency. In the following block « Is the termination criterion reached? » the received decision is analyzed. If the decision is optimal, the following step of design comes. Otherwise on the basis of the evolutionary adaptation block a new population of alternative decisions is formed, and search process proceeds iteratively until the reception of effective decisions of a problem of ECE schemes elements placement.

4 Experimental Research

The software product in the object-oriented designing environment Borland C ++ Builder 6.0 is developed. Testing of the developed algorithms was carried out on computer AMD FX (tm)-8121 Eight-Core Processor 3.10 GHz, the RAM 4,00 GB.

Series of experiments for a different set of the test examples differing with quantity of elements in the scheme [8, 9] have been carried. Average results of experiments are given in table 1 and on fig. 2.

Table 1. Dependence of the developed algorithms operating time on the scheme size

	100	1000	2500	5000	7500	10000
EA	1,5 s	56,3 s	122 s	214 s	276 s	341 s
GA	3,6 s	87,7 s	154 s	371 s	531 s	774 s
BS	8,5 s	108,8 s	186 s	401 s	679 s	1021 s

Results of researches allow drawing a conclusion that the developed genetic algorithms running time does not fall outside the limits of polynomial dependence, and can be expressed by the formula: O (n log n) - O (n^3), where n is a number of the scheme elements.

For the developed algorithms efficiency definition the researches of the decision quality on test examples [8, 9] have been carried out. Here, as the objective function is considered expression (4), i.e. the total length of all connections is calculated in conventional units. Results of the researches are presented in table 2 and on fig.3.

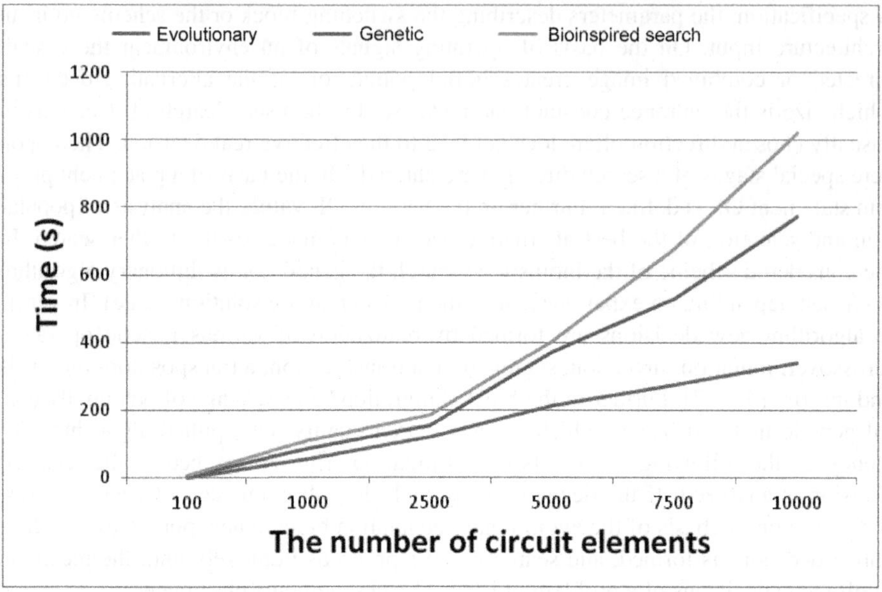

Fig. 2. Schedules of dependence of the developed algorithms operating time on quantity of the scheme elements

Table 2. Definition of algorithms efficiency

	100	1000	2500	5000	7500	10000
EA	50,1	76,3	102,7	124,6	144,9	193,6
GA	60,3	71,2	82,6	104,1	136,7	155,8
BS	58,9	67,4	80,2	95,7	123,1	145,1

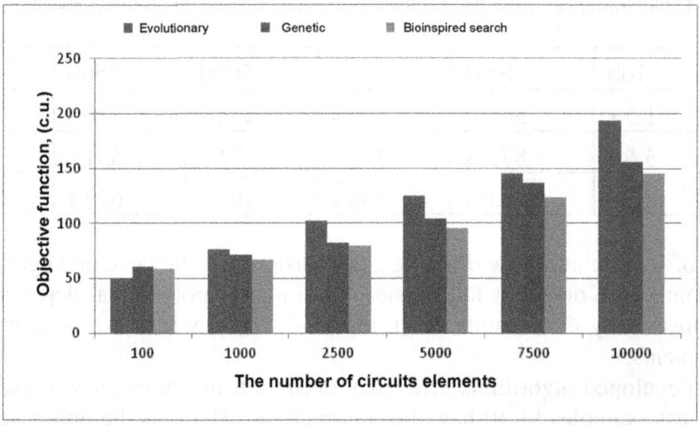

Fig. 3. The histogram of comparison of quality of the decision received by developed algorithms

On the basis of the analysis of the researches for problems of small dimension (to 1000 elements) the evolutionary algorithm is effective. And at the decision of a placement problem in the dimension approached to industrial volumes (more than 5000 elements), the bioinspired search is considered to be effective.

The important stage of research was the analysis on the basis of comparison of the developed algorithm with benchmarks. As benchmarks are used widely known algorithms [10] and calculated the wires lengths in micron. Results of testing are presented in table 3 and on fig. 4.

Table 3. The results of experiments

Circuit	Num. of el.	Capo 8.6, (micron)	Feng Shui 2.0, (micron)	Dragon 2.23, (micron)	BS, (micron)
ibm01	12752	4,97	4,87	4,42	4,28
ibm02	19601	15,23	14,38	13,57	13,05
ibm03	23136	14,06	12,84	12,33	11,96
ibm04	27507	18,13	16,69	15,41	15,58
ibm05	29347	44,73	37,3	36,38	35,96
ibm06	32498	21,96	20,27	20,38	20,52
ibm07	45926	36,06	31,5	29,97	31,82
ibm08	51309	37,89	34,14	32,2	33,8
ibm09	53395	30,28	29,86	28,1	27,46
ibm10	69429	61,25	57,99	57,2	53,04
ibm11	70558	46,45	43,28	40,77	40,66
ibm12	71076	81,55	75,91	71,03	73,17
ibm13	84199	56,47	54,09	50,57	48,37

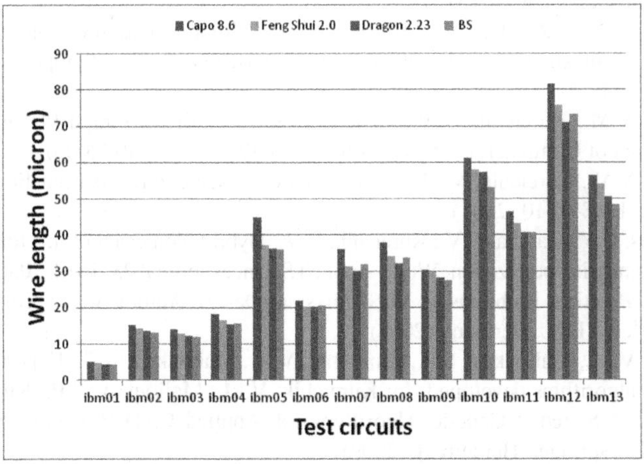

Fig. 4. The histogram of wires lengths comparison at placement of test schemes ibm01-imb13 by various algorithms

5 Conclusion

The modified hybrid architecture of the bioinspired search using multilevel evolution and the mechanism of migration, allowing threading decision process is offered, making an exchange of a genetic material and partially eliminating a problem of algorithms preliminary convergence. Genetic and evolutionary algorithms were developed, allowing receiving sets of quasi-optimum solutions for polynomial time. Basic difference of the offered architecture is division of search process into two stages and application of various algorithms into each of these stages. The program environment in language C ++ has been developed. Computing experiment has been made. During the computing experiment empirical dependences, ranges of change of entrance parameters have been established and a number of recommendations about their optimum choice have been developed. The series of tests and experiments have allowed specifying theoretical estimations of time complexity of designing algorithms and their behavior for schemes of various structures. The best running time of algorithms is ~O (n log n); the worst one is O (n^3).

Acknowledgment. This research is supported by grants of the Ministry of Education and Science of the Russian Federation. The project #8.823.2014.

References

1. Kureichik, V.M., Malioukov, S.P., Kureichik, V.V., Malioukov, A.S.: General questions of automated design and engineering. In: Kureichik, V.M., Malioukov, S.P., Kureichik, V.V., Malioukov, A.S. (eds.) Genetic Algorithms for Applied CAD Problems. SCI, vol. 212, pp. 1–22. Springer, Heidelberg (2009)
2. Alpert, C.J., Dinesh, P.M., Sachin, S.S.: Handbook of Algorithms for Physical design Automation. Auerbach Publications Taylor & Francis Group, USA (2009)
3. Sherwani, N.A.: Algorithms For VLSI Physical Design Automation, 3rd edn. Kluwer Academic Publisher, USA (2013)
4. Zaporozhets, D.U., Zaruba, D.V., Kureichik, V.V.: Representation of solutions in genetic VLSI placement algorithms. In: 2014 East-West Design & Test Symposium (EWDTS), pp. 1–4 (2014)
5. Kureichik, V.M., Kureichik, V.V.: Genetic Algorithm for the Graph Placement. International Journal of Computer and Systems Sciences 39(5), 733–740 (2000)
6. Kureichik, V.V., Kureichik, V.M.: Genetic search-based control. Automation and Remote Control 62, 1698–1710 (2001)
7. Zaporozhets, D.Y., Zaruba, D.V., Kureichik, V.V.: Hybrid bionic algorithms for solving problems of parametric optimization. World Applied Sciences Journal 23, 1032–1036 (2013)
8. Lim, S.K.: Practical Problems in VLSI Physical Design Automation. Springer Science + Business Media B.V., Germany (2008)
9. Kureichik, V.M., Malioukov, S.P., Kureichik, V.V., Malioukov, A.S.: Experimental investigation of algorithms developed. In: Kureichik, V.M., Malioukov, S.P., Kureichik, V.V., Malioukov, A.S. (eds.) Genetic Algorithms for Applied CAD Problems. SCI, vol. 212, pp. 211–223. Springer, Heidelberg (2009)
10. IBM-PLACE 2.0 benchmark suits, `http://er.cs.ucla.edu/benchmarks/ibm-place2/bookshelf/ibm-place2-all-bookshelf-nopad.tar.gz`

Informational System to Support the Design Process of Complex Equipment Based on the Mechanism of Manipulation and Management for Three-Dimensional Objects Models

A.N. Dukkardt, A.A. Lezhebokov, and D.Yu. Zaporozhets

Southern Federal University, Rostov-on-Don, Russia
{a.duckardt,legebokov,elpilasgsm}@gmail.com

Abstract. The paper describes the construction of an information system to support the design process of complex equipment based on mechanisms of manipulation and control three-dimensional models of objects. Problem creating interactive three-dimensional models of objects with reference to specific points on the marker image is also placed in the topic. The authors analyze the existing and proposed new ways and mechanisms of manipulation and control of three-dimensional models of objects based on augmented reality technology. The part of current research is developing of algorithms that demonstrate how to work with the information system. The authors propose the architecture of an information system adapted for solving problems of support decision-making processes in computer-aided design. Information system to support the design process of complex packaging equipment in the form of applications for mobile devices based on augmented reality technology was developed. The effectiveness of the approach was confirmed by the appropriate computational experiments.

Keywords: Three-dimensional simulation, Informational systems, Design, Augmented reality, Mobile application.

1 Introduction

Nowadays, a constructor, as well as a designer, uses the various methods for visualizing the information analysis process and reporting the results. The most common of them are the following: the creation of a full-scale layout, creating a three-dimensional model through the use of specialized software, preparation of illustrations or video of the object, prototyping the 3D-printer, the use of augmented reality technology [1-2].

Creation of a layout is a time consuming way of visualization and requires a lot of overhead. The reason is that the layout should be a large-scale and detailed image of the object of design, in other words, the future of the real object of construction or production. The layout construction may contain inaccuracies or be quite fragile, that makes its transportation for demonstration in scope of the exhibition or presentation

R. Silhavy et al. (eds.), *Artificial Intelligence Perspectives and Applications*,
Advances in Intelligent Systems and Computing 347, DOI: 10.1007/978-3-319-18476-0_7

as a big problem. Disadvantages of the method of prototyping are a lack of operational changes, scaling and replication.

Three-dimensional visualization method allows solving some of the problems of modeling. Three-dimensional model with a given degree of detail is developed by the special software. Special software or pre-prepared visual image of the model as a set of images or video in the format is used for the demonstration model. The disadvantage of this solution is that the user needs the special software that cannot be efficiently distributed, as well as to track change versions and the individual characteristics of the user's environment. Result of model visualization in this approach shows only a one-sided view of the model. Footage requires detailed study of the script and also makes it impossible to complete a comprehensive study or analysis of the object [2-3].

Printing method of three-dimensional model of the 3D-printer allows you to bypass some of the limitations of the computer analysis of the test or the designed object. The accuracy of the printed model will be limited to the technical capabilities of a particular printer model. Production of printed copies model is an expensive process, the cost of which depends on the size of the model, the material used and equipment. Print dynamic elements such as moving parts, requires a high-end equipment accuracy and a high enough level of detail of three-dimensional model [3].

Thus, the actual task is to develop new models, methods, and approaches to manipulate and control objects in three-dimensional models of intelligent information systems to support decision-making processes in the design and management.

2 New Methods and Mechanisms of Manipulation and Control Three-Dimensional Models of Objects

Augmented reality is a technology that enables you to impose a three-dimensional computer graphics, animation, videos, or text information on a real-time objects. In contrast, virtual reality, AR-interface allows the user to see the real world and the virtual objects are embedded manipulated in real time [4-5].

Augmented reality allows us to construct and visualize the three-dimensional model of any complexity via mobile camera devices. In this case, the binding of the model using control points to a prepared image marker. Due to the capabilities of the technology decision maker is the ability to view, manipulate and manage three-dimensional models. Used a three-dimensional model can be animated and include various interactive elements, which opens up new possibilities to manipulate and control the object being studied. Augmented reality allows you to view a virtual object against a background of real world in real time. The company IKEA uses augmented reality in the product catalog to inspire their potential customers, designers and architects. Three-dimensional visualization of objects makes it possible to view them in the real world of the future interior [5-6].

The use of augmented reality technology is a topical approach to the problem of visualization of the design object. The technology allows you to create new methods of manipulation and control three-dimensional models of objects with the help of special markers. The user interacts with the information system, manages to find the camera lens and image recognition marker. At this stage, a precise positioning of the marker in the space. To a certain position marker is bound three-dimensional model of the proposed facility.

The model is created in a specialized program editor on the conceptual image or design drawings. Three-dimensional model has animations or perform physical simulation. For maximum realism obeying rules or laws of physics animation. For the description of interface and control interactive elements are used interactive rules defined in the original terms of reference.

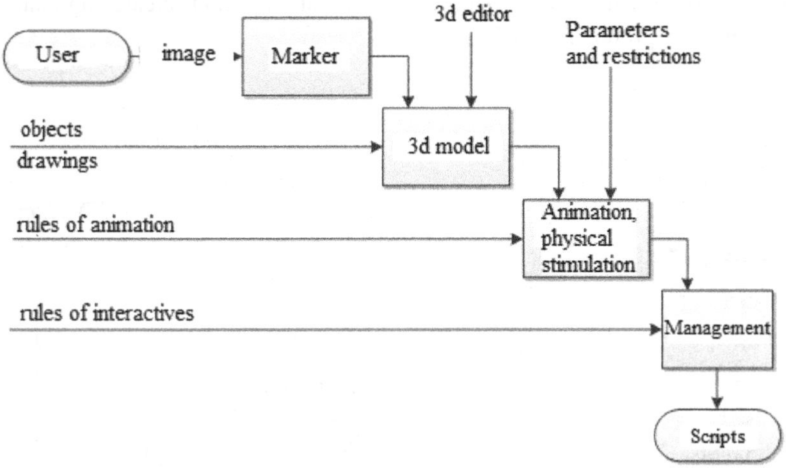

Fig. 1. Functions of information system

The result of the decomposition of the structure and definition of the functions of information system is a set of requirements for the control script, shown in an object-oriented high-level programming language. Obtained at this stage of the source code can be used in various systems design and programming, for descriptions of methods of manipulation and management.

3 Information System Architecture

Let us consider the information system architecture decisions shown in Fig. 2. The architecture includes two databases (DB): base of objects and primitives, and database of finished projects. Database of objects and primitives contains all ever created by designers three-dimensional models, which may be reused in new projects.

Projects database is responsible for storing collections of data, such as the position of objects in space, a set of scripts, rules, animation, process logic control and manipulation, additional elements (such as lights, sound and video elements, textures).

Databases interact with the kernel information system on the fly kernel can change the data. For example, moving an object in space, i.e. change the project database, but the database is not primitive.

3D-editor is an external tool for creating and working with three-dimensional models. Call the editor comes from the control unit primitives, after saving for each primitive creates a new version. This unit, as the name implies, manages and setting up a model created in the editor or database objects and primitives.

The model is validated according to predefined rules specific project in the unit of analysis and statistics. If there are any inaccuracies, defects or desired changes, the model will sent back to the control unit and finalized in an external editor. As models of control proceeds to block programming animation, which creates animation settings for one or a group of models.

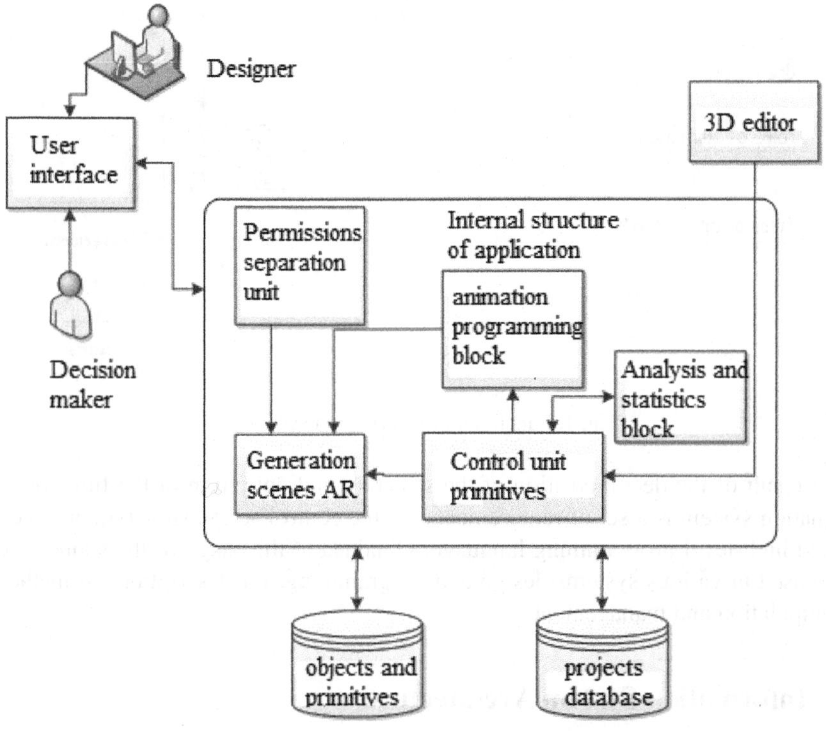

Fig. 2. Architecture of information system

All models are prepared and their animation settings are sent to the scene generation unit augmented reality, as well as added data from the database project if the system should support decision-making in several subject areas. At the scene generation unit augmented reality affects separation unit access rights determining the level of access to the primitives, animation or individual projects.

Designer develops the user interface, in other words, determines its appearance and functionality. User interface itself is the outer shell of the application that receives data from the indoor unit (core) of information system.

The decision maker is the primary user-consumer who interacts with the application via the user interface. It uses its functionality and sends control commands to the core information system.

4 Users' Operations in the Information System

Let us consider the algorithms for the two main groups of users who have access to functional information system. The developer's algorithm shows the process of creating a library of three-dimensional primitives based on the terms of reference. After you create each entity approved by the customer (or the decision maker) or sent to the correction and revision. Animation and interaction between primitives created and configured. Further, the process of compilation of scenes decision-making for specific technical information system project. At this stage, there is an association of primitives in the model, the integration of the markers with the models, the description of the interface and add interactive elements such as menus, event handlers. After that there is a check on the correctness of the rules work, ie the search for errors and optional decision maker makes a small change. If everything works according to the original job, and no changes are needed, a mobile application is generated for selected mobile platform.

The algorithm works by a linear structure. The first printed marker or other document containing the image marker. After starting the application, the user gets access to the full or limited current privilege level functionality of the information system. The principle of operation augmented reality is to build a camera lens image marker. After the desired location marker in space and implementation of procedures to manipulate the user performs a task analysis of design decisions. The result of solving design problems determines the need for repetitive actions with three-dimensional models.

5 Effectiveness of the Practical Implementation

New models, algorithms and architecture are the result of the work. All of these elements are implemented in the project to build an information system to support the development of complex packaging equipment. Let us consider in detail the process of developing a mobile app with augmented reality as an example demonstrating the packing equipment. According to the source drawing (Figure 3) Evaluation is the appearance of the equipment determined by its constituent components and dimensions. Used for programming animation videos, recording the work process of an actual sample of the equipment.

The process of developing a three-dimensional model begins with a rapid prototyping model partitioning the moving parts [3, 7]. Next, the developed model is smoothed (Figure 4) and imported into the programming environment and animation settings. Programming scene augmented reality is to write scripts that describe the behavior of each individual part of the model.

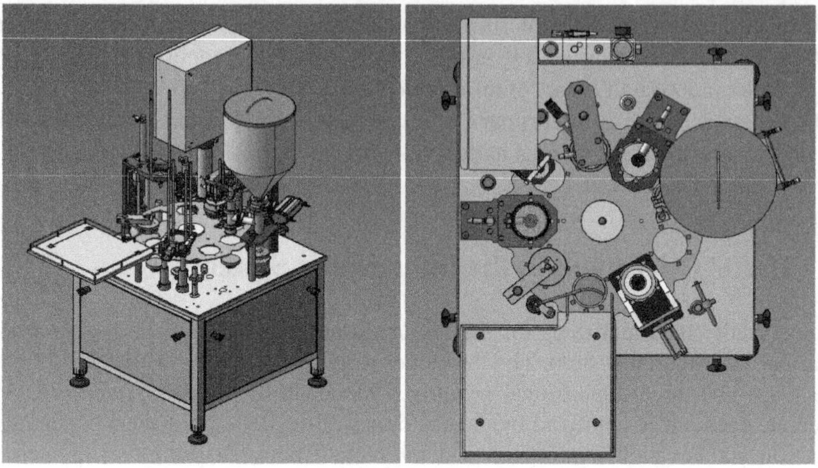

Fig. 3. Drawing of packaging equipment

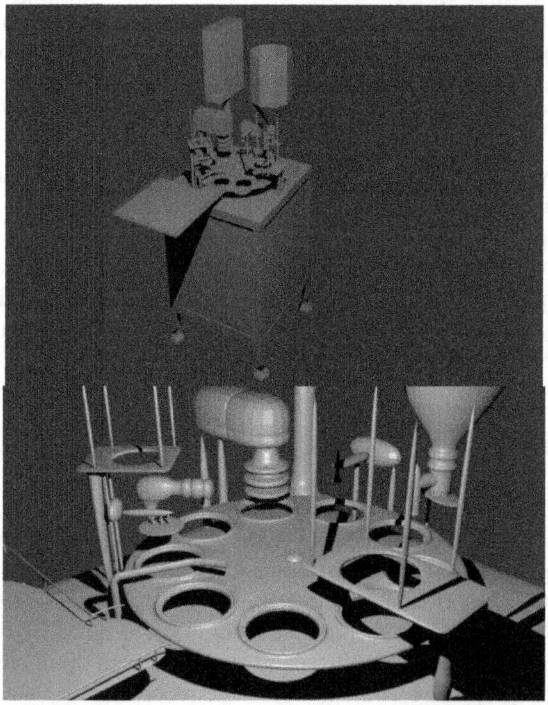

Fig. 4. Three-dimensional model of equipment

After programming the animation scene, there is export the project in the format of mobile applications for Android or iOS. The mobile application is not demanding to resources and works on most devices that are running the most common version of the mobile operating systems [10] (Figure 5).

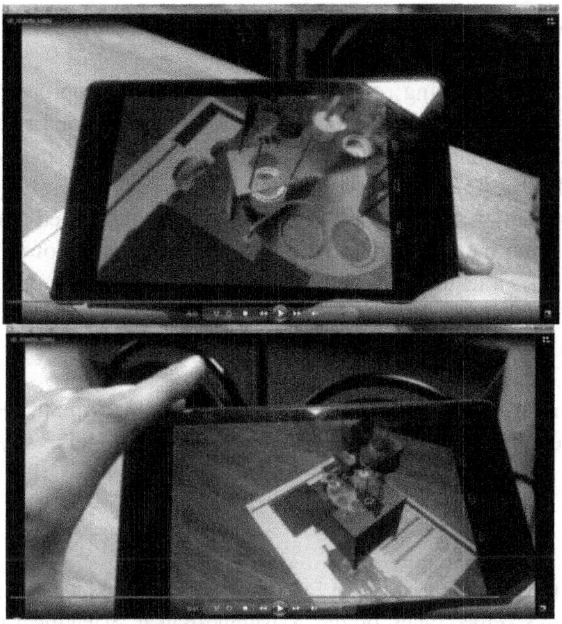

Fig. 5. Mobile application with augmented reality

The application of intelligent information system with augmented reality has reduced engineering costs by 3-5%. Reducing costs, primarily due to an increase of efficiency of processes of visualization, the presence of new management tools and manipulate three-dimensional models of objects for quick and rapid acceptance of design solutions [7-9].

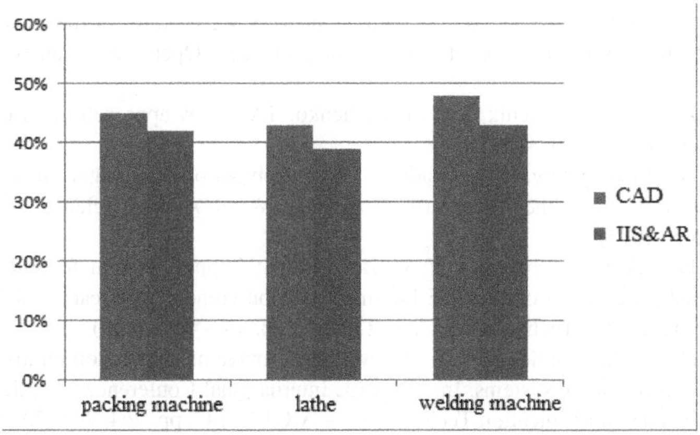

Fig. 6. Designing costs

6 Conclusion

Authors obtained the following results of the research: the approach to building information model, the decomposition of functions, algorithms and rules of operation, system architecture, based on the mechanisms of manipulation and control of three-dimensional models of objects. Proposed mechanisms can successfully solve tasks in the design of information systems through the usage of augmented reality technology based on markers and mechanisms of manipulation and control of three-dimensional models of objects. As a result, the practical implementation of the developed software and information platform for creating mobile applications that demonstrate examples of solving design problems.

Acknowledgment. The study was performed by the grant from the Russian Science Foundation (project # 14-11-00242) in the Southern Federal University.

References

1. Zaporozhets, D.Y., Kravchenko, Y.A., Lezhebokov, A.A.: Methods of data mining in complex systems. Proceedings of the Kabardino-Balkar Scientific Centre of Russian Academy of Sciences 3, 52–57 (2012)
2. Lezhebokov, A.A., Shkalenko, B.I.: Augmented Reality and "markers" in the course "Information Technology". In: Proceedings of the Congress on Intelligent Systems and Information Technology - IT 2013, pp. 353–358. FIZMATLIT, Moscow (2013)
3. Lezhebokov, A.A., Bova, V.V., Shugushkhov, K.H.M.: Tools and technologies to support virtual prototyping processes aided design. Proceedings of the Kabardino-Balkar Scientific Centre of Russian Academy of Sciences 5, 38–43 (2013)
4. Lezhebokov, A.A., Gladkov, L.A.: Automated workplace instructor with intellectual support. Software Products and Systems 4, 12 (2005)
5. Lezhebokov, A.A., Kravchenko, Y.A., Pashchenko, S.V.: Features of the use of augmented reality technology to support educational processes. Open Education 3(104), 49–54 (2014)
6. Lezhebokov, A.A., Kureichik, V.V., Pashchenko, S.V.: New approach to virtual learning. Open Education 3(104), 4–9 (2014)
7. Bova, V.V., Lezhebokov, A.A., Gladkov, L.A.: Problem-oriented algorithms of solutions search based on the methods of swarm intelligence. World Applied Sciences Journal 27(9), 1201–1205 (2013)
8. Lezhebokov, A.A., Kravchenko, Y.A., Bova, V.V.: Support system for QR-code-based educational processes. In: 8th IEEE International Conference "Application of Information and Communication Technologies – AICT 2014", pp. 482–485 (2014)
9. Kravchenko, Y.A., Kureichik, V.V.: Knowledge management based on multi-agent simulation in informational systems. In: 8th IEEE International Conference "Application of Information and Communication Technologies – AICT 2014", pp. 264–267 (2014)
10. Rodzina, L., Kristoffersen, S.: Context-dependent car navigation as kind of human-machine collaborative interaction. In: Proc. of the International Conference on Collaboration Technologies and Systems, San-Diego, pp. 253–259 (2013)

Combined Method of Analyzing Anaphoric Pronouns and Inter-sentential Relationships between Transitive Verbs for Enhancing Pairs of Sentences Summarization

Trung Tran and Dang Tuan Nguyen

Faculty of Computer Science, University of Information Technology,
Vietnam National University - Ho Chi Minh City, Vietnam
ttrung@nlke-group.net,
dangnt@uit.edu.vn

Abstract. The main content of this paper is to present a solution based on the approach which we proposed in previous researches for summarizing the meaning of pairs of simple sentences having general characteristics: (a) both of two sentences have the transitive verb of action; (b) the second sentence has only one anaphoric pronoun taking the object role of the transitive verb of action. The general approach in our researches is to generate a new reducing sentence having the content which summarizes the meaning of the original pair of sentences. However, to be suitable for pairs of sentences having above characteristics and the new reducing sentence have to satisfy the naturalism for the cognition of native speakers, our approach in this research is based on the main idea: (i) resolve the anaphoric pronoun and identify that the antecedent is human or thing; (iii) analyze inter-sentential relationships of transitive verb of action; (iii) propose the new generating algorithm. These points help for better understanding the meaning of the original pair of sentences and generating the new reducing sentence. One more important point is the proposed solution in this research can be applied for many different natural languages.

Keywords: Anaphoric Pronoun, Discourse Representation, Sentence Generation, Inter-sentential Action Relationship, Meaning Summarization.

1 Introduction

To summarize the meaning of a given document, there were a lot of researches [5], [7], [8], [9], [11], [12], [15]. Base on analyzing the summarizing results, D. Shen et al. [15] classified approaches into two different dimensions: (1) "abstract-based" when inside the content of the result there are lexicons and constituents which are not contained in the original paragraph; (2) "extract-based" when the content of the result contains sentences holding the main content of the original paragraph.

Another main research field is to generate a new information form which can be understandable for human from original information contents which were gathered from many different sources. In traditional researches (Cf. [13], [14]), authors presented models, techniques which help for generating information contents with

R. Silhavy et al. (eds.), *Artificial Intelligence Perspectives and Applications*,
Advances in Intelligent Systems and Computing 347, DOI: 10.1007/978-3-319-18476-0_8

different purposes. The common features of these researches: (i) the input is information gathered from different sources, which are not complete paragraph; (ii) depending on the objective, the output is represented in the form which can be understandable for user, and often are not complete paragraph; (iii) the set of lexicons which is used for generating the output is independent with lexicons in the input.

Combining the idea about generating new information content and summarizing paragraph in "abstract-based" approach, as an initial research, in [17] we proposed a model consisting of methods and techniques for "generating the new reducing Vietnamese sentence (NRVS)" for "summarizing the meaning of the source pair of simple Vietnamese sentences". The main approach consists of two processing threads:

- Apply the method, techniques in [16] to resolve anaphoric pronouns and build the structure DRS – Discourse Representation Structure (Cf. [1], [3], [4], [10]) as the mechanism which help for understanding and representing the semantic of the original pair of simple Vietnamese sentences;
- Establish the mechanism for generating the NRVS consisting of three steps: (i) determine the main content in the source pair of Vietnamese sentences based on analyzing the structure DRS; (ii) generate the syntactic structure of the NRVS based on analyzing inter-sentential relationships; (iii) complete the NRVS.

With the new approach, we studied four types of Vietnamese paragraphs in which each paragraph composes two simple sentences. These paragraphs have the general characteristics: the first sentence has a transitive verb which relates to two human objects; the second sentence has a pronoun indicating person, standing with demonstrative adjective ["ta" / "ấy" / "này"]. After finding the antecedent for each anaphoric pronoun, we visualized the relationship between two sentences by a diagram according to the main predicates in the DRS structure of this pair.

Follow the idea about sentence-generation approach of summarization in [17], the main content in this paper is presenting a solution for satisfying three following requirements:

- Summarize pairs of simple sentences having the common characteristic: (a) both two sentences have a transitive verb of action; and (b) the second sentence has only one anaphoric pronoun taking the object role of transitive verb of action.
- The generated NRVS has the naturalism for the cognition of native speakers.
- Can be applied for many different natural languages.

Table 1. Characteristics of pairs of sentences which are considered in this research

Form	Characteristics
1	• The first sentence has two objects: the human object takes the subject role of the action; the thing object takes the object role of the action. • The second sentence has one pronoun indicating human, standing alone and taking the object role of the action. **Example 1**: *Nhân học toán. Lễ dạy em.* (English: Nhân learns math. Lễ teaches him.)

Table 1. (*continued*)

2	• The first sentence has two objects: the thing object takes the subject role of the action; the human object takes the object role of the action. • The second sentence has one pronoun "nó" taking the object role of the action. **Example 2**: *Con chó chơi với Nhân. Lễ nhìn nó.* (English: The dog plays with Nhân. Lễ looks at it.)
3	• The first sentence has two objects: the human object takes the subject role of the action; the thing object takes the object role of the action. • The second sentence has one pronoun "nó" taking the object role of the action. **Example 3**: *Nhân mua máy tính. Lễ mượn nó.* (English: Nhân buys a calculator. Lễ borrows it.)
4	• The first sentence has two objects: one human object takes the subject role of the action; one human object takes the object role of the action. • The second sentence has one pronoun indicating human, standing with demonstrative adjective ["ta" / "ấy" / "này"] and taking the object role of the action. **Example 4**: *Nhân gặp Lễ. Nghĩa giới thiệu anh ta.* (English: Nhân meets Lễ. Nghĩa introduces him.)

According to four forms of pair of Vietnamese sentences in Table 1, in this research, we generate the NRVS which satisfies the above second requirement by applying the Vietnamese linguistic foundations: sentence in passive voice; and relative clause. Besides, our solution in this research based on these foundations also can be applied to summarize pairs of sentences in several other languages.

2 The Solution for Sentence Pair Summarization

In this section, we present our solution to summarize the meaning of pairs of sentences in Table 1 with four main phases:

- First: improve the solution of resolving inter-sentential anaphoric pronouns for pairs of simple sentences and building structure DRS presented in [16].
- Second: analyze the structure DRS to determine some inter-sentential relationships.
- Third: propose new algorithm for generating the syntactic structure of the NRVS.
- Forth: combine the syntactic structure with the lexicon set to complete the NRVS.

2.1 Improving Inter-sentential Anaphoric Pronouns Resolution

As presented in [16], the solution of resolving anaphoric pronouns for pair of simple Vietnamese sentences is performed through main steps: (i) analyze the structure of the paragraph into two separate sentences; (ii) analyze the syntactic structure of each sentence and describe appropriate grammatical characteristics for each constituent; (iii) describe appropriate grammatical characteristics for each lexicon and build the DRS structure based on these characteristics; (iv) determine the antecedent for each anaphoric pronoun based on appropriate finding strategy.

In this research, we still keep processing steps in the solution [16], only perform some improvements with following meanings:

- We do not classify into groups as in [16] instead integrate general processing, therefore there will be adjustments about finding antecedent strategies as well as appropriate finding algorithm.

- Add some appropriate information into information representation predicate of lexicons in the step describing grammatical characteristics of lexicons: (i) category: `[noun]` – indicate noun / `[action]` – indicate action; (ii) sub category: `[proper]` – indicate proper noun / `[common]` – indicate common noun / `[transitive]` – indicate transitive verb of action; (iii) species: `[human]` – indicate human object / `[thing]` – indicate thing in general (animate or non-animate object).
  - ⇨ Consider proper noun "Nhân" in Example 1, the information representation predicate of proper noun "Nhân" in the structure DRS: `named(I, [nhân], [human], [noun], [proper])`.
  - ⇨ Consider transitive verb of action "nhìn" (English: look at) in Example 2, the information representation predicate of transitive verb of action "nhìn" in the structure DRS: `nhìn(Arg1, Arg2, [nhìn], [action], [transitive])`.

The finding antecedent strategy:

- For anaphoric pronoun indicating human and standing alone: Determine characteristics of the antecedent: (i) Appear at the first sentence; (ii) Is human object; (iii) Take the subject role of action.
- For anaphoric pronoun indicating human and standing with demonstrative adjective ["ta" / "ấy" / "này"]: Determine characteristics of the antecedent: (i) Appear at the first sentence; (ii) Is human object; (iii) Take the object role of action.
- For anaphoric pronoun "nó": Determine characteristics of the antecedent: (i) Appear at the first sentence; (ii) Is thing object.

```
While (index I is in list U of DRS) Do
   While (predicate associated with I is in list Con of DRS) Do
      If (Pronoun standing alone) Then
         If ((position(I) is [first]) And (species(I) is [human]) And
            (role(I) is [subject])) Then
               Index of the antecedent = I;
         End If
      Else If (Pronoun standing with "ấy" / "ta" / "này") Then
         If ((position(I) is [first]) And (species(I) is [human]) And
            (role(I) is [object])) Then
               Index of the antecedent = I;
         End If
      Else If (Pronoun "nó") Then
         If ((position(I) is [first]) And (species(I) is [thing])) Then
            Index of the antecedent = I;
         End If
      End If
   End While
End While
```

Fig. 1. The algorithm of finding antecedent

2.2 Analyze the DRS Structure and Determine Relationships

In [17] and in this research, we determine following components in the structure DRS showing the main content of the source pair of sentences:

- List U contains unique indexes which associate with objects.
- Main predicates which represent information in list Con.

We analyze the DRS structure for determining some inter-sentential relationships:

- Inter-sentential anaphoric pronoun relationship.
- Relationship of actions: between the action at the first sentence and the action at the second sentence. In this research, we consider two actions are performed independently and in parallel, and analyze relationships with the same object to determine the performing context in the NRVS.

As an example, consider the pair of sentences in Example 3, the DRS structure with main predicates of this pair of sentences:

```
[1,2,3]
-------------------------------------------------------
named(1,[nhân],[human],[noun],[proper])
máy_tính(2,[máy,tính],[thing],[noun],[common])
mua(1,2,[mua],[action],[transitive])
named(3,[lễ],[human],[noun],[proper])
mượn(3,2,[mượn],[action],[transitive])
```

Fig. 2. The DRS structure with main predicates of pair of sentences: "Nhân mua máy tính. Lễ mượn nó."

The inter-sentential relationships according to this DRS structure as follows:

- Inter-sentential anaphoric pronoun: object 1 has relationship with both first and second action.
- Inter-sentential relationship of transitive verb of action: the first action – the second action. In which: object 1 has the subject role relationship with the first action, simultaneously has the object role relationship with the second action.
  - ⇨ Two actions are performed independently and in parallel.
  - ⇨ Active context for the first action.
  - ⇨ Passive context for the second action.

2.3 Generate the Syntactic Structure of the NRVS

To generate the syntactic structure of the NRVS, we base on the DRS structure with main predicates and determination of inter-sentential relationships, propose the new algorithm with the main idea consisting of following points:

- Combine the active context for the first action and the passive context for the second action. In Vietnamese, with the passive context, apply structure "được….bởi". In English, with the passive context, apply structure "is….by".
- Apply structure "relative clause" for object 2: In Vietnamese, use only lexicon "mà" taking the pronoun role; in English, use relative pronoun "who" for human, use relative pronoun "which" for thing.

```
Add [predicate object 1] into structure;
Add [first predicate action] into structure;
Add [predicate object 2] into structure;
If (object 1 has the object relationship with second action) Then
   Add "và" (Vietnamese) / "and" (English) into structure;
Else If (object 2 has the object relationship with second action) Then
   If (object 2 is human) Then
      Add "mà" (Vietnamese) / "who" (English) into structure;
   Else If (object 2 is thing) Then
      Add "mà" (Vietnamese) / "which" (English) into structure;
   End If
End If
Add "được" (Vietnamese) / "is" (English - passive voice) into structure;
Add [second predicate action] into syntactic structure;
Add "bởi" (Vietnamese) / "by" (English - passive voice) into structure;
Add [predicate object 3] into syntactic structure;
```

Fig. 3. The general algorithm for generating the syntactic structure of the NRVS for four forms of pair of sentences in Table 1

Implementing the above algorithm, we have the syntactic structure of the NRVS according to the DRS structure in Figure 2: `named(1) + mua(1,2) + máy_tính(2) + "mà" (which) + "được" (is) + mượn(3,2) + "bởi" (by) + named(3)`

2.4 Complete the NRVS

In this research, we perform an improvement in comparison with [17] when building the set of lexicons with the main idea: define only one lexical class, in which attributes taking values are corresponding to values of information in information representation predicate of lexicons. Class Lexicon is defined with attributes as follows: (i) `semantic` → take the value is corresponding to name of information representation predicate of lexicon; (ii) `arg1` → take the value is corresponding to index I in information representation predicate of noun or Arg1 in information representation predicate of transitive verb of action; (iii) `arg2` → take the value is corresponding to information Arg2 in information representation predicate of transitive verb of action; (iv) `content` → take the value is corresponding to information content in information representation predicate of transitive verb of lexicon; (v) `species` → take the value is corresponding to information species in information representation predicate of transitive verb of noun; (vi) `category` → take the value is corresponding to information category in information representation predicate of transitive verb of lexicon; (vii) `class` → take the value is corresponding to information sub-category in information representation predicate of transitive verb of lexicon; (viii) `morphology` → take the value is an actual form of lexicon.

⇨ Consider information representation predicate of proper noun "Nhân" in section 2.1, lexical object npNhan is defined in the set of lexicons with attributes and values: `{semantic → [named]}; {arg1 → I}; {arg2}; {content → [nhân]}; {species → [human]}; {category → [noun]}; {class → [proper]}; {morphology → [Nhân]}.`

⇨ Consider information representation predicate of transitive verb of action "nhìn" (English: look at) in section 2.1, lexical object atNhin is defined in the set of lexicons with attributes and values: `{semantic → [nhìn]}; {arg1 → Arg1}; {arg2 → Arg2}; {content → [nhìn]}; {species}; {category → [action]}; {class → [transitive]}; {morphology → [nhìn]}.`

We combine the set of lexicons with the syntactic structure of the NRVS with following general algorithm:

```
For (element i in syntactic structure) Do
   If (element i is predicate) Then
      For (lexical object j in set of lexicons) Do
         If (properties of lexical object j is corresponding with infor-
mation in element i) Then
            Replace (element i) by (property morphology of lexical ob-
ject j);
         End If
      End For
   Else If (element i is linking word)
      Keep (element i);
   End If
End For
```

Fig. 4. The general algorithm for combining the set of lexicons with the syntactic structure of the NRVS

In Table 2, we present the result when performing the algorithm of combining the set of lexicons and the syntactic structure to complete the NRVS:

Table 2. Complete the NRVS

Form	The complete NRVS
1	The syntactic structure of the new reducing sentence when using morphology of lexicons: "Nhân" + "học" + "toán" + "và" + "được" + "dạy" + "bởi" + "Lễ" ⇨ The complete new reducing sentence: *Nhân học toán và được dạy bởi Lễ.* (English: Nhân learns math and is taught by Lễ.)

Table 2. (*continued*)

2	The syntactic structure of the new reducing sentence when using morphology of lexicons: "con chó" + "chơi với" + "Nhân" + "và" + "được" + "nhìn" + "bởi" + "Lễ" ⇨ The complete new reducing sentence: *Con chó chơi với Nhân và được nhìn bởi Lễ.* (English: The dog plays with Nhân and is seen by Lễ.)
3	The syntactic structure of the new reducing sentence when using morphology of lexicons: "Nhân" + "mua" + "máy tính" + "mà" + "được" + "mượn" + "bởi" + "Lễ" ⇨ The complete new reducing sentence: *Nhân mua máy tính mà được mượn bởi Lễ.* (English: Nhân buys the computer which is borrowed by Lễ.)
4	The syntactic structure of the new reducing sentence when using morphology of lexicons: "Nhân" + "gặp" + "Lễ" + "mà" + "được" + "giới thiệu" + "bởi" + "Nghĩa" ⇨ The complete new reducing sentence: *Nhân gặp Lễ mà được giới thiệu bởi Nghĩa.* (English: Nhân meets Lễ who is introduced by Nghĩa.)

3 Experiment and Discussion

We proceed to build a new data set suitable for testing in this research according to the following rule:

- The first sentence: (i) There is one transitive verb of action; (ii) Taking the subject and object role can be human or thing.
- The second sentence: (i) There is one transitive verb of action; (ii) Taking the subject role can be human or thing; (iii) Taking the object role can be one of three types: pronoun indicating human and standing alone, pronoun indicating human and standing with demonstrative adjective ["ta" / "ấy" / "này"], pronoun "nó".

Follow this rule, we collected 41 pairs of Vietnamese sentences suitable for four forms in Table 1. Testing these 41 pairs of sentences is performed through two steps:

- **Step 1:** test resolving anaphoric pronouns and building the structure DRS;
- **Step 2:** test generating the new reducing Vietnamese sentence with two main requirements: (i) summarize the meaning of the source pair of Vietnamese sentences; and (2) have the naturalism for the cognition of Vietnamese speakers.

Performing step 1, resolve anaphoric pronouns and build the structure DRS for 33 pairs of Vietnamese sentences, the successful rate is 0.8. Performing step 2 for 33 built structure DRS, the result is presented in Table 4 as follows:

Table 3. The result when testing generating the new reducing vietnamese sentence

Form	The number of structure DRS	The number of generated NRVSs	The number of NRVSs satisfying requirements	The rate of satisfying requirements
1	15	15	10	0.67
2	4	4	3	0.75
3	2	2	2	1
4	11	11	9	0.82

Analyze the result of testing, we see that there are some points:

- Generate the new reducing Vietnamese sentence for all source pairs of Vietnamese sentences which can be built the structure DRS.
- In comparison with methods, techniques in [17], in this research we also consider the relationship between the transitive verbs of action at the first sentence with the transitive verb of action at the second sentence. This help for the new generated Vietnamese sentence better satisfying two given requirements, especially the naturalism for the cognition of Vietnamese speakers.
- The cause of not correctly resolving anaphoric pronouns for some source pairs of Vietnamese sentences as well as some new generated Vietnamese sentences do not satisfy two given requirements:
  - Because of presupposition in [16] and in this research about the characteristic of anaphoric pronouns: pronoun indicating human only relates to human, pronoun "nó" only relates to thing. In some pairs of sentences, occur one of two situations: (a) the second sentence has pronoun indicating human but two objects at the first sentence are both thing; (b) the second sentence has pronoun "nó" but two objects at the first sentence are both human.

Example 5: *Ông Nhân nói chuyện với Lễ. Nghĩa hỏi nó.*
(English: Mr Nhân talks to Lễ. Nghĩa asks it.)
➔ In Vietnamese, pronoun "nó" can indicate human having lower age. In this case [nó = Lễ] and should use "him" in English.

- With some pairs of sentences belonging to Form 1 or Form 2 in Table 1, due to the true meaning of transitive verbs of action at the first and second sentence as well as the reality context, the actual relationship is "cause – effect".

Example 6: *Nghĩa bán chiếc xe đạp. Lễ trách em.*
(English: Nghĩa sells the bicycle. Lễ blames him.)
➔ In this case, the new reducing sentence should be "Vì Nghĩa bán chiếc xe đạp nên bị Lễ trách" (English: Because Nghĩa sells the bicycle so is blamed by Lễ).

- With some pairs of sentences belonging to Form 3 or Form 4 in Table 1, in this research we use structure "relative clause" in positive side – in the sense that the second object achieves something. However, in many pairs of sentences, due to the true meaning of the transitive verb of action at the second sentence as well as the reality context, need to use structure "relative clause" in negative side – in the sense that the second object is affected.

Example 7: *Nghĩa an ủi đứa bé. Nhân la rầy em.*
(English: Nghĩa comforts the child. Nhân scolds him.)
➔ In this case, the new reducing sentence should be "Nghĩa an ủi đứa bé mà bị Nhân la rầy." (English: Nghĩa comforts the child who is scolded by Nhân.)

4 Conclusion

In this paper, we presented a solution based on approaches in [17] for summarizing the meaning of pairs of sentences having common features: (a) both sentences have transitive verbs of action; and (b) the second sentence has only one anaphoric pronoun and this pronoun takes the object role of the transitive verb of action. The main idea of this solution consists of two processing threads: (1) improve methods, techniques in [16] for resolving anaphoric pronouns and building the structure DRS of the original pair of sentences; (ii) analyze relationships based on this structure DRS and generate the NRVS.

We also presented testing, made analysis, evaluation and gave some limitations in the current approach in section 3. Besides given limitations, one important point is need to consider many more forms of pairs of sentences, inter-sentential relationships as well as paragraphs having more than two sentences.

These limitations will be considered as objectives in our next researches.

References

1. Blackburn, P., Bos, J.: Representation and Inference for Natural Language - Volume II: Working with Discourse Representation Structures. Department of Computational Linguistics, University of Saarland, Germany (1999)
2. Cao, H.X.: Tiếng Việt: Sơ thảo ngữ pháp chức năng (Vietnamese: Brief of Functional Grammar). Nhà xuất bản giáo dục (2006)
3. Covington, M.A., Schmitz, N.: An Implementation of Discourse Representation Theory. Advanced Computational Methods Center, The University of Georgia (1989)
4. Covington, M.A., Nute, D., Schmitz, N., Goodman, D.: From English to Prolog via Discourse Representation Theory. ACMC Research Report 01-0024, The University of Georgia (1988)
5. Das, D., Martins, A.F.T.: A survey on automatic text summarization. Language Technologies Institute, Carnegie Mellon University (2007)
6. Halliday, M.A.K., Matthiessen, C.M.I.M.: An Introduction to Functional Grammar, 3rd edn. Hodder Arnold (2004)
7. Jezek, K., Steinberger, J.: Automatic Text summarization. In: Snasel, V. (ed.) Znalosti 2008, FIIT STU Brarislava, Ustav Informatiky a Softveroveho Inzinierstva, pp. 1–12 (2008) ISBN 978-80-227-2827-0
8. Jones, K.S.: Automatic summarising: a review and discussion of the state of the art. Technical Report 679. Computer Laboratory, University of Cambridge (2007)
9. Jones, K.S.: Automatic summarizing: factors and directions. In: Mani, I., Marbury, M. (eds.) Advances in Automatic Text Summarization. MIT Press (1999)
10. Kamp, H.: A theory of truth and semantic representation. In: Groenendijk, J., Janssen, T.M.V., Stokhof, M. (eds.) Formal Methods in the Study of Language, Part 1, pp. 277–322. Mathematical Centre Tracts (1981)
11. Lloret, E.: Text summarization: an overview. Paper supported by the Spanish Government under the project TEXT-MESS (TIN2006-15265-C06-01) (2008)
12. Mani, I., Maybury, M.T.: Advances in Automatic Text Summarization. MIT Press (1999)
13. Reiter, E., Dale, R.: Building Natural Language Generation System. Cambridge University Press (1997)

14. Reiter, E., Dale, R.: Building Applied Natural Language Generation Systems. Natural Language Engineering 3(1), 57–87 (1997)
15. Shen, D., Sun, J.T., Li, H., Yang, Q., Chen, Z.: Document summarization using conditional random fields. In: 20th International Joint Conference on Artificial Intelligence (IJCAI 2007), pp. 2862–2867 (2007)
16. Tran, T., Nguyen, D.T.: A Solution for Resolving Inter-sentential Anaphoric Pronouns for Vietnamese Paragraphs Composing Two Single Sentences. In: 5th International Conference of Soft Computing and Pattern Recognition (SoCPaR 2013), Hanoi, Vietnam, pp. 172–177 (2013)
17. Tran, T., Nguyen, D.T.: Merging Two Vietnamese Sentences Related by Inter-sentential Anaphoric Pronouns for Summarizing. In: 1st NAFOSTED Conference on Information and Computer Science (NICS 2014), Hanoi, Vietnam, pp. 371–381 (2014)

Hierarchical Approach for VLSI Components Placement

D.Yu. Zaporozhets, D.V. Zaruba, and V.V. Kureichik

Southern Federal University, Taganrog, Russia
{elpilasgsm,daria.zaruba}@gmail.com

Abstract. To solve the problem of VLSI components' placement a modified hierarchical approach is proposed. This approach consists of three levels. At the first level a preliminary decomposition of search space with the use of evolutionary algorithms is performed. Geometric parameters of each group are determined by the total area of its constituent components. At the second level VLSI components are placed within decomposition groups on the basis of the modified genetic algorithm. At the third level decomposition groups are placed within a connection field using genetic search methods. The suggested methods of the encoding and decoding of alternative solutions enable the authors to perform genetic procedures. These methods consist in using the reverse Polish notation. A computational experiment, which confirmed the theoretical estimates of the performance and efficiency of the developed algorithms, was carried out. The hierarchical algorithm was compared with classical methods and the bioinspired search. The time complexity of the algorithm is represented as O (n log n).

Keywords: VLSI, Computer-aided design, Placement, layout, Multi-level optimization, Genetic algorithm, Bioinspired search.

1 Introduction

At present the VLSI manufacturing technology allows placing more than a few millions of elements on a chip. So it is necessary to develop the relevant computer-aided design methods and means. The technological development on a submicrometer scale caused the emergence of new physical phenomena. Therefore, the design of new heuristics algorithms to solve engineering problems is a topical issue of modern engineering [1-3]. In this paper the authors suggest a solution of the VLSI components' placement. It should be pointed out that this problem is NP-full. So, to obtain results in the polynomial time it is necessary to develop the heuristics allowing to find quasi-optimal solutions. Evolutionary and bioinspired search mechanisms proved to be effective in terms of computer aided design [2, 4, 5]. To solve the VLSI components placement problem a hierarchical approach on the basis of the bioinspired search is designed.

R. Silhavy et al. (eds.), *Artificial Intelligence Perspectives and Applications*,
Advances in Intelligent Systems and Computing 347, DOI: 10.1007/978-3-319-18476-0_9

2 Problem Definition

A connection field, a circuit layout which contains a predetermined number of components, and a net list comprise the initial data to solve the placement problem.

Let us formulate the placement problem as follows. At the predetermined connection field it is necessary to place components with respect to each other so that the system performance F tends to obtain the optimum value. Performance F is a total time delay during the signal transmission. The optimization of this parameter comes to minimization of the critical connection length. The critical connection is an electrical circuit with a maximum connection length [1-3,5-8].

The problem of the VLSI components placement is formalized as follows. Let B_1, …, B_n be the components that must be placed within the connection field. Each component $B_i | 1 \leq i \leq N$ is defined by geometrical dimensions, namely height h_i and width w_i. Let $N = \{N_i \mid i = 1, m\}$ be the net list and L_i be the length of net $N_i \mid i = 1,m$. To solve the placement problem one needs to find a rectangular region within the connection field for each element from B. The rectangular region is defined by $R = \{R_i | \, i = 1, n\}$ so that:

1. Each block of components is placed inside the corresponding rectangular region in such a way that R_i has height h_i and width w_i.
2. All the regions must be disjoint.
3. The total area of the bounding rectangle tends to obtain the minimum value.
4. The total length of the connections is minimized so that

$$F = \sum_{i=1}^{m} L_i . F \rightarrow min. \qquad (1)$$

To find length L_i of the i-th net it is suggested that one use the bounding rectangular of this circuit since the algorithm's time complexity takes a linear value.

3 Hierarchical Search Structure

In regard to the components placement the key problem is the processing of huge amounts of information. For the effective solution of this problem, we suggest applying the modified hierarchical approach consisting of three levels. Due to the use of the proposed method a significant amount of computer resources and search time are reduced. Also the researchers had an opportunity to obtain optimal and quasi-optimal results in the polynomial time. The modified search structure is shown in Fig 1.

At the first level the VLSI components are broken up into decomposition groups by a evolutionary algorithm. At the second level components within the decomposition groups are place by a genetic algorithm. At the third level a similar genetic algorithm is performed to place the decomposition groups within the connection field. At the second and third levels the reverse Polish notation is used to encode the solution. The reverse Polish notation is a notation in which the operands preceded the operators. Encoding and decoding techniques are described in detail in [9]. The authors suggest the following operators:

- "right" - the subsequent fragment is located on the right of the previous ones along Axis X;
- "up" - the subsequent fragment is located above the previous ones along Axis Y.

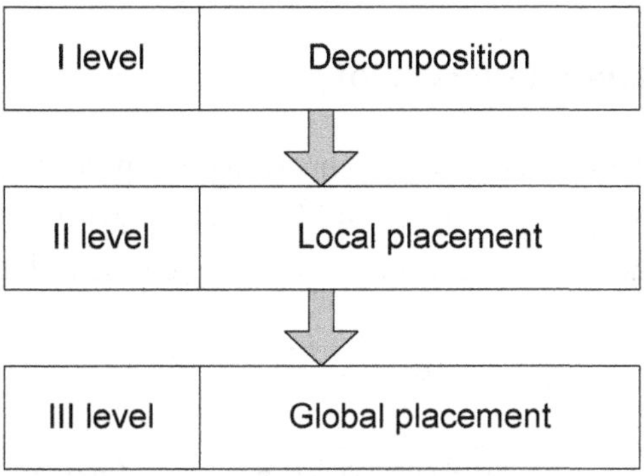

Fig. 1. The hierarchical search structure

Let operator "up" be represented as "+" and operator "right" be represented as "*" for the notational convenience [9].

4 Circuit Decomposition

As for the large-scale problems, the algorithm of the initial data decomposition on undersize subsets is generally used. However, the decomposition problem is NP-complete and requires particular attention. The authors suggest using a decomposition evolutionary algorithm. The time complexity of this algorithm is represented by O (n log n) [10, 11]. In response to the placement problem the authors introduce a criterion that allows to adequately place the components and decomposition groups (DG) within the connection fields. Let W(c) be the weight of the net equal to the number of elements (vertices) in Net $c \in C$. Therefore, the net total weight criterion F_d is calculated as follows

$$F_d = \sum_{j=1}^{N_c} W(j). F_d \rightarrow \min. \quad (2)$$

Here N_c is a number of circuits. In this case the restriction is represented as follows:

$$\sum_{p=1}^{M_k} S(p) \leq S/_K. \quad (3)$$

Here M_k is a number of components in the k-th decomposition group; S(p) is a p-th component area; S is a connection field area; K is a number of decomposition groups.

The decomposition group of components is represented as a tuple

$$db_k =< EM_k, LC_k, CC_k >. \quad (4)$$

Here EM_k is a subset of elements from set B in the k block; LC_k is a net list in the block, CC_k is an external connection list in the block. As a result, we obtain the set of decomposition group $DB = \{db_k\}, k = \overline{1, K}$, where k is defined by the decision maker.

5 Local and Global Placement

At the second level let us place the decomposition group db_k within the connection field with dimensions $\frac{a}{\sqrt{K}} \times \frac{b}{\sqrt{K}}$ by means of the modified genetic algorithm [4,5]. The general structure of the modified genetic algorithm is represented in Fig.2.

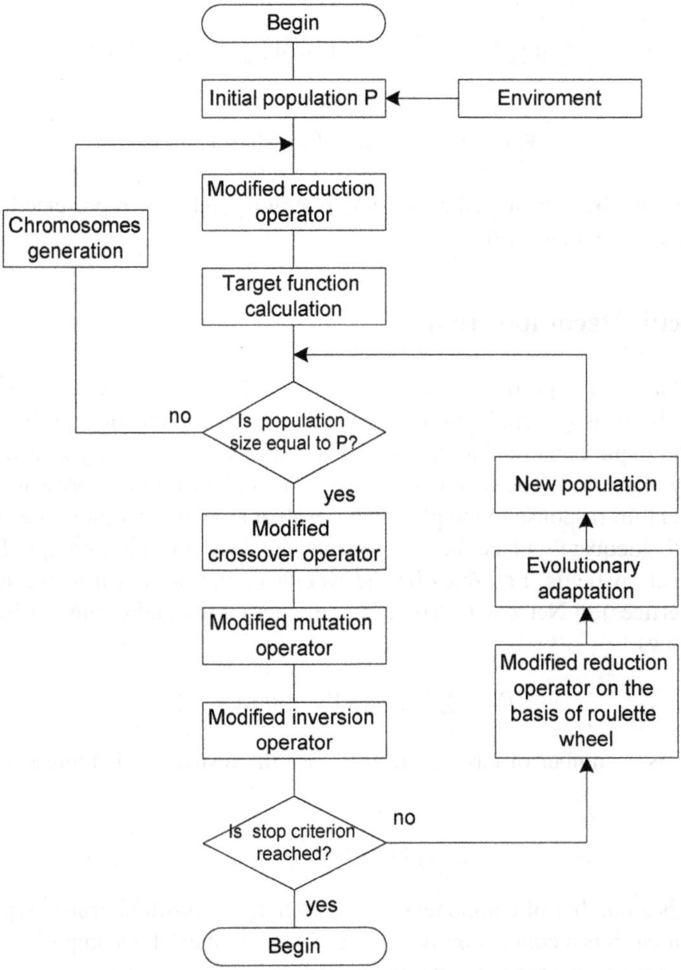

Fig. 2. The general structure of the modified genetic algorithm

Let us describe the algorithm in detail. At the zero iteration the population with P alternative solutions (chromosomes) is initialized in accordance with the rules described above. The length of each chromosome is represented as 2n-1 where n is a number of elements. Then, to remove the chromosomes that do not take the optimal solution ("the worst chromosomes") a modified reduction operator is used. At this step "the worst chromosomes" can be corrected. If the correction is impossible, "the worst chromosomes" are removed from the population.

The authors suggest the following procedure of the chromosome correction.

1. The turning of the component set about the axes.
2. The singling out of frequently used operators [5,12]. So, the size of the new population may differ from the initial one. Therefore, it is possible to generate additional chromosomes and estimate them. Three crossover operators (CO1, CO2 and CO3) are designed to generate descendants.

The operation principle of CO1 is represented in Fig. 3.

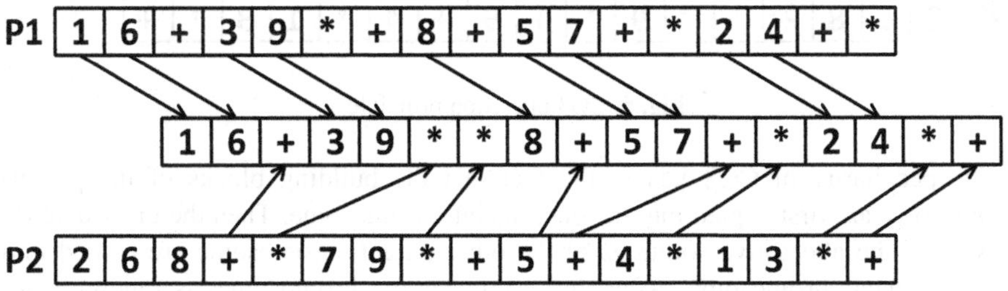

Fig. 3. CO1 operation principle

At the first step the gene-operands of the parent-chromosome P1 are copied into the respective genes of the descendant-chromosome. Then the gene-operators of the parent-chromosome P2 are copied into the descendant-chromosome from the left to the right, filling the remaining positions. The Fig. 4 shows the operation principle of CO2.

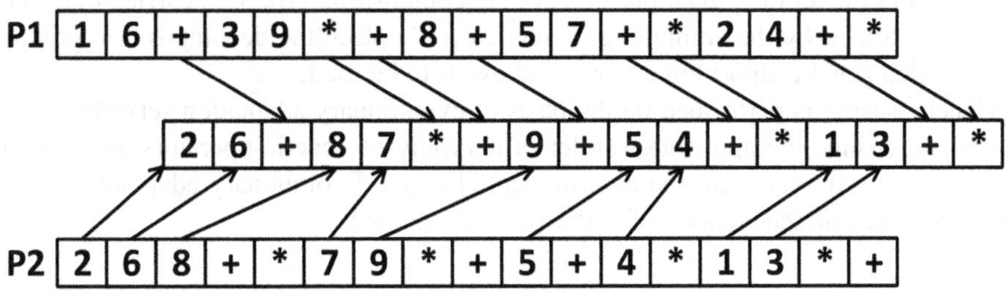

Fig. 4. CO2 operation principle

Primarily the gene-operators of the parent-chromosome P2 are copied into the genes of the descendant-chromosome, and then the gene-operands of the parent-chromosome P1 are copied into the descendant-chromosome from the left to the right. The operation principle of the CO3 is based on the organization of building blocks. The building blocks connect individual genes or groups of genes (parts of alternative solutions) whose modification is undesirable during the search [12]. The operation principle of CO3 is illustrated in Fig.5.

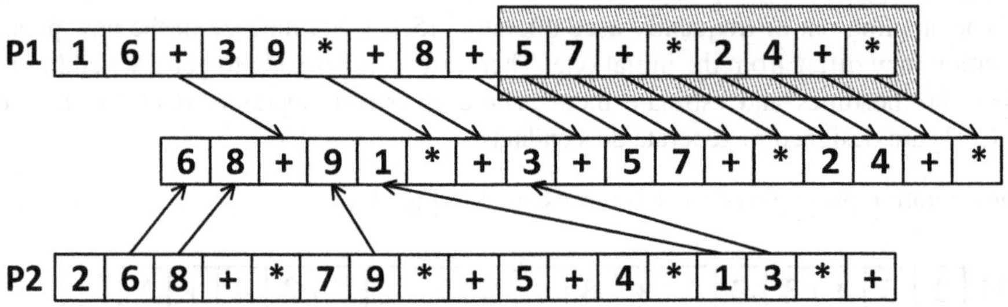

Fig. 5. CO3 operation principle

The peculiarity of CO3 lies in the fact that the building blocks of the parent-chromosome are first copied into the descendant-chromosome. Then the unused genes or parts of the alternative solutions of the other parent-chromosome are copied into the descendant-chromosome. After that a number of chromosomes is randomly selected for the implementation of a modified mutation operator.

The modified mutation operator implements the paired permutation of two same-type genes, i.e. a gene-operand can be moved only to the place of a gene-operand, whereas a gene-operator can be moved only to the place of a gene- operator. Next, a modified inversion operator is used. The sequence of gene-operands or gene-operands with the maximum Hamming distance is recorded in reverse [4,6]. Then we apply a modified selection operator on the basis of roulette wheel to correct “the worst chromosomes” or delete them.

At the third level the global placement of decomposition groups is carried out. The net list is formed by the combining of sets CC_k. To place the decomposition groups the modified genetic algorithm described above is to be used.

Genetic search is performed on the basis of evolutionary adaptation served to configure and change the application order of a variety of genetic operators and search patterns taking into account the environment changes. Evolutionary adaptation has a direct influence on the creation of a new population [4-6].

6 Experiments

To carry out computational experiments a software environment that implemented the placement algorithms was developed. A series of experiments with various sets of test examples was conducted. Average experiment results are shown in Fig. 6. Time complexity is represented by O (n log n).

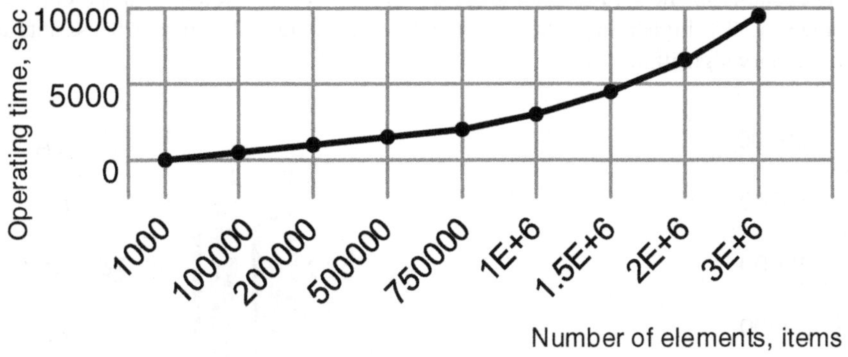

Fig. 6. Dependence of the developed algorithms' operating time on the number of the circuit elements

Next, we carried out a series of tests with different numbers of decomposition groups for the circuit with a number of elements equal to 524,288 items. Fig. 7 shows the dependence of the solution quality on the number of decomposition groups.

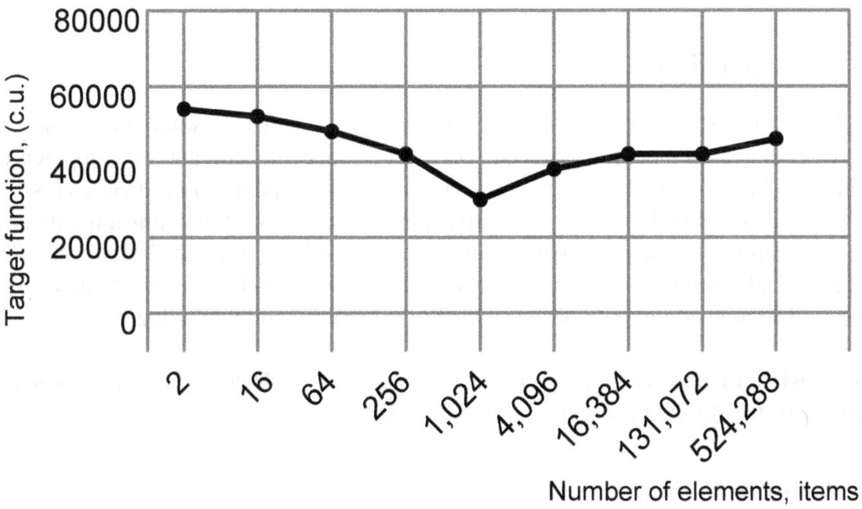

Fig. 7. Dependence of the quality of the solutions on the number of decomposition groups

On the basis of the obtained results the optimal number of decomposition groups is equal to 1024. In this case the size of the decomposition group is approximately 500 elements.

Fig. 8 shows a bar chart comparing the solution quality of the proposed hierarchical algorithm (HA), the bees algorithm (BA), the ant colony algorithm (ACA), the genetic algorithm (GA) and the evolutionary algorithm (EA).

The quality of the solutions obtained on the basis of the hierarchical search, are on average, 4.38% higher than the placement results obtained with the bees, ants and evolutionary algorithms.

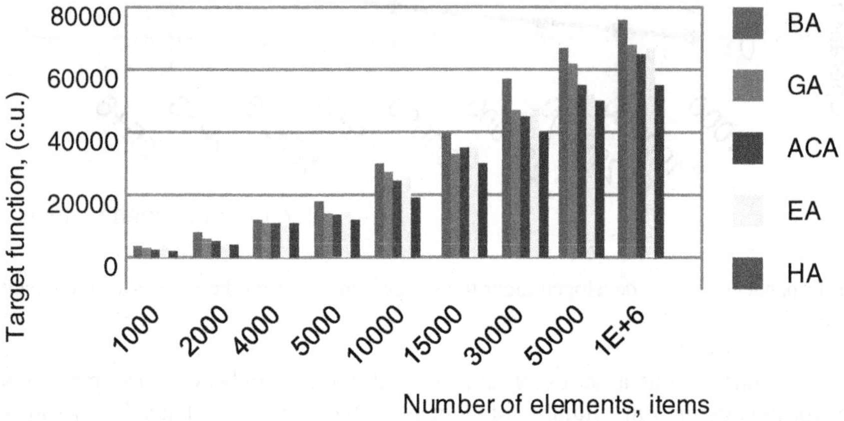

Fig. 8. The comparison of the solutions quality of hierarchical algorithm (HA), bees algorithm (BA), ant colony algorithm (ACA), genetic algorithm (GA) and evolutionary algorithm (EA)

7 Conclusion

In the paper the authors suggested the hierarchical approach to solve the problem of VLSI circuit components placement that can reduce the amount of computer resources to be used and the search time to obtain optimal and quasi-optimal results.

Computational experiments were carried out which allowed to confirm the theoretical estimations of the performance and efficiency of the developed algorithm. The suggested algorithm was compared with bioinspired and the classical search algorithms. The time complexity of the algorithm is represented by O (n log n).

Acknowledgments. This research is sponsored by the Ministry of Education and Science of the Russian Federation. Project #8.823.2014.

References

1. Sherwani, N.A.: Algorithms for VLSI Physical Design Automation, 3rd edn. Kluwer Academic Publisher, USA (2013)
2. Kureichik, V.M., Malioukov, S.P., Kureichik, V.V., Malioukov, A.S.: General Questions of automated design and engineering. In: Kureichik, V.M., Malioukov, S.P., Kureichik, V.V., Malioukov, A.S. (eds.) Genetic Algorithms for Applied CAD Problems. SCI, vol. 212, pp. 1–22. Springer, Heidelberg (2009)

3. Alpert, C.J., Dinesh, P.M., Sachin, S.S.: Handbook of Algorithms for Physical design Automation. Auerbach Publications Taylor & Francis Group, USA (2009)
4. Goldberg, D.E.: Genetic Algorithms in Search, Optimization, and Machine Learning. Addison-Wesley Professional, USA (1989)
5. Kureichik, V.V., Kureichik, V.M.: Genetic search-based control. Automation and Remote Control 62(10), 1698–1710
6. Kureichik, V.V., Zaporozhets, D.Y.: Modern problems of placing VLSI elements. Izvestiya SFedU. Engineering Sciences 120, 68–73 (2011)
7. Cong, J., Shinnerl, J.R., Xie, M., Kong, T., Yuan, X.: Large-scale circuit placement. ACM Transactions on Design Automation of Electronic Systems, 389–430 (2005)
8. Kureichik, V.M., Kureichik, V.V.: Genetic Algorithm for the Graph Placement. International Journal of Computer and Systems Sciences 39(5), 733–740 (2000)
9. Zaporozhets, D.Y., Zaruba, D.V., Lezhebokov, A.A.: A method of coding solutions for solving problems placement. Izvestiya SFedU. Engineering Sciences 136, 183–188 (2012)
10. Bova, V.V., Kureichik, V.V.: Integrated hybrid and combined search in the problems of design and management. Izvestiya SFedU. Engineering Sciences 113, 37–42 (2010)
11. Kureichik, V.V., Kureichik, Vl.Vl.: Hybrid search in design. Izvestiya SFedU. Engineering Sciences 132, 22–27 (2012)
12. Zaporozhets, D.Y., Zaruba, D.V., Kureichik, V.V.: Representation of solutions in genetic VLSI placement algorithms. In: 2014 East-West Design & Test Symposium (EWDTS), pp. 1–4 (2014)

Pre-processing, Repairing and Transfer Functions Can Help Binary Electromagnetism-Like Algorithms

Ricardo Soto[1,2,3], Broderick Crawford[1,4,5], Alexis Muñoz[1], Franklin Johnson[6], and Fernando Paredes[7]

[1] Pontificia Universidad Católica de Valparaíso, Chile
[2] Universidad Autónoma de Chile, Chile
[3] Universidad Central de Chile, Chile
[4] Universidad Finis Terrae, Chile
[5] Universidad San Sebastián, Chile
[6] Universidad de Playa Ancha, Chile
[7] Escuela de Ingeniería Industrial, Universidad Diego Portales, Chile
{ricardo.soto,broderick.crawford}@ucv.cl,
alexis.munoz.c@mail.pucv.cl, franklin.johnson@upla.cl,
fernando.paredes@udp.cl

Abstract. The Electromagnetism-like algorithm is a relatively modern metaheuristic based on the attraction-repulsion mechanism of particles in the context of electromagnetism theory. This paper focuses on improving performance of this metaheuristic when solving binary problems. To this end, we incorporate three elements: pre-processing, repairing, and transfers functions. The pre-processing allows to reduce the size of instances, while repairing eliminates those potential solutions that violate the constraints. Finally, the incorporation of a transfer function adapts the solutions to a binary domains. We illustrate experimental results where the incorporation of these elements improve the resolution phase, when solving a set of 65 non-unicost set covering problems.

Keywords: Pre-processing, Metaheuristics, Electromagnetism-like Algorithm.

1 Introduction

Metaheuristics are approximate algorithms, whose purpose is to provide good enough solutions by exploring the promising regions of the search space, however they are unable to always reach the global optimum of the problem. On the contrary, exact algorithms explore the complete search space, guaranteeing the global optimum if exists, but they are limited when the search space exploration requires huge amount of memory and time. In this paper, we focus on the improvement of a relatively modern metaheuristic called Electromagnetism-like algorithm by incorporating three interesting elements: pre-processing, repairing, and transfers functions. The pre-processing allows to reduce the size of instances,

R. Silhavy et al. (eds.), *Artificial Intelligence Perspectives and Applications*,
Advances in Intelligent Systems and Computing 347, DOI: 10.1007/978-3-319-18476-0_10

while repairing eliminates those potential solutions that violate the constraints. The incorporation of a transfer function efficiently adapts the solutions to binary domains. We test the proposed approach in a well-known binary benchmark called Set Covering Problem (SCP). The SCP belongs to the NP-Hard class of problems [7,5,4] and has wide real life application. Some examples are airline crew scheduling [12], plant location [8], network discovery [6], and service allocation [3] among others. We present interesting results on 65 non-unicost instances of this problem where the incorporation of the aforementioned elements clearly improve the obtained results.

The remainder of this paper is organized as follows: Section 2 introduces the Electromagnetism-like algorithm. Next section presents the pre-processing, repairing, and transfer function. Section 3 briefly describes the set covering problem and provides the experimental results. Finally, conclusions and future work are given.

2 Electromagnetism-Like Algorithm

The Electromagnetism-like algorithm (EM) was developed by Ilker Birbil and Shu-Cherng Fang [2] to resolve incomplete search problems for optimization, inspired in the electromagnetism theory where particles can attract or repels others particles. The magnitude of attraction or repulsion of a particle w.r.t the population is determined by its charge. The higher is the charge of a particle, the stronger is its attraction. After charge calculations, the movement direction of each particle is determined, which will be subjected to the forces exerted among particles.

The EM algorithm consist of four phases: initialization of EM, local search procedure, calculation of the charge and force exerted on each particle and finally the movement of the particle as shown in Algorithm 1.

Algorithm 1 - EM

Initialize(Pop_size)
$It \leftarrow 1$
While $It < MAXITER$ **do**
  Local Search$(LSITER, \delta)$
  Calculate Forces
  Move
  $It{+}{+}$
End While

Initialization: The initialization procedure is used to find an initial population of particles, randomly constructed. In this case, the values of the solution vector lie between 0 and 1. After the initial population construction, we compute the fitness to each particle and the best one is selected. The EM algorithm evaluates all the solutions w.r.t the objective function and selects the best one at each iteration, in such a way to obtain a result that can be used as a reference point to obtain best optimal solutions. The best particle is stored to be compared to other possible solutions.

Local Search: This procedure is employed to gather local and relevant information for a given particle. Two input parameters are needed: LSITER defines the number of local iterations and δ represents the multiplier for the neighborhood search.

Calculation of Charge and Force: The calculation of charge and force is carried out via Eq. 1 and 2, respectively:

$$q^i = \exp\left(-n\frac{f\left(x^i\right) - f\left(x^{best}\right)}{\sum_{k=1}^{pop\_size}\left(f\left(x^k\right) - f\left(x^{best}\right)\right)}\right), \quad \forall i \in pop\_size \tag{1}$$

$$F_d^i = \sum_{k=1,k\neq i}^{pop\_size} \left\{ \begin{array}{l} (x_d^k - x_d^i)\frac{q^i q^k}{\|x^i - x^k\|^2} \; if \; f(x^k) < f(x^i) \\ (x_d^i - x_d^k)\frac{q^i q^k}{\|x^k - x^i\|^2} \; if \; f(x^k) \geq f(x^i) \end{array} \right\}, \forall i \tag{2}$$

where q^i is the charge of particle x^i, n represents the dimension, x^{best} is the best particle found, and F_d^i is the force of component d of particle x^i. After charge computation, some particles will hold higher charges (better fitness) than others, which determines the attraction or repulsion magnitude exerted. A good particle generates attraction helping to converge to better populations, while bad particles produce repulsion.

Movement: After force and charge computation, the movement of solutions is governed by Eq. 3, where λ is a random step length, which we uniformly distribute between 0 and 1. Finally, RNG is a vector controlling that movements do not escape from the domain of variables.

$$x^i = x^j + \lambda\frac{F^i}{\| F^i \|}(RNG) \quad i = 1, 2, ..., pop\_size \tag{3}$$

3 The Set Covering Problem

This paper focuses on the improvement of the Electromagnetism-like algorithm when solving binary problems. To this end we test the proposed approach by using a widely employed binary benchmark, namely, the set covering problem [11]. This problem consist in finding the lowest number of sets that can contain every element from a given set. As previously mentioned, this problem has several practical applications such as airline crew scheduling [12], plant location [8], network discovery [6], and service allocation in general [3]. The formal definition for SCP is as follows:

$$min\,(z) = \sum_{j=1}^{n} c_j X_j \tag{4}$$

Subject to:

$$\sum_{j=1}^{n} a_{ij} x_j \geq 1 \qquad \forall i \in M \tag{5}$$

$$x_j = \begin{cases} 1 & j \in S \\ 0 & otherwise \end{cases} \qquad \forall j \in N \tag{6}$$

where $A = (a_{ij})$ is a $m \times n$ binary matrix, $C = (c_{ij})$ is a n-dimensional integer array, $M = \{1, ..., m\}$, and $N = \{1, ..., n\}$. c_j $(j \in N)$ represents the cost of column j and we can assume, without generality loss, that $c_j > 0$ for $(j \in N)$. So we can say that a column $(j \in N)$ cover a row i that already exists in M if $a_{ij} = 1$. The SCP looks for a subset of columns called S with minimal cost, such that each existent row i in M is covered by at least one existent column j in S.

4 Pre-processing, Repairing, and Transfer Function

In this section, we describe the elements we introduce in order to improve performance of the EM algorithm: Pre-processing, repairing, and a transfer function.

- ***Pre-process:*** A popular way to accelerate problem solving is to reduce the instance sizes by applying pre-processing. Then, to effectively tackle SCPs we employ *Column Domination* and *Column Inclusion* [10]. Column Domination allows to eliminate columns where their rows are covered by others columns with lowest *ratio*, we compute the *ratio* as follows:

$$ratio = \frac{Cost\ of\ a\ column\ c}{number\ of\ covered\ rows\ by\ column\ c} \tag{7}$$

  The Column inclusion forces to include in the solution the columns that are unique in covering a given row after applying column domination.
- ***Repairing:*** EM as other metaheuristics use random generation of initial populations, which commonly produce solutions that violate the instance constraints. Repairing methods are used to improve the solving process by turning those unfeasible solutions into feasible ones. In this work, we repair solutions by replacing the values that lead to constraint violations by values that satisfy the constraint but at the lowest cost.
- ***Binarization:*** In the EM movement phase, Eq. 3 produces a real number which must be transformed to a binary one due to the nature of the problem treated. To this end we employ the following transfer function, which was the best performing one from the eight transfer functions proposed in [9].

$$T(x) = \frac{1}{1 + e^{-2x}}$$

  Finally, the resulting value is discretized via the standard method as follows:

$$x(t+1) = \begin{cases} 1 & \text{if } rand < T(x(t+1)) \\ 0 & otherwise \end{cases}$$

5 Experimental Results

The proposed algorithm has been implemented using Java and the experiments have been launched on an 2.53 Ghz Intel Core I3 with 4GB RAM DDR3 1333 MHz machine running Windows 7. We have tested 65 (organized in sets: 4, 5, 6, A, B, C, D, NRE, NRF, NRG, NRH) non-unicost instances of the SCP, which have been taken from the OR-Library by J.E. Beasley[1]. Details about instances are given in Table 1.

Table 1. SCP instances

Instance set	No. of instances	n	m	Cost Range	Density(%)	Optimal solution
4	10	200	1000	[1,100]	2	Known
5	10	200	2000	[1,100]	2	Known
6	5	200	1000	[1,100]	5	Known
A	5	300	3000	[1,100]	2	Known
B	5	300	3000	[1,100]	5	Known
C	5	400	4000	[1,100]	2	Known
D	5	400	4000	[1,100]	5	Known
NRE	5	500	5000	[1,100]	10	Unknown
NRF	5	500	5000	[1,100]	20	Unknown
NRG	5	1000	10000	[1,100]	2	Unknown
NRH	5	1000	10000	[1,100]	5	Unknown

For the EM algorithm, we employ $MAXITER = 25$, $LSITER = 1000$, and 25 particles, which was the best configuration obtained after training. Each experiment was launched 30 times. Tables 2 and 3 depict the results obtained, where column 1 represent the instance number, followed by the best known optimal solution, the best optimal solution found, the average (new Avg) for the 30 executions and the relative percentage deviation (new RPD) of the proposed approach. Finally, the average (Avg) for the 30 executions and the relative percentage deviation (RPD) of the classic EM (without pre-processing, repairing and transfer functions) are depicted. The computation of the RPD is as follows:

$$RDP = \frac{(Z - Z_{opt})}{Z_{opt}} \times 100$$

The results clearly show the positive impact of the three integrated elements: pre-processing, repairing, and transfers functions. The new approach significantly outperforms the classic EM in all cases. This is produced mainly by the size instance reduction achieved by the pre-processing, the acceleration produced in the solving process by the repairing, and the suitable adaptation to binary domain performed by the employed transfer function.

Table 2. Results obtained using EM for groups 4, 5, 6, A, B, C, and D

Instance	Opt	Best	new Avg	new RPD	Avg	RPD
4.1	429	447	448	4.20	453	5.36
4.2	512	559	568	9.18	564	10.16
4.3	516	537	543	4.07	560	8.33
4.4	494	527	528	6.68	532	7.69
4.5	512	527	530	2.93	538	4.69
4.6	560	607	607	8.39	611	8.93
4.7	430	448	448	4.19	455	5.81
4.8	492	509	509	3.46	626	26.83
4.9	641	682	684	8.74	715	11.39
4.10	514	571	573	11.09	578	12.26
5.1	253	280	281	10.67	289	13.83
5.2	302	318	319	5.30	335	10.93
5.3	226	242	242	7.08	266	16.81
5.4	242	251	251	3.72	264	8.68
5.5	211	225	225	6.64	235	11.37
5.6	213	247	248	15.96	262	23.00
5.7	293	316	316	7.85	335	13.99
5.8	288	315	316	9.38	337	16.32
5.9	279	314	315	12.54	330	17.92
5.10	265	280	282	5.66	296	10.94
6.1	138	152	152	10.14	166	20.29
6.2	146	160	160	9.59	164	11.64
6.3	145	160	160	10.34	162	11.72
6.4	131	140	140	6.87	142	8.40
6.5	161	184	184	14.29	194	20.50
A.1	253	261	262	3.16	276	8.70
A.2	252	279	279	10.71	293	15.48
A.3	232	252	253	8.62	275	18.10
A.4	234	250	251	6.84	268	14.10
A.5	236	241	241	2.12	258	9.32
B.1	69	86	87	24.64	92	33.33
B.2	76	88	89	15.79	94	23.68
B.3	80	85	85	6.25	92	13.75
B.4	79	84	85	6.33	89	11.39
B.5	72	78	79	8.33	86	19.44
C.1	227	237	237	4.41	245	7.93
C.2	219	237	237	8.22	247	12.33
C.3	243	271	272	11.52	286	16.87
C.4	219	246	247	12.33	257	16.89
C.5	215	224	224	4.19	238	10.23
D.1	60	62	63	3.33	67	11.67
D.2	66	73	74	10.61	79	18.18
D.3	72	79	80	9.72	83	15.28
D.4	62	67	67	8.06	72	16.13
D.5	61	66	66	8.20	73	18.03

Table 3. Results obtained for groups NRE, NRF, NRG, and NRH

Instance	Opt	Best	new Avg	new RPD	Avg	RPD
NRE.1	29	30	31	3.45	35	17.24
NRE.2	30	35	35	16.67	38	26.67
NRE.3	27	34	34	25.93	36	33.33
NRE.4	28	33	34	17.86	37	32.14
NRE.5	28	30	31	7.14	34	21.43
NRF.1	14	17	17	21.43	21	50.00
NRF.2	15	18	19	20.00	23	46.67
NRF.3	14	17	18	21.43	21	42.86
NRF.4	14	17	17	21.43	21	50.00
NRF.5	13	16	16	23.08	20	53.85
NRG.1	176	194	194	10.23	211	19.89
NRG.2	154	176	176	14.29	187	20.13
NRG.3	166	184	185	10.84	189	13.25
NRG.4	168	196	196	16.67	212	26.19
NRG.5	168	198	198	17.86	210	24.40
NRH.1	63	70	70	11.11	74	17.46
NRH.2	63	71	71	12.70	76	19.05
NRH.3	59	68	69	15.25	74	25.42
NRH.4	58	70	71	20.69	76	31.03
NRH.5	55	69	70	25.45	73	32.73

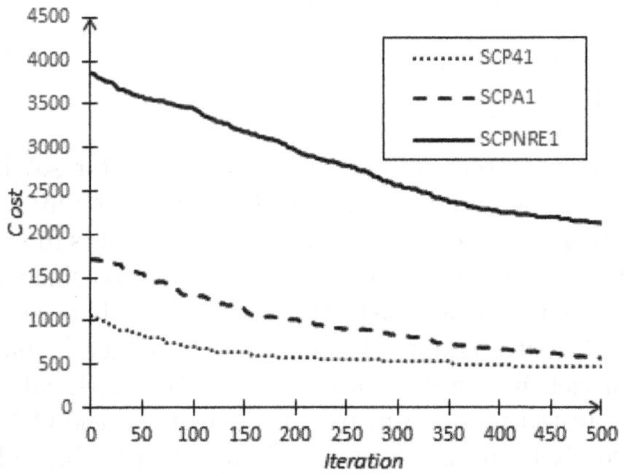

Fig. 1. LSITER convergence for instances scp41, scpa1 and scpnre1

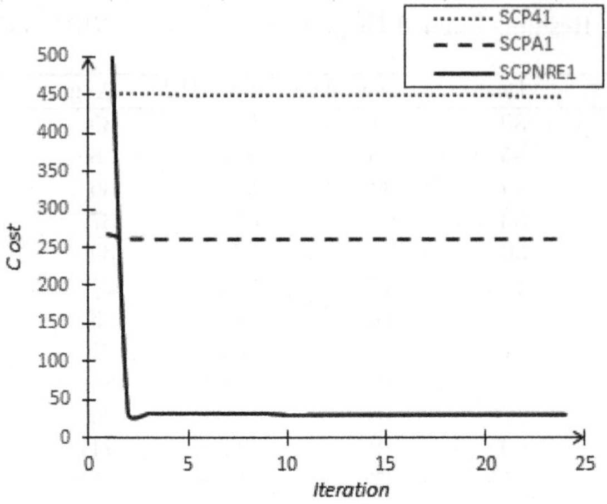

Fig. 2. MAXITER convergence for instances scp41, scpa1 and scpnre1

Fig. 1 and 2 show the evolution of three representative instances w.r.t the LSITER and MAXITER values. The fast convergence of these instances demonstrates that EM can be competitive with others metaheuristics to solve the SCP. LSITER is able to rapidly converge before the iteration 500, helping to the MAXITER value to converge before iteration 25. The experimental results also exhibit the robustness of the approach, which is able to reach reasonable good results by keeping the same parameter configuration.

6 Conclusions

In this paper we have presented a new EM algorithm for solving SCPs. The main idea was to improve performance of this algorithm by incorporating three interesting elements: pre-processing, repairing, and transfers functions. The pre-processing helps to reduce the size of instances, while repairing eliminates those potential solutions that do not satisfy the constraints. The incorporation of a transfer function efficiently adapts the solutions to binary domains. We tested the proposed approach in a well-known binary benchmark called set covering problem. We have tested 65 non-unicost instances from the OR-Library where the results have clearly been improved with respect to the classic EM algorithm. The results have also exhibited the rapid convergence and robustness of the proposed algorithm which is capable to obtain reasonable good results by keeping its configuration. As future work, we plan to hybridize EM with another metaheuristics as well as to incorporate filtering techniques from the constraint programming domain to improve performance as illustrated in [13,14].

Acknowledgements. Ricardo Soto is supported by Grant CONICYT/ FONDECYT/INICIACION/ 11130459, Broderick Crawford is supported by Grant CONICYT/FONDECYT/ 1140897, and Fernando Paredes is supported by Grant CONICYT/FONDECYT/ 1130455.

References

1. Beasley, J.E.: A lagrangian heuristic for set-covering problems. Naval Research Logistics (NRL) 37(1), 151–164 (1990)
2. Birbil, S.I., Fang, S.-C.: An electromagnetism-like mechanism for global optimization. Journal of Global Optimization 25(3), 263–282 (2003)
3. Ceria, S., Nobili, P., Sassano, A.: Annotated Bibliographies in Combinatorial Optimization. J. Wiley and Sons (1997)
4. Ceria, S., Nobili, P., Sassano, A.: A Lagrangian-based heuristic for large-scale set covering problems. Mathematical Programming 81(2), 215–228 (1998)
5. Chvatal, V.: A greedy heuristic for the set-covering problem. Mathematics of Operations Research 4(3), 233–235 (1979)
6. Grossman, T., Wool, A.: Computational experience with approximation algorithms for the set covering problem. European Journal of Operational Research 101(1), 81–92 (1997)
7. Hochba, D.S.: Approximation algorithms for NP-hard problems. ACM SIGACT News 28(2), 40–52 (1997)
8. Krarup, J., Bilde, O.: Plant Location, Set Covering and Economic Lot Size: An 0 (mn)-Algorithm for Structured Problems. Springer (1977)
9. Mirjalili, S., Hashim, S., Taherzadeh, G., Mirjalili, S.Z., Salehi, S.: A Study of Different Transfer Functions for Binary Version of Particle Swarm Optimization. In: GEM 2011. CSREA Press (2011)
10. Pezzella, F., Faggioli, E.: Solving large set covering problems for crew scheduling. Top 5(1), 41–59 (1997)
11. Ren, Z.-G., Feng, Z.-R., Ke, L.-J., Zhang, Z.-J.: New ideas for applying ant colony optimization to the set covering problem. Computers & Industrial Engineering 58(4), 774–784 (2010)
12. Rubin, J.: A technique for the solution of massive set covering problems, with application to airline crew scheduling. Transportation Science 7(1), 34–48 (1973)
13. Soto, R., Crawford, B., Galleguillos, C., Monfroy, E., Paredes, F.: A hybrid AC3-tabu search algorithm for solving Sudoku puzzles. Expert Systems with Applications 40(15), 5817–5821 (2013)
14. Soto, R., Crawford, B., Galleguillos, C., Monfroy, E., Paredes, F.: A Pre-filtered Cuckoo Search Algorithm with Geometric Operators for Solving Sudoku Problems. The Scientific World Journal, Article ID 465359, 12 pages (2014)

Heuristic Feasibility and Preprocessing for a Set Covering Solver Based on Firefly Optimization

Ricardo Soto[1,2,3], Broderick Crawford[1,4,5], José Vilches[1], Franklin Johnson[6], and Fernando Paredes[7]

[1] Pontificia Universidad Católica de Valparaíso, Chile
[2] Universidad Autónoma de Chile, Chile
[3] Universidad Central de Chile, Chile
[4] Universidad Finis Terrae, Chile
[5] Universidad San Sebastián, Chile
[6] Universidad de Playa Ancha, Chile
[7] Escuela de Ingeniería Industrial, Universidad Diego Portales, Chile
{ricardo.soto,broderick.crawford}@ucv.cl,
jose.vilches.f@mail.pucv.cl, franklin.johnson@upla.cl,
fernando.paredes@udp.cl

Abstract. The set covering problem is a classic benchmark that has many real applications such as positioning of communications systems, logical analysis, steel production, vehicle routing, and service allocation in general. In this paper, we present an improved firefly algorithm to the efficient resolution of this problem. The firefly algorithm is a recent metaheuristic based on the flashing characteristics of fireflies that attract each other by using their brightness. We improve this approach by incorporating pre-processing and an heuristic feasibility operator resulting in an interesting solver able to clearly outperform the previously reported results obtained from firefly algorithms.

Keywords: Set Covering Problem, Firefly Algorithm, Metaheuristic.

1 Introduction

The Set Covering Problem (SCP) is a classic benchmark belonging to the NP-complete class [15] of problems. The idea is to find a set of solutions that satisfy a range of needs at the minimum possible cost. This problem has several real applications such as positioning of communications systems, logical analysis, steel production, vehicle routing, and service allocation in general [17,18].

During the last two decades, various techniques has been reported to tackle this problem. For instance, metaheuristics perform an incomplete exploration of the search space, analyzing only its promising regions. Some examples in this context are genetic algorithms [4], simulated annealing [5], tabu search [8], and the firefly algorithm [12]. On the contrary, exact methods explore the complete search space guaranteeing the global optimum if exist, but they are limited when the search space exploration requires huge amount of memory and time.

R. Silhavy et al. (eds.), *Artificial Intelligence Perspectives and Applications*,
Advances in Intelligent Systems and Computing 347, DOI: 10.1007/978-3-319-18476-0_11

Some exact methods that efficiently tackle the set covering problem are reported in [1,11,16,9,7,3,6]

In this paper we present an improved firefly algorithm to the efficient resolution of this problem. The firefly algorithm is a recent metaheuristic based on the flashing characteristics of fireflies that attract each other by using their brightness. We improve this classic approach by incorporating pre-processing and an heuristic feasibility operator. The proposed approach is able to clearly outperform results obtained by the best reported firefly algorithm for set covering problems on a set of 65 instances from the Beasley's OR-Library.

The remainder of this paper is organized as follows: Section 2 introduces the set covering problem and the preprocessing. An overview of the firefly algorithm is given in the next section. Section 4 presents the proposed approach including the heuristic feasibility operator. Finally, we provide the experimental results, conclusions, and future work.

2 Problem Description

The Set Covering Problem (SCP) can be formally defined as follows:

Let $A = (a_{ij})$ be an n-row, m-column, zero-one matrix. We say that a column j covers a row i if $a_{ij} = 1$. Each column j is associated with a nonnegative real cost c_j. Let $I = \{1, ..., n\}$ and $J = \{1, ..., m\}$ be the row set and column set, respectively. The SCP calls for a minimum cost subset $S \subseteq J$, such that each row $i \in I$ is covered by at least one column $j \in S$. A mathematical model for the SCP is as follows:

$$Minimize\, f(x) = \sum_{j=1}^{m} c_j x_j \tag{1}$$

subject to

$$\sum_{j=1}^{m} a_{ij} x_j \geq 1, \quad \forall i \in I \tag{2}$$

$$x_j \in \{0, 1\}, \quad \forall j \in J \tag{3}$$

The purpose is to minimize the sum of the costs of the selected columns, where $x_j = 1$ if the column j is in the solution, 0 otherwise. Each row i must be covered by at least one column j, which is guaranteed by the SCP constraints.

2.1 Preprocessing

With the purpose of speeding resolution times, preprocessing is applied to instances so as to reduce its size. We employ two types of preprocessing, which are reported in [14].

Column Domination: When a column j, whose rows I_j can be covered by another column with lower cost c_j, it is said that column j is dominated and can be deleted.

Column Inclusion: When a row is covered by a unique column after column domination, this column must be included in the optimal solution.

3 Overview of Firefly Algorithm

The firefly algorithm is a modern metaheuristic [19,20] inspired on the firefly behavior, whose movements are mainly guided by the brightness that they naturally emit. Three main rules govern the implementation of this metaheuristic:

1. All fireflies are unisex, which means that they are attracted to other fireflies regardless of their sex.
2. The degree of attractiveness of a firefly is proportional to its brightness, and for any two flashing fireflies, the less brighter one will move towards the brighter one. More brightness means less distance between two fireflies. If there is no brighter one than a particular firefly, a random movement is performed.
3. Finally, the brightness of a firefly is determined by the value of the objective function. For a maximization problem, the brightness of each firefly is proportional to the value of the objective function. In case of minimization problem, brightness of each firefly is inversely proportional to the value of the objective function.

As the attractiveness of a firefly is proportional to the light intensity seen by adjacent fireflies, it is now possible to define the variation of attractiveness β with respect to the distance r as follows:

$$\beta(r) = \beta_0 e^{-\gamma r^m}, m \geq 1 \tag{4}$$

where γ is an absorption coefficient and the distance r_{pq} between firefly p and q is determined by Eq. 5.

$$r_{pq} = \| X_{p,s} - X_{q,s} \| = \sqrt{\sum_{i=1}^{d} (X_{p,s} - X_{q,s})^2} \tag{5}$$

where d is the dimension, and $X_{p,s}$ is the s_{th} component of the spatial coordinate of the p_{th} firefly [19,20]. Finally, the movement determined by the brightness, from firefly p to firefly q, is computed by Eq 6.

$$X'_p = X_p + \beta(r)(X_p - X_q) + \alpha(rand - \frac{1}{2}) \tag{6}$$

where X_p is the current position of the p_{th} firefly and X'_p is the next generation position. α represents a randomly move that is calculated in each of iteration in order to decrease its value and to expect a more bounded search. The α value is computed as follows:

$$\alpha = 1 - \Delta\alpha^t \tag{7}$$

$$\Delta = 1 - \left(\frac{10^{-4}}{0.9}\right)^{\frac{1}{numIterations}} \tag{8}$$

4 Description of the Proposed Approach

In this section, we present the approach for solving SCPs and the associated heuristic feasibility operator.

1. Initialize the parameters: β, γ, number of fireflies, and maximum number of generation.
2. Initialize the random position in the multidimensional space $M = [X_1, X_2, X_3, X_n]$, where n is the number of fireflies, and X represents a solution, which in turn is modeled as a d-dimensional binary vector.
3. Each solution has its fitness and it is calculated by using the objective function of the SCP (Eq. 1).
4. The firefly move is computed by using Eq 6. The result is a real number, which needs to be transformed to a binary value. To this end, we employ the following transfer function, which was the best performing one from the functions proposed in [10].

$$S(x) = \left|\frac{2}{\pi} arctan(\frac{\pi}{2}x)\right|$$

5. The new solution found needs to pass for an evaluation process in order to check whether it satisfy the constraint. If so, the solution is kept, otherwise the solution is repaired by using the heuristic feasibility operator described in Sect 4.1.
6. Store the best solution and pass to the next iteration.
7. Stop, if termination criterion has been met (max. generation number). Otherwise, return to the step 3.

4.1 Heuristic Feasibility Operator

Firefly, as various metaheuristics, may provide solutions that violate the constraints of the problem. For instance, in the SCP, a new solution that has not all his rows covered, clearly violates a subset of constraints. In order to provide feasible solutions and to improve performance, we incorporate a heuristic operator that achieves the generation of feasible solutions, and additionally eliminates column redundancy [2].

To make all feasible solutions we compute a ratio based on the sum of all the constraint matrix rows covered by a column j.

$$\frac{Cost\ of\ a\ column\ j}{number\ of\ uncovered\ rows\ which\ it\ covers} \quad (9)$$

The unfeasible solution are repaired by covering the columns of the solution that had the lower ratio. After this, a local optimization step is applied, where column redundancy is eliminated. A column is redundant when it can be deleted and the feasibility of the solution is not affected.

Let:

I = The set of all rows,
J = The set of all columns,
α_i = The set of columns that cover row i, $i \in I$,
β_j = The set of rows covered by column j, $j \in J$,
S = The set of columns in a solution,
U = The set of uncovered rows,
w_i = The number of columns that cover row i, $i \in I$.

(i) Initialize $w_i := |S \cap \alpha_i|, \forall i \in I$
(ii) Initialize $U := \{i | w_i = 0, \forall i \in I\}$
(iii) For each row i in U(in increasing order of i):
  (a) Find the first column j in increasing order of j in α_i that minimizes $c_j/|U \cap b_j|$
  (b) Add j to S and set $w_i := w_i + 1, \forall i \in b_j$. Set $U := U - b_j$.
(iv) For each column j in S (in decreasing order of j), if $w_i \geq 2$, $\forall i \in \beta_j$, set $S := S - j$ and set $w_i := w_i - 1$, $\forall i \in \beta_j$.
(v) S is now a feasible solution for the SCP that contains no redundant columns.

Steps (i) and (ii) identify the uncovered rows. Steps (iii) and (iv) are "greedy" heuristics in the sense that in step (iii) columns with lower cost ratio are being considered first and in step (iv) the columns with higher cost are dropped whenever possible.

5 Experiments and Results

The firefly algorithm for solving SCP was implemented in Java and launched in a 1.8 GHz Intel i7 with 8 GB RAM machine running MS Windows 7. We use a population of 20 fireflies and $\gamma = 0.0002$, $\beta_0 = 1$, and 7000 iterations. The parameters values have empirically been defined after a large tuning phase. We test 65 SCP non-unicost instances (organized in sets: 4, 5, 6, A, B, C, D, NRE, NRF, NRG, NRH) from OR-Library (available at `http://www.brunel.ac.uk/~mastjjb/jeb/info.html`), all of them were executed 30 times. Each instance was pre-processed by using rules defined in Sect. 2. Table 1 shows detailed information about tested instances, where "Density" is the percentage of non-zero entries in the SCP matrix.

Table 1. Tested instances

Instance set	No. of instances	n	m	Cost Range	Density(%)	Optimal solution
4	10	200	1000	[1,100]	2	Known
5	10	200	2000	[1,100]	2	Known
6	5	200	1000	[1,100]	5	Known
A	5	300	3000	[1,100]	2	Known
B	5	300	3000	[1,100]	5	Known
C	5	400	4000	[1,100]	2	Known
D	5	400	4000	[1,100]	5	Known
NRE	5	500	5000	[1,100]	10	Unknown
NRF	5	500	5000	[1,100]	20	Unknown
NRG	5	1000	10000	[1,100]	2	Unknown
NRH	5	1000	10000	[1,100]	5	Unknown

Tables 2 and 3 show the results obtained using the proposed firefly algorithm. Column 1 depicts the instance number, m' represents the number of constraint matrix columns of the preprocessed instance. Column 3 depicts the best known optimum value for the instance. Columns 4 provides the minimum optimum value reached by the proposed firefly algorithm ($Min(impFF)$). Columns 5 depicts the minimum optimum value reached by the best firefly algorithm reported ($Min(FF)$) [12]. Finally, $RPD(impFF)$ shows the difference between Opt and $Min(impFF)$ in terms of percentage, which is computed as follows:

$$RPD(impFF) = \frac{(Min(impFF) - Opt)}{Opt} \times 100$$

The results illustrate the effectiveness of the improvements incorporated to the proposed approach (pre-processing and the heuristic feasibility operator), where the fitness reached by the improved firefly algorithm outperforms for the whole set of instances the results obtained by the best reported firefly algorithm [12]. The proposed approach is also able to provide RPDs equal (or very near) to 0 for various instances.

Table 2. Computational results on preprocessed instances

Instance	m'	*Opt*	$Min(impFF)$	$Min(FF)$ [12]	$RPD(impFF)$
4.1	169	429	430	481	0.23
4.2	212	512	515	580	0.58
4.3	225	516	520	619	0.77
4.4	200	494	500	537	1.21
4.5	215	512	514	609	0.39
4.6	229	560	561	653	0.17
4.7	194	430	431	491	0.23
4.8	215	492	497	565	1.016
4.9	243	641	643	749	0.31
4.10	200	514	523	550	1.75
5.1	220	253	257	296	1.58
5.2	262	302	307	372	1.65
5.3	215	226	229	250	1.32
5.4	225	242	242	277	0
5.5	185	211	212	253	0.47
5.6	211	213	214	264	0.46
5.7	220	293	297	337	1.36
5.8	245	288	291	326	1.04
5.9	230	279	285	350	2.15
5.10	229	265	269	321	1.50
6.1	212	138	141	173	2.17
6.2	243	146	148	180	1.36
6.3	237	145	148	160	2.06
6.4	200	131	131	161	0
6.5	249	161	164	186	1.86
A.1	383	253	256	285	1.18
A.2	387	252	257	285	1.98
A.3	391	232	238	272	2.58
A.4	378	234	237	297	1.28
A.5	387	236	238	262	0.84
B.1	453	69	70	80	1.44
B.2	459	76	76	92	0
B.3	498	80	80	93	0
B.4	488	79	80	98	1.26
B.5	460	72	72	87	0

Table 3. Computational results on preprocessed instances

Instance	m'	Opt	$Min(impFF)$	$Min(FF)$ [12]	$RPD(impFF)$
C.1	519	227	231	279	1.76
C.2	561	219	223	272	1.82
C.3	588	243	250	288	2.88
C.4	550	219	227	262	3.65
C.5	541	215	216	262	0.46
D.1	639	60	60	71	0
D.2	679	66	67	75	1.51
D.3	693	72	74	88	2.77
D.4	652	62	63	71	1.61
D.5	627	61	61	71	0
NRE.1	831	29	31	32	6.89
NRE.2	953	30	31	36	3.33
NRE.3	878	27	29	35	7.40
NRE.4	927	28	31	34	10.71
NRE.5	970	28	28	34	0
NRF.1	726	14	16	17	14.28
NRF.2	672	15	15	17	0
NRF.3	752	14	16	21	14.28
NRF.4	698	14	16	19	14.28
NRF.5	659	13	14	16	7.69
NRG.1	2076	176	176	230	0
NRG.2	1942	154	155	191	0.64
NRG.3	2003	166	168	198	1.20
NRG.4	1974	168	171	214	1.78
NRG.5	2022	168	171	223	1.78
NRH.1	2796	63	65	85	3.17
NRH.2	2745	63	66	81	4.76
NRH.3	2765	59	63	76	6.77
NRH.4	2779	58	61	75	5.17
NRH.5	2703	55	56	68	1.81

Fig. 1 shows the evolution of three representative SCP instances. The fast convergence of these instances demonstrates that the proposed approach can be competitive with others metaheuristics to solve the SCP, where instances rapidly converge in about 60 iterations. The experimental results also exhibit the robustness of the algorithm, which is able to reach reasonable good results by keeping the same parameter configuration.

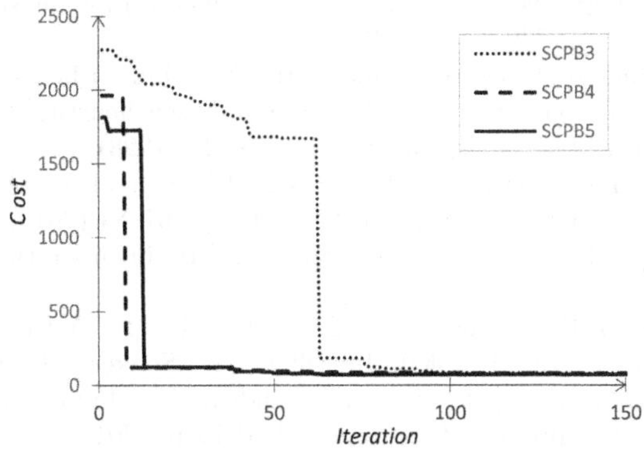

Fig. 1. Convergence for instances B.3, B.4 and B.5

6 Conclusions

In this paper we have presented an improved firefly algorithm for solving SCPs. We have incorporated pre-processing and an heuristic feasibility operator in order to improve performance. We have tested 65 non-unicost instances from the Beasley's OR-Library where the quality of results have clearly been increased for all instances w.r.t to the previous work. Additionaly, the proposed approach is also able to provide RPDs equal (or very near) to 0 for various instances. As future work, we plan to hybridize the firefly algorithm with another metaheuristics as well as to incorporate filtering techniques from the constraint programming domain to improve performance as illustrated in [13].

Acknowledgements. Ricardo Soto is supported by Grant CONICYT/ FONDECYT/INICIACION/ 11130459, Broderick Crawford is supported by Grant CONICYT/FONDECYT/ 1140897, and Fernando Paredes is supported by Grant CONICYT/FONDECYT/ 1130455.

References

1. Balas, E., Carrera, M.C.: A dynamic subgradient-based branch-and-bound procedure for set covering. Oper. Res. 44(6), 875–890 (1996)
2. Beasley, J.E., Chu, P.C.: A genetic algorithm for the set covering problem. Eur. J. Oper. Res. 94(2) (1996)
3. Beasley, J.E.: A Lagrangian heuristic for set covering problems. Naval Res. Logistics 37(1), 151–164 (1990)
4. Beasley, J.E., Chu, P.C.: A genetic algorithm for the set covering problem. Eur. J. Oper. Res. 94(2), 392–404 (1996)
5. Brusco, M.J., Jacobs, L.W., Thompson, G.M.: A morphing procedure to supplement a simulated annealing heuristic for cost- and coverage-correlated set-covering problems. Ann. Oper. Res. 86, 611–627 (1999)
6. Caprara, A., Toth, P., Fischetti, M.: Algorithms for the set covering problem. Ann. Oper. Res. 98, 353–371 (2000)

7. Caprara, A., Fischetti, M., Toth, P.: A heuristic method for the set covering problem. Oper. Res. 47(5), 730–743 (1999)
8. Caserta, M.: Tabu search-based metaheuristic algorithm for large-scale set covering
9. Ceria, S., Nobili, P., Sassano, A.: A Lagrangian-based heuristic for large-scale set covering problems. Math. Program. 81(2), 215–228 (1998)
10. Chandrasekaran, K., Sishaj, P.S., Padhy, N.P.: Binary real coded firefly algorithm for solving unit commitment problem. Inf. Sci. 249, 67–84 (2013)
11. Chvatal, V.: A greedy heuristic for the set-covering problem. Math. Oper. Res. 4(3), 233–235 (1979)
12. Crawford, B., Soto, R., Olivares-Suárez, M., Paredes, F.: A Binary Firefly Algorithm for the Set Covering Problem. In: Silhavy, R., Senkerik, R., Oplatkova, Z.K., Silhavy, P., Prokopova, Z. (eds.) Modern Trends and Techniques in Computer Science. AISC, vol. 285, pp. 65–73. Springer, Heidelberg (2014)
13. Crawford, B., Castro, C., Monfroy, E., Soto, R., Palma, W., Paredes, F.: Dynamic Selection of Enumeration Strategies for Solving Constraint Satisfaction Problems. Romanian Journal of Information Science and Technology 15(2), 106–128 (2013)
14. Fisher, M.L., Kedia, P.: Optimal solution of set covering/partitioning problems using dual heuristics. Management Science 36(6) (1990)
15. Garey, M.R., Johnson, D.S.: Computers and Intractability: A Guide to the Theory of NP-Completeness. W. H. Freeman & Co., New York (1990)
16. Lan, G., DePuy, G.W.: On the effectiveness of incorporating randomness and memory into a multi-start metaheuristic with application to the set covering problem. Comput. Ind. Eng. 51(3), 362–374 (2006)
17. Vasko, F.J., Wolf, F.E.: Optimal selection of ingot sizes via set covering. Oper. Res. 35(3), 346–353 (1987)
18. Vasko, F.J., Wilson, G.R.: Using a facility location algorithm to solve large set covering problems. Operations Research Letters 3(2), 85–90 (1984)
19. Yang, X.S.: Nature-Inspired Metaheuristic Algorithms. Luniver Press, UK (2008), Inspired Computing (NaBIC 2009), India, p. 210. IEEE Publications, USA (December 2009)
20. Yang, X.-S.: Firefly algorithms for multimodal optimization. In: Watanabe, O., Zeugmann, T. (eds.) SAGA 2009. LNCS, vol. 5792, pp. 169–178. Springer, Heidelberg (2009)

Self-learning of the Containers Service Coordinator Agent in Multi-agent Automation Environment of Transit Cargo Terminal

M.V. Lutsan, E.V. Nuzhnov, and V.V. Kureichik

Southern Federal University, Rostov-on-Don, Russia
nev@sfedu.ru

Abstract. The article deals with some problems of transport logistics, concerning the improvement of the organization and automation of basic processes of transit cargo terminal. The terminal operates with three-dimensional blocks which contain packaged goods: it receives arriving blocks, temporarily stores and sends them to the customers. Blocks are transported on trucks in a receptacle of limited size, conventionally called a container. When the resources of loading and unloading of containers, transport and storage are limited, and there are not enough some ordered blocks, queues of containers waiting for loading and unloading may occur. The authors applied multi-agent approach to the terminal management: the work is distributed among the four agents: containers unloading agent, warehouse agent, containers loading agent and the main coordinator agent. A new function of coordinator agent – self-learning based on the results of its previous work – is presented and described in the article. Self-learning is an important property of intelligent agent. This property can contribute to increasing the effectiveness of using the agents for the organization and automation of transit cargo terminal.

Keywords: Self-learning, Coordinator agent, Transit terminal, Multi-agent approach, Container.

1 Introduction

Development and growth of commodity production is associated with the improvement of processes organization of goods transportation. Continental, national and regional trucking of goods is of great importance. Usually items are packed into three-dimensional receptacles of different sizes (named blocks), blocks are transported in receptacles of trucks that can be called containers. A key role in the transport network of blocks in freight containers is played by transit cargo terminals, where the following processes take place: the blocks unloading from containers of cars arriving; blocks temporary warehousing, as well as their loading (three-dimensional packaging) into containers delivered on order. Articles [1-2] describe the basic tasks for such transit terminals and approaches to their solutions. Improving the functioning effectiveness of the transit automobile transshipment terminal by automating planning and scheduling processes of loading and unloading containers with three-dimensional rectangular

R. Silhavy et al. (eds.), *Artificial Intelligence Perspectives and Applications,*
Advances in Intelligent Systems and Computing 347, DOI: 10.1007/978-3-319-18476-0_12

blocks of different sizes and processes of temporary blocks storage is the actual problem. The contribution made by authors of the article to solving this problem consists in solving the management tasks of queues of unloading arriving containers and loading outgoing ones, three-dimension blocks packing into container, as well as the management of warehouse of temporary arriving blocks storage [3-7]. To improve the organization and automation of containers and blocks processing in transit terminal the authors of the article applied multi-agent approach (described by Tarasov V.B. in [8]) to the transit cargo terminal management and described it in the articles [9-13]. The terminal processes are distributed among the four agents on the base of two level scheme of their functioning and interaction. At the lower level the following agents work: the agent of unloading arriving containers; the temporary storage warehouse agent of arriving blocks; the agent of loading blocks into container at the request of customers. At the top level the coordinator agent works. As it's known, learning is an important feature of the intelligent system [14-16]. This article regards the intellectual component of the coordinator agent activities associated with its self-learning on the base of its previous work results.

2 Tasks of the Coordinator Agent

1. Selection of a filled container for unloading from the arriving containers list based on their priorities, storage requirements and orders for containers.
2. Selection of order for loading a container from the list of orders and selection of an empty container, taking into account the orders and containers priorities, as well as the individual types of blocks priorities.
3. Warehouse management of temporary blocks storage to prevent shortage of blocks required for execution of orders, as well as to prevent the warehouse filling.
4. Control of containers queues and orders processing to prevent containers or containers order delay exceeding the processing time limit.
5. Saving of terminal characteristics and containers priorities after the decision-making and using this information for future decisions.

Unlike the lower-level warehouse agent and agents of loading and unloading, the coordinator agent has feedback mechanisms and the ability to accumulate experience. Every decision made by the coordinator agent is stored in the database of states, and each step is based on the states in the previous steps [9-13].

3 The Implementation of a Feedback Mechanism

Every decision of the coordinator agent is estimated by a set of negative characteristics that the terminal can have after the decision has been made by the agent. These negative characteristics take only the values {yes, no}. The presence of the characteristics indicates the negative consequences of the decision. The following set of negative characteristics is used:

- H_1 – there is a blocks deficiency in the terminal;
- H_2 – blocks temporary storage warehouse is filled;
- H_3 – waiting period for any container is exceeded;
- H_4 – waiting period for any order is exceeded.

A set of characteristics defines the vector of the current state of the terminal:

$$H = (H_1, H_2, H_3, H_4) \quad (1)$$

The best decision is the one that leads to a state of the terminal, described by the vector of zero length H $\|H\| = 0$, and the worst results in the state described by the vector of Euclidean length $\|H\| = 2$. We introduce the factor $\omega, \omega \in [0,1]$, determining the quality of the decision made:

$$\omega = 1 - \|H\|/2 \quad (2)$$

Thus, $\omega = 0$ means the worst decision, $\omega = 1$ is the best one.

Since each time the terminal may have a different number of containers in the unloading list, and accordingly, the agent of unloading can send to the coordinator agent vector W_u of varying length, the coordinator agent normalizes vector W_u length so that it is the maximum length of vector W'_{ui}, where $i = \overline{1, h}$, h is the number of states in the state database of the coordinator agent. When the length D of vector Wu is D > D', where D' is the maximum vector length in the states base, vector Wu is complemented with zeros to match the length D = D'. Otherwise, all vectors in the state database are complemented with zeros to the length D so that D' = D.

4 Calculation of Priorities for Containers Arriving for Unloading

Initially, the adjustment of estimates provided by the agent of unloading is made. To do this every possible outcome of the agent coordinator's decision-making is estimated, and ω_i is calculated for each W_{ui}. Then each W_{ui} is adjusted:

$$W_{ui} = W_{ui} - W_{ui} * (1 - \omega_i) * (1 - W_{ui}) \quad (3)$$

The equation (2) shows that the adjustment is inversely proportional to the initial priority of the container. The higher the initial priority of the container, the less the adjustment influences it. After applying corrections to the vector Wu, the coordinator agent using the previous state chooses the next container for unloading. To do this the coordinator agent takes the information in the state s-b, b= $\overline{1, S}$, where s is the current state of the coordinator agent, S is maximum number of states which is used to adjust the current step while s-1>S. The function of priorities adjusting is additive, the value of correction of previous states is inversely proportional to the distance from the current state. The aim is to reduce the influence of states, which are very far from the current state, on decision-making. The task of the coordinator agent also includes the search for past states of containers having the trade (block) composition the most corresponding to the minimum threshold μ, below which the containers are considered to be different and previous decisions made for these containers do not affect the containers in the next iterations. The search of the most appropriate containers in a particular state is carried out in the following way:

$$\forall c_i \in U, max\Delta_i^b = 1 - \left\| \frac{U_i^b - c_i}{c_i} \right\| \quad (4)$$

where $i = \overline{1, D}$, U is the set of all block compositions of containers to be unloaded at the current moment; U_i^b is a set of all block compositions of a container, for the state b and the container i.

If $\Delta_i^b < \mu$, then it is believed that for the container c_i in the state b there were no similar containers, and the state b is not taken into account when adjusting the decision, so we set $\Delta_i^b = 0$. Adjusted importance (priority) of the container i is described by the formula:

$$P_{ui} = W_{ui} + \sum_{b=1}^{S} (W_{ui}^b - \partial^b) \Delta_i^b (1 - \partial'^b) \quad (5)$$

where ∂^b for b=1 takes the initial value W_{ui} for the current container, and for b>1 takes calculated value W_{ui} for the previous step (i.e. for the step b-1), for b=1 $\partial'^b = 0$, and for b>1 $\partial'^b = \partial^b$.

In the formula (4) the additive correction function of value W_{ui} based on the values W_{ui} in the previous iterations is used. The advantages of this function are the following. The previous decision affects the result of the current decision at most; the impact of previous decisions on the current decision is reduced in proportion to the distance between the current step to the previous one.

5 Example

Let us consider the example of calculating the value of W_{ui} by formula (5), when three filled containers with the following composition of the blocks are done: U_1=<100; 140; 200>, U_2=<110; 150; 200>, U_3=<115; 160; 205>. Let us calculate the importance of the container U_3, to be unloaded in step 3. The container U_1 was unloaded in step 1 with W_{u1} = 0.4, and the container U_2 was unloaded in step 2 with W_{u2} = 0.8. Before the adjustment application in accordance with formula (5) the importance of the container U_3 in step 3 W_{u3} = 0.6. Calculate the correspondence container U_3 to other containers according to formula (5) (where for Δ_1 for convenience take appropriate the container U_3 to the container U_1, and for Δ_2 – appropriate the container U_3 to the container U_2):

$$\Delta_1 = 1 - \left\| \frac{\langle 110; 140; 200 \rangle - \langle 115; 160; 205 \rangle}{\langle 115; 160; 205 \rangle} \right\| = \quad (6)$$

$$= 1 - \left\| \frac{\langle -5; -20; -5 \rangle}{\langle 115; 160; 205 \rangle} \right\| = 1 - \| \langle -0.05; -0.125; 0.024 \rangle \| =$$

$$= 1 - 0.137 = 0.863$$

$$\Delta_2 = 1 - \left\| \frac{\langle 110;\ 150;\ 200\rangle - \langle 115;\ 160;\ 205\rangle}{\langle 115;\ 160;\ 205\rangle} \right\| = \tag{7}$$

$$= 1 - \left\| \frac{\langle -5; -10; -5\rangle}{\langle 115;\ 160;\ 205\rangle} \right\| = 1 - \|\langle -0.05; -0.125;\ 0.024\rangle\| = 1 - 0.08 = 0.92$$

Let us calculate the corrected value Wu3 for the container U3 in step 3: W_{u3}=0.6+(0.4–0.6)*0.863+(0.8–(0.6+(0.4–0.6)*0.863*(1–0.6–(0.4–0.6)*0.863))*0.92 =0.4274+0.275=0.702.

Thus, the final value of the importance of the container U3 is equal to 0.702. The dependence of corrective action on the iteration step number is shown in Figure 1, where the dashed line shows the extrapolated values for the above example.

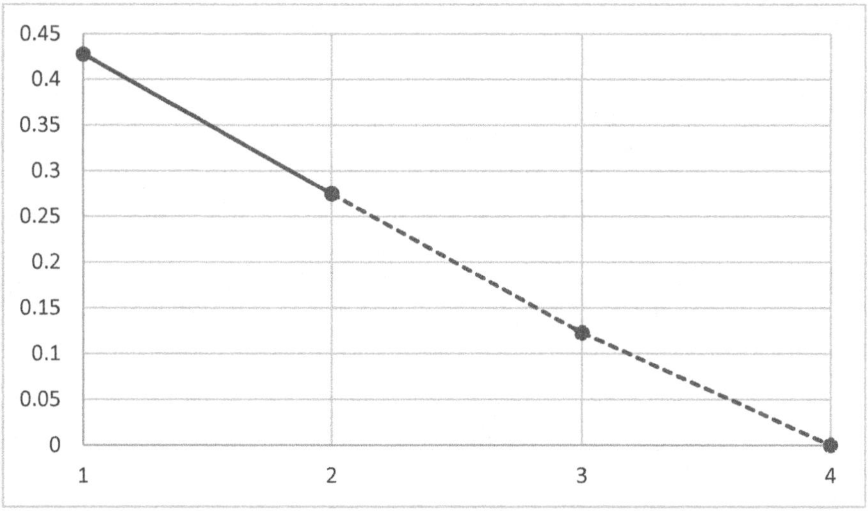

Fig. 1. The dependence of corrective action on the iteration step number

Let us compare the effect of compliance each container to other containers from previous iterations (Δ) on the correction value for the current container W_{u1} and various W^b_u the value of the adjusted W^b_{u1} in each step b, $b = \overline{1,4}$ in Table 1 and in Table 2.

Table 1. Dependence of corrective action on various parameters

b	$W_{ul} = 0.6$	$\Delta = 0.5$	$\Delta = 0.8$	$\Delta = 0.9$
1	0.8	0.65	0.74	0.77
2	0.6	0.61	0.66	0.68
3	0.4	0.58	0.61	0.62
4	0.2	0.55	0.57	0.59

Table 2. Dependence of corrective action on various parameters

b	$W_{u1} = 0.3$	$\Delta = 0.5$	$\Delta = 0.8$	$\Delta = 0.9$
1	0.6	0.57	0.54	0.57
2	0.8	0.66	0.62	0.66
3	0.7	0.62	0.58	0.62
4	0.6	0.59	0.56	0.59

In Table 1 column b reflects the step b number; column $W_{u1} = 0.6$ shows priority container according W_{u1} in step b; the remaining columns present value W'_{u1} adjusted to take this step. Diagram of corrective actions is shown in Figure 2.

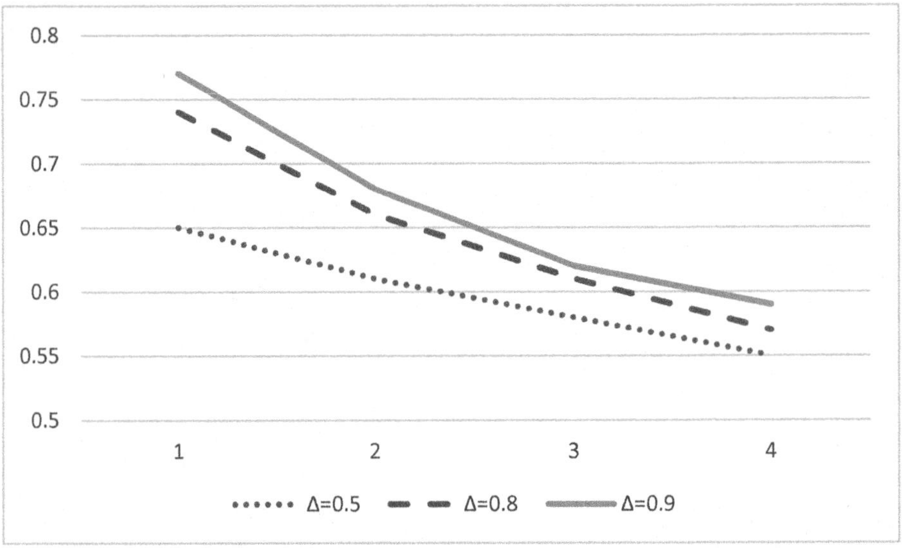

Fig. 2. Diagram of corrective actions

The dependence of the corrective action on various parameters for positive adjustment of the initial decision is shown in Table 2. Diagram of corrective actions is shown in Figure 3.

On the base of analysis of Figure 2 and Figure 3 it can be concluded that with increasing the iteration number (horizontal axis) the value of change W_{u1} decreases, which is consistent with the formula (4). With increasing of the number of steps, in each next step the value of change W_{u1} decreases, which is consistent with the principles of neural networks training and Q-learning [14-16].

To improve the efficiency of the arrival and departure containers areas, if technically feasible and necessary resources (containers) exist, the coordinator agent can assess the feasibility of increasing the number of platforms (for parallelizing processes) unloading / loading of containers or reducing their number at overstocking (filling and lack of space) stock.

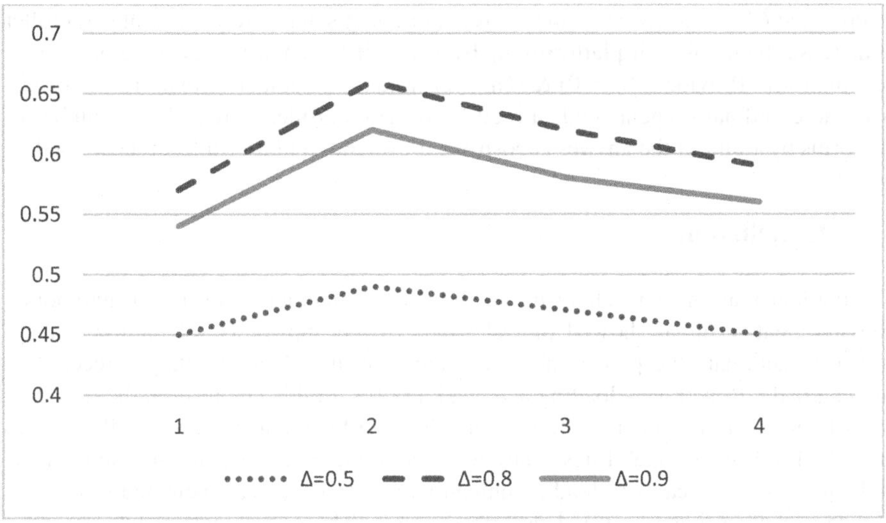

Fig. 3. Diagram of corrective actions

6 Optimizing the Number of Platforms for Unloading

Let the number of platforms for unloading be $N_U = \overline{1,p}$, where p is maximum number of platforms for unloading. Let us enter logical state variables $S_i = \{0,1\} \mid (i=\overline{1,4})$: there are delivered and not unloaded containers (S_1); there is a free platform for unloading (S_2); the warehouse has free space (S_3); an empty container is needed for loading (S_4). Then the condition of reasonability for increasing the number of platforms for unloading is:

$$S^U = S_1 \,\&\, S_2 \,\&\, S_3 \,\&\, S_4 \tag{8}$$

Provided ($S^U = 1$) & ($N_U \neq p$) the coordinator agent sends a request to the controller to increase the number of platforms N_U by 1 and if the answer is positive, it opens a new platform. Provided ($S^U = 0$) & ($N_U \neq 1$), when the number of platforms is excessive, the coordinator agent sends a request to the controller to reduce the number of platforms by 1 and if the answer is positive, it excludes a platform from those used.

7 Optimizing the Number of Platforms for Loading

Let the number of platforms for loading be $N_L = \overline{1,q}$, where q is maximum number of platforms for loading. Let us enter logical state variables $S_i = \{0,1\} \mid (i=\overline{5,7})$: there are containers waiting for being loaded (S_5); there is a free platform for loading (S_6); the warehouse has all blocks for at least one plan of loading (S_7). Then the condition of reasonability for increasing the number of platforms for loading is as follows:

$$S^L = S_5 \,\&\, S_6 \,\&\, S_7 \tag{9}$$

Provided (S^L=1) & ($N_L \neq q$) the coordinator agent sends a request to the controller to increase the number of platforms N_L by 1 and, if the answer is positive, it opens a new platform. Provided ($S^L = 0$) & ($N_L \neq 1$), when the number of platforms is excessive, the coordinator agent sends a request to the controller to reduce the number of platforms by 1 and, if the answer is positive, it excludes a platform from those used.

8 Conclusion

The implementation of mechanisms of the agents' interaction in transit automobile transshipment terminal showed the relevance of the application of multi-agent approach to automate the processing of containers in the chain «waiting – reception» and «unloading - storage - loading and sending off». In this connection, the introduction of self-learning function offered into the coordinator agent work will allow to raise the level of its capabilities. They will determine its actions and decisions made in the most difficult cases of loading/unloading processes management and warehouse service processes identified and described above. The coordinator agent can make operational key decisions to optimize the number of platforms for containers loading and unloading that increase the overall efficiency of the transit automobile cargo terminal.

Acknowledgments. This research is sponsored by the Ministry of Education and Science of the Russian Federation. Project #8.823.2014.

References

1. Lee, Y.H., Jung, J.W., Lee, K.M.: Vehicle routing scheduling for cross-docking in supply chain. Computers & Industrial Engineering 51, 247–256 (2006)
2. Yu, W., Egbelu, P.J.: Scheduling of inbound and outbound trucks in cross docking systems with temporary storage. European Journal of Operational Research 184, 377–396 (2008)
3. Lutsan, M.V., Nuzhnov, E.V.: Methods and facilities of containers queues processing in automated cargo terminal. In: Proceedings of SFU. Technical Sciences. Thematic Issue "Intelligent CAD", vol. (7), pp. 179–184. Publishing House of SFU, Taganrog (2013)
4. Lutsan, M.V., Nuzhnov, E.V.: Information system architecture to support the operation of automated cargo terminal. In: Proceedings of the Congress on Intelligent Systems and Information Technology "IS & IT 2013". Scientific publication in 4 volumes, vol. 2, pp. 192–199. FIZMATLIT, Moscow (2013)
5. Lutsan, M.V., Nuzhnov, E.V.: Intelligent information system supporting the activities of the cargo terminal. In: Proceedings of the Kabardino-Balkar Scientific Centre of Russian Academy of Sciences, vol. (4), pp. 48–55. Publishing KBSC Russian Academy of Sciences, Nalchik (2013)
6. Lutsan, M.V., Nuzhnov, E.V.: Three-dimensional packaging of rectangular objects with determination of their loading sequence. In: Proceedings of the Congress on Intelligent Systems and Information Technology "IS & IT 2011". Scientific publication in 4 volumes, vol. 3, pp. 285–291. FIZMATLIT, Moscow (2011)

7. Lutsan, M.V., Nuzhnov, E.V.: Solution of the problem of three-dimensional packaging with palletizing of containers. In: Proceedings of SFU. Technical Sciences. Thematic Issue "Intelligent CAD", vol. (7), pp. 196–204. Publishing House of SFU, Taganrog (2014)
8. Tarasov, V.B.: Of multi-agent systems to intelligent organizations: philosophy, psychology, informatics, p. 352. Editorial URSS, Moscow (2002)
9. Lutsan, M.V., Nuzhnov, E.V.: Automation cargo terminal using intelligent agents. In: Proceedings of the Third International Conference "Automation of Management and Intelligent Systems and the Environment". Scientific publication, vol. 2, pp. 67–70. Publishing KBSC RAS, Nalchik (2012)
10. Lutsan, M.V., Nuzhnov, E.V.: The use of intelligent agents for automated cargo terminal. In: Proceedings of SFU. Technical Sciences. Thematic Issue "Intelligent CAD", vol. (7), pp. 174–180. Publishing house of SFU, Taganrog (2012)
11. Lutsan, M.V., Nuzhnov, E.V.: Heuristics of intelligent agents in automated cargo terminal. In: Proceedings of SFU. Technical Sciences. Thematic Issue "Intelligent CAD", vol. (11), pp. 232–237. Publishing house of SFU, Taganrog (2012)
12. Lutsan, M.V., Nuzhnov, E.V.: Using the multi-agent approach for automation of cargo container terminal. In: Information Technologies, Systems Analysis and Administration/Proceedings of X Russian Scientific Conference of Young Scientists and Students, vol. 2, pp. 5–13. Publishing House of SFU, Taganrog (2012)
13. Nuzhnov, E.V., Lutsan, M.V.: Information environment to support automated cargo terminal through the use of intelligent agents. In: Kureichik, V.M. (ed.) Intelligent Systems. Collective Monograph, vol. (6), pp. 227–242. FIZMATLIT, Moscow (2013)
14. Panait, L., Luke, S.: Cooperative multi-agent learning: The state of the art. Autonomous Agents and Multi-Agent Systems 11(3), 387–434 (2005)
15. Nielsen, M.: Neural Networks and Deep Learning, `http://neuralnetworksanddeeplearning.com/` (visited: November 01, 2014)
16. Ollington, R.B., Vamplew, P.W.: Concurrent Q-Learning: Reinforcement Learning for Dynamic Goals and Environments. International Journal of Intelligent Systems (20), 1037–1052 (2005)

On the Performance of Ensemble Learning for Automated Diagnosis of Breast Cancer

Aytuğ Onan

Celal Bayar University, Faculty of Engineering,
Department of Computer Engineering,
Manisa, 45140, Turkey
aytug.onan@cbu.edu.tr

Abstract. The automated diagnosis of diseases with high accuracy rate is one of the most crucial problems in medical informatics. Machine learning algorithms are widely utilized for automatic detection of illnesses. Breast cancer is one of the most common cancer types in females and the second most common cause of death from cancer in females. Hence, developing an efficient classifier for automated diagnosis of breast cancer is essential to improve the chance of diagnosing the disease at the earlier stages and treating it more properly. Ensemble learning is a branch of machine learning that seeks to use multiple learning algorithms so that better predictive performance acquired. Ensemble learning is a promising field for improving the performance of base classifiers. This paper is concerned with the comparative assessment of the performance of six popular ensemble methods (Bagging, Dagging, Ada Boost, Multi Boost, Decorate, and Random Subspace) based on fourteen base learners (Bayes Net, FURIA, K-nearest Neighbors, C4.5, RIPPER, Kernel Logistic Regression, K-star, Logistic Regression, Multilayer Perceptron, Naïve Bayes, Random Forest, Simple Cart, Support Vector Machine, and LMT) for automatic detection of breast cancer. The empirical results indicate that ensemble learning can improve the predictive performance of base learners on medical domain. The best results for comparative experiments are acquired with Random Subspace ensemble method. The experiments show that ensemble learning methods are appropriate methods to improve the performance of classifiers for medical diagnosis.

Keywords: Ensemble learning, Breast cancer diagnosis, Classification.

1 Introduction

The automated diagnosis of diseases is a promising field in medical informatics which involves developing models that should perform as accurate and as efficient as possible. Cancer is a class of diseases characterized by abnormal cell growth. There are many different types of cancer each of which is classified by the type of cell that is originated from. Breast cancer is the most commonly diagnosed cancer (not taking skin cancer into account) and the second most common cause of cancer death among U.S. women [1]. Moreover, the diagnosis and treatment of breast cancer in its earliest phases as possible enables dramatic improvements in the outcomes of cancer patient

R. Silhavy et al. (eds.), *Artificial Intelligence Perspectives and Applications*,
Advances in Intelligent Systems and Computing 347, DOI: 10.1007/978-3-319-18476-0_13

[2]. The early detection and accurate diagnosis of breast cancer is crucial, since a long-term survival rate can be acquired for females with not metastasized breast cancer [3].

Machine learning algorithms are widely employed for automated diagnosis of breast cancer with high classification accuracies. In [4], artificial neural network was utilized to predict 5-, 10- and 15-year breast cancer specific survival rates. The results indicated that neural networks can be used as a viable tool for cancer survival prediction due to their high accuracy and good predictive performance.

In [5], an empirical analysis of three data mining techniques (Naive Bayes, the back-propagated neural network and decision tree algorithm) on the survivability rate prediction of breast cancer patients was conducted. The results reported that decision tree algorithm (C4.5) has a better predictive performance. Akay [6] proposed a support vector machine based model combined with feature selection to apply on diagnosing breast cancer with 99.51% classification accuracy. Delen et al. [7] utilized artificial neural networks (ANN) and decision trees along with logistic regression (LR) to develop the prediction model. The experimental results indicated that the decision tree (C5) outperforms ANN and LR in terms of accuracy, sensitivity and specificity. In [8], adaptive neuro-fuzzy inference system (ANFIS) was integrated with the neural network adaptive capabilities and the fuzzy logic to diagnose breast cancer and the reported accuracy was 99.08%. In [9], a breast cancer survivability prediction model based on bagging with Random Tree is proposed. In [10], decision tree (C5) algorithm was integrated with bagging on breast cancer diagnosis. The experimental results indicated that with the use of bagging algorithm, the predictive performance of the model was increased. In [11], CART (Classification and Regression Tree) classifier combined with the bagging algorithm with feature selection was applied to breast cancer diagnosis. The comparative study highlighted the importance of pre-processing and ensemble learning to enhance classification accuracy. A detailed survey on application of machine learning techniques in cancer prediction and prognosis was reported in [12-13].

Recent years have seen a growing interest in using ensemble learning techniques, which trains a set of base classifiers and combines their outputs with a fusion strategy [14]. To our knowledge, however, there is no an extensive empirical comparative study about the ensemble learning methods on medical diagnosis. The basic motivation for this study stems from this involvement.

In this study, a comparative analysis of the performance of seven popular ensemble methods (Ada Boost, Bagging, Dagging, Multi Boost, Decorate, Logit Boost, and Random Subspace) based on fourteen base learners (Bayes Net, FURIA, K-nearest Neighbors, C4.5, RIPPER, Kernel Logistic Regression, K-star, Logistic Regression, Multilayer Perceptron, Naïve Bayes, Random Forest, Simple Cart, Support Vector Machine, and LMT) for automatic detection of breast cancer is presented.

The outline of this study is as follows. In "Classification Algorithms" section, a high-level description for machine learning classifiers is given. In "Ensemble Learning Methods" section, ensemble methods are briefly explained. In "Experimental Study" section, the details for data set, evaluation metrics, experimental process and the empirical results are presented. Finally, "Conclusions" section presents the concluding remarks of the study.

2 Classification Algorithms

The following presents the short introduction of various machine learning classifiers used in the experimental study.

RIPPER (Repeated Incremental Pruning to Produce Error Reduction) is a rule-based classification method which extends IREP rule learning algorithm in terms of error rates [15]. The algorithm mainly consists of two phases: building phase and optimization stage. The algorithm builds up a rule set greedily by adding one rule at a time. Initially, the training data is sorted in the order of less prevalent class to the more frequent class based on corresponding class frequencies. The instances associated with the least prevalent class are divided into positive subset whereas the remaining instances are assigned to negative subset [16]. Then, the rule set that splits the least prevalent class from the other classes is found with IREP algorithm. Afterwards, an empty rule set is initialized and positive and negative subsets are divided into four groups, namely growing positive, growing negative, pruning positive and pruning negative subsets such that positive subsets consist of instances associated with the least prevalent class. Growing one rule is done by growing positive and growing negative subsets by greedily adding conditions to the rule. This is done by iteratively adding conditions that maximize the information gain criterion until the rule cannot cover any negative instances from the growing dataset. After rule growing, rule pruning is applied to eliminate overfitting problem. Then, rule optimization is applied to the rule set.

FURIA (Fuzzy Unordered Rule Induction Algorithm) is a fuzzy rule-based classification method which modifies and extends the well-known rule learner RIPPER algorithm [17]. FURIA preserves simple rule set structure of RIPPER algorithm, but instead of conventional rules and rule lists, FURIA algorithm uses fuzzy rules and unordered rule sets. Moreover, the pruning step of RIPPER algorithm was omitted due to the negative effect on the performance. In addition, it uses an efficient rule stretching method to deal with uncovered examples.

IBk algorithm [18] is an instance based learning algorithm which is a slightly modified version of K-nearest neighbour algorithm. The algorithm can determine the appropriate value for K based on cross-validation. It normalizes the ranges of attributes, processes instances incrementally and has a policy for tolerating missing values. Moreover, the algorithm employs a "wait and see" evidence-gathering method to determine which of the saved instances are expected to perform well during classification. In addition, the algorithm enables users to specify several choices of distance weighting [19].

K-star algorithm [20] is an instance based learning algorithm which uses an entropy based distance function. It handles with symbolic attributes, real-valued attributes and missing values properly owing to the use of entropy as a distance function. The technique of summing probabilities over all possible paths overcomes the problem of smoothness.

Naïve Bayes classifier [21] is a statistical classification algorithm which is based on Bayes' theorem. Naïve Bayes classifier gives comparable results with decision tree and neural network algorithms and have high accuracy and speed on large datasets [22]. Naïve Bayes classifier has a clear semantics in representing, using and learning probabilistic knowledge. The assumptions of accepting that the predictive attributes are conditionally independent given the class and no hidden or latent attributes influence the predictive process make the algorithm a viable tool for classification and learning [21].

Bayes Net algorithm is a Bayesian classification algorithm which specify joint conditional probability distributions [22]. Bayes Net algorithm consists of a directed acyclic graph and a set of conditional probability tables. Each random variable is represented by a node in the directed acyclic graph. The conditional probability table for the values of the variables indicate each possible combination of the values of its parent nodes [22]. Training process of Bayesian networks consists of two stages, namely learning a network structure and learning the probability tables [23]. There are several different ways of structure learning, such as local score metrics, conditional independence tests, global score metrics and fixed structure. Based on these ways, a number of search algorithms, such as hill climbing, simulated annealing and tabu search are implemented in Weka [23].

Logistic regression is a statistical model that predicts the probability of some event occurring as a linear function of a set of predictor variables. Linear regression suffers from two problems: the membership values are not proper probability values and the least-squares regression takes errors as both statistically independent and normally distributed with the same standard deviation. In order to get rid of the aforementioned problems, logistic regression generates a linear model based on a transformed target variable [24]. Logistic regression with a ridge estimator improves the parameter estimation and diminish the error made by following predictions [25].

Kernel logistic regression model is a statistical classifier that generates a fit model by minimizing the negative log-likelihood with a quadratic penalty using BFGS optimization [24].

Multilayer perceptron (MLP) is a feed forward classifier with one or more hidden layers that uses back-propagation to classify instances. The architecture of a multilayer perceptron typically consists of input layer, hidden layers and output layer, where the input signals are propagated in forward direction [26]. Multilayer perceptron generally uses back-propagation as a learning method. In back-propagation schema, input signals are propagated through the network from left to right and error signals are propagated in reverse direction. When there is a difference between actual and desired outputs, the weights are adjusted so that error is reduced.

SMO algorithm is support vector machine classifier using sequential minimal optimization for training. Support vector machine is a classification method for classification of linear and nonlinear data which uses a nonlinear mapping for transforming the original data into a higher dimension [22]. Support vector machines have highly accurate predictive performance. Besides, they are less vulnerable to overfitting problems. On the other hand, the training time of the algorithm is extremely slow and training algorithms are too complex. To fill this gap, Platt (1998) [27] proposed a sequential minimal optimization for training that breaks large quadratic programming problem required for training into smaller subsets as possible.

J48 is a slightly modified version of C4.5 in WEKA. C4.5 is a successor of ID3 algorithm [28]. The test attribute selection criteria of the algorithm is information gain to overcome the attribute bias problem of ID3. For a given set, each time the algorithm selects an attribute with the highest information gain. The algorithm can deal with continuous and default attribute values properly. The pruning mechanism of the algorithm overcomes the problem of overfitting, eliminates the exceptions and noise in the training set [29].

Random Forest algorithm is an ensemble of classification or regression trees, induced from bootstrap samples of the training data [30]. In this model, the generalization error of the classifier depends on power of the individual trees and the association between the trees. Random feature selection is used in the tree induction process. This enables algorithm to perform comparable to Adaboost algorithm and to be tolerable with noisy data.

LMT (Logistic Model Trees) algorithm is a classification algorithm that integrates decision tree induction with logistic regression [31]. The tree structure of the algorithm is grown in a similar manner to the C4.5 algorithm. Here, an iterative training of additive logistic regression models is performed. By splitting, the logistic regressions of the parent node are passed to the child nodes. This provides to have all parent models and probability estimates for each class at the leaf nodes of the final model [32].

CART (Classification and Regression Trees) algorithm is a classification model that is generated by recursively partitioning the data space and fitting a simple prediction model for each partition [33]. The algorithm can deal with highly skewed data, multi-state numerical data and the order or disorder data. The test attribute selection criteria of the algorithm is Gini coefficient [29]. CART employs 10-fold cross validation to estimate errors.

3 Ensemble Learning Methods

The following explains the well-known ensemble learning methods used in the experimental study.

Bagging (bootstrap aggregating) is a popular ensemble learning method proposed by Breiman [34]. The method aims to obtain a single prediction with increased accuracy by combining the outputs of individual classifiers. Bootstrapping is done by random sampling with replacement. The sizes of each sample is identical to the original training set's size. The composite bagged classifier returns the class prediction based on the majority voting. Hence, bagging can eliminate the instability of individual base learners [35]. In bagging, each instance is chosen based on equal probability.

Dagging is a similar ensemble learning method to Bagging, but instead of bootstrapping, it uses disjoint samples [36]. Thus, it is an effective method when individual classifiers have bad time complexity. The outputs of individual classifiers are amalgamated via voting.

Boosting [37] encompasses a family of methods to obtain a series of classifiers. The training sets for each classifier are adopted based on the classification performance of the classifiers in the earlier cycle so that classifiers in newer cycles can perform better on the instances that are incorrectly classified by the earlier classifiers [38]. Among the many boosting algorithms in the literature, AdaBoost algorithm is one of the most widely used one owing to its speed, robustness, simplicity, not requiring parameter tuning and prior knowledge about weak classifiers [39].

Multi Boosting aims to use the two methods with different mechanisms, namely AdaBoost boosting method and a variant of bagging, together to outperform either in isolation [40]. While the method preserves AdaBoost algorithm's high bias and variance reduction, it also benefits from wagging's superior variance reduction property.

Decorate is an ensemble learning method for generating ensembles that directly constructs diverse classifiers by using artificially-constructed training examples [41]. The larger the ensemble, the more accurate the produced model. However, this may require greater training time and complexity [24].

Random Subspace is an ensemble learning method proposed by Ho [42]. It combines multiple classifiers trained on randomly selected feature subspaces. This method aims to avoid overfitting while providing high accuracy rate.

4 Experimental Results

4.1 Dataset

The effectiveness of the ensemble learning methods for medical diagnosis is verified by using a well-known Wisconsin breast cancer data set [43, 44]. The data set was collected from the patients of University of Wisconsin-Madison Hospitals. The data consists of 699 records. However, in order to be consistent, we removed the 16 records with missing values, so a data set of 239 malignant and 444 benign instances is obtained. The attribute information of the data set is summarized in Table 1.

Table 1. Description of attributes of Wisconsin breast cancer data *(N=683 observations)*

Attribute number	Attribute description	Minimum	Maximum	Mean	Standard deviation
1	Clump thickness	1	10	4.442	2.821
2	Uniformity of cell size	1	10	3.151	3.065
3	Uniformity of cell shape	1	10	3.215	2.989
4	Marginal adhesion	1	10	2.830	2.865
5	Single epithelial cell size	1	10	3.234	2.223
6	Bare nuclei	1	10	3.545	3.644
7	Bland chromatin	1	10	3.445	2.450
8	Normal nucleoli	1	10	2.870	3.053
9	Mitoses	1	10	1.603	1.733

4.2 Performance Evaluation

In order to evaluate the performance of the compared base learning and ensemble learning methods, two well-known performance measures in classification were used. These are classification accuracy and F-measure.

Classification accuracy (ACC) is the ration of true positives and true negatives obtained by the classifier over the total number of instances in the test dataset, as given by Equation 1.

$$ACC = \frac{TN + TP}{TP + FP + FN + TN} \tag{1}$$

For Equations 1-3, TN, TP, FP, and FN denotes number of true negatives, true positives, false positives and false negatives, respectively.

Precision (PRE) is the proportion of the true positives against the true positives and false positives as given by Equation 2:

$$PRE = \frac{TP}{TP + FP} \tag{2}$$

Recall (REC) is the proportion of the true positives against the true positives and false negatives as given by Equation 3:

$$REC = \frac{TP}{TP + FN} \tag{3}$$

F-measure is the harmonic mean of precision and recall which is given by Equation 4. F-measure takes values in [0, 1] interval and values of F-measure closer to 1 indicate better classification performance.

$$F - measure = \frac{2 * PRE * REC}{PRE + REC} \tag{4}$$

4.3 Experimental Procedure

In experimental analysis, 10-fold cross validation method is used to divide the original data set into randomly partitioned 10 mutually exclusive folds. Training and testing process is performed ten times, so that each time one fold is used for testing the model as a validation data and the rest folds are used for training. The experimental results indicate the average results of the 10-fold cross validation process. In experiments, two rule-based learning methods (RIPPER and FURIA algorithms), two lazy learning methods (IBk and K-star algorithms), two Bayesian classifiers (Naïve Bayes and Bayes Net algorithms), four function classifiers (logistic regression, kernel logistic regression, multilayer perceptron and SMO algorithms) and four decision tree classifiers (J48, Random Forest, LMT, CART algorithms) are used as base learners. In order to analyse the effect of ensemble learning, six ensemble methods (Bagging, Dagging, AdaBoost, Decorate and Random Subspace) are utilized with the fourteen base learners mentioned in advance.

We conduct a set of experiments on a PC with a Intel Core i7 CPU 3.40 GHz with 8.00 GB RAM. The experiments are performed with the machine learning toolkit WEKA (Waikato Environment for Knowledge Analysis) version 3.7.11. It is an open-source platform that contains many machine learning algorithms implemented in JAVA. For the base learners and ensemble learning methods, the default parameters given in WEKA are used.

4.4 Results and Discussion

In Tables 2-3, classification accuracy and F-measure values for the experimental results of base learners and ensemble learning methods are presented, respectively. The highest (the best) results for each row and column are highlighted with boldface.

As it can be observed from Table 2, the use of ensemble learning methods with base classifiers improves generally the classification accuracy of the classifiers. The highest accuracy rates for learners are generally achieved by Random Subspace method. It obtains ten highest average accuracy rates (RIPPER, FURIA, IBk, K-star, logistic regression, kernel logistic regression, MLP, Random Forest, and LMT). This is followed by Multi Boosting method, where the highest average accuracy rates for five classifiers are obtained over fourteen classifiers. Compared to the results achieved by the base learner, Dagging method also performs well. It obtains results better predictive performance than the base learners for eleven cases in total. Bagging method also generally obtains better results than the base learners. Among the ensemble learning methods compared, the predictive performance of AdaBoost is relatively low. It generally receives the same or lower accuracy rates than combined the base learners.

In Table 3, F-measure results for experimental comparisons are presented. As mentioned in advance, higher F-measure results indicate better classification performance of a learning method. The F-measure results listed here are parallel to the results listed in Table 2. Again, the highest (the best) F-measure values among comparisons are generally achieved with the use of Random Subspace ensemble learning method.

Among the all empirical study, the best (the highest) classification accuracy is 97.6574%. This is achieved with the use of Dagging method with Bayes Net classifier. The best F-measure result is also achieved by the same combination. These findings emphasize that ensemble learners can improve the performance of base classifiers for automatic diagnosis of diseases.

Table 2. Classification accuracies

Classifier	Base Learner	Ensemble Learning Methods					
		Bagging	Dagging	AdaBoost	Multi Boosting	Decorate	Random Subspace
RIPPER	95.1684	96.7789	96.3397	96.3397	**96.9253**	96.6325	**96.9253**
FURIA	96.1933	96.6325	96.3397	96.1933	96.1933	95.6076	**96.9253**
IBK	95.9004	96.4861	96.1933	95.9004	95.9004	95.1684	**96.9253**
K-star	95.4612	95.7540	94.8755	94.8755	95.4612	95.3148	**96.0469**
Naive Bayes	96.0469	96.1933	**96.3397**	96.0469	**96.3397**	95.9004	96.1933
BayesNet	97.2182	97.3646	**97.6574**	96.3397	96.9253	96.6325	97.2182
Logistic Regression	96.7789	**96.9253**	**96.9253**	96.7789	**96.9253**	**96.9253**	**96.9253**
Kernel Logistic Regression	96.7789	**96.9253**	96.7789	96.1933	96.6325	**96.9253**	**96.9253**
MLP	96.0469	96.4861	96.6325	95.7540	96.1933	96.1933	**97.2182**
SMO	97.0717	96.6325	96.1933	97.0717	**97.2182**	97.0717	96.6325
J48	96.0469	96.0469	96.3397	96.1933	96.6325	**96.7789**	**96.7789**
Random Forest	96.1933	**96.9253**	96.4861	96.7789	96.4861	96.4861	**96.9253**
LMT	95.6325	96.7789	96.0469	96.4861	96.4861	96.3397	**97.0717**
Simple Cart	95.1684	95.9004	95.9004	**96.9253**	96.3397	96.7789	95.1684

Table 3. F-measure results

Classifier	Base Learner	Ensemble Learning Methods					
		Bagging	Dagging	AdaBoost	Multi Boosting	Decorate	Random Subspace
RIPPER	0.952	0.968	0.963	0.964	**0.969**	0.966	**0.969**
FURIA	0.962	0.966	0.963	0.962	0.962	0.956	**0.969**
IBK	0.959	0.965	0.962	0.959	0.959	0.951	**0.969**
K-star	0.954	0.957	0.948	0.949	0.954	0.953	**0.960**
Naive Bayes	0.961	0.962	**0.964**	0.960	**0.964**	0.959	0.962
BayesNet	0.972	0.974	**0.977**	0.963	0.969	0.966	0.972
Logistic Regression	0.968	**0.969**	**0.969**	0.968	**0.969**	**0.969**	**0.969**
Kernel Logistic Regression	0.968	**0.969**	0.968	0.962	0.966	**0.969**	**0.969**
MLP	0.961	0.965	0.966	0.958	0.962	0.962	**0.972**
SMO	0.971	0.966	0.962	0.971	**0.972**	0.971	0.966
J48	0.961	0.961	0.963	0.962	0.966	**0.968**	**0.968**
Random Forest	0.962	**0.969**	0.965	0.968	0.965	0.965	**0.969**
LMT	0.966	0.968	0.960	0.965	0.965	0.963	**0.971**
Simple Cart	0.952	0.959	0.959	**0.969**	0.963	0.968	0.952

5 Conclusion

The automated diagnosis of diseases with high predictive performance is an important topic in medical informatics domain. In this study, an empirical analysis and evaluation of six well-known ensemble learning methods (Bagging, Dagging, Ada Boost, Multi Boost, Decorate and Random Subspace) based on fourteen base learners (Bayes Net, FURIA, K-nearest Neighbors, C4.5, RIPPER, Kernel Logistic Regression, K-star, Logistic Regression, Multilayer Perceptron, Naïve Bayes, Random Forest, Simple Cart, Support Vector Machine, and LMT) is performed for breast cancer data. Among the experimental results, the highest classification accuracy is achieved by the use of Dagging method with Bayes Net. Experimental results also indicate that Random Subspace, Multi Boosting and Dagging are useful ensemble learners for medical domain. The experiments illustrate that ensemble learning methods can be used as a viable method for improving the performance of medical diagnosis classifiers.

References

1. Ahmad, A.: Breast Cancer Metastasis and Drug Resistance Progress and Prospects. Springer, Berlin (2013)
2. Tabar, L., Tot, T., Dean, P.B.: Breast Cancer-The Art and Science of Early Detection with Mammography: Perception, Interpretation, Histopathologic Correlation. Thieme, New York (2004)

3. Westa, D., Mangiamelib, P., Rampalc, R., Westd, V.: Ensemble strategies for a medical diagnostic decision support system: A breast cancer diagnosis application. European Journal of Operational Research 162(2), 532–551 (2005)
4. Lundin, M., Lundin, J., Burke, H.B., Toikkanen, S., Pylkkanen, L., Joensuu, H.: Artificial Neural Networks Applied to Survival Prediction in Breast Cancer. Oncology 57, 281–286 (1999)
5. Bellaachia, A., Guven, E.: Predicting Breast Cancer Survivability using Data Mining Techniques. In: Proceedings of the Sixth SIAM International Conference on Data Mining, pp. 1–4. SIAM, Maryland (2006)
6. Akay, M.F.: Support vector machines combined with feature selection for breast cancer diagnosis. Expert Systems with Applications 36(2), 3240–3247 (2009)
7. Delen, D., Walker, G., Kadam, A.: Predicting breast cancer survivability: a comparison of three data mining methods. Artificial Intelligence in Medicine 34(2), 113–127 (2005)
8. Ubeyli, E.D.: Adaptive neuro-fuzzy inference systems for automatic detection of breast cancer. Journal of Medical Systems 33(5), 353–358 (2009)
9. Thongkam, J., Sukmak, V.: Bagging Random Tree for Analyzing Breast Cancer Survival. KKU Res. J. 17(1), 1–13 (2012)
10. Ya-Qin, L., Cheng, W.: Decision Tree Based Predictive Models for Breast Cancer Survivability on Imbalanced Data. In: Proc. 3rd International Conference on Bioinformatics and Biomedical Engineering, pp. 1–4. IEEE Press, New York (2009)
11. Lavanya, D., Rani, K.U.: Ensemble Decision Tree Classifier for Breast Cancer Data. International Journal of Information Technology Convergence and Services (IJITCS) 2(1), 17–24 (2012)
12. Cruz, J.A., Wishart, D.S.: Application of Machine Learning in Cancer Prediction and Prognosis. Cancer Informatics 2006(2), 59–77 (2006)
13. Gayathri, B.M., Sumathi, C.P., Santhanam, T.: Breast Cancer Diagnosis Using Machine Learning Algorithm- A Survey. International Journal of Distributed and Parallel Systems 4(3), 105–112 (2013)
14. Li, L., Hu, Q., Wu, X., Yu, D.: Exploration of classification confidence in ensemble learning. Pattern Recognition 47, 3120–3131 (2014)
15. Cohen, W.W.: Fast Effective Rule Induction. In: Proc. Twelfth International Conference on Machine Learning, pp. 115–123. Morgan Kaufmann, San Francisco (1995)
16. Duma, M., Twala, B., Marwala, T., Nelwamondo, F.V.: Improving the Performance of the Ripper in Insurance Risk Classification- A Comparative Study using Feature Selection. In: Ferrier, J.-L., Bernard, A., Yu, O., Gusikin, K.M. (eds.) Proceedings of the 8th International Conference on Informatics in Control, Automation and Robotics, vol. 1, pp. 203–210. SciTePress, Netherlands (2011)
17. Hühn, J., Hüllermeier, E.: FURIA: an algorithm for unordered fuzzy rule induction. Data Mining and Knowledge Discovery 19(3), 293–319 (2009)
18. Aha, D.W., Kibler, D., Albert, M.K.: Instance-Based Learning Algorithms. Machine Learning 6, 37–66 (1991)
19. Wu, X., Kumar, V.: The Top Ten Algorithms in Data Mining. Taylor & Francis Group, New York (2009)
20. Clearly, J.G., Trigg, L.E.: K*: An Instance-based learner using and entropic distance measure. In: Proc. Twelfth International Conference on Machine Learning, pp. 108–114. Morgan Kaufmann, San Francisco (1995)
21. John, G.H., Langley, P.: Estimating continuous distributions in Bayesian classifiers. In: Proc. of the Eleventh Conference on Uncertainty in Artificial Intelligence, pp. 338–345. Morgan Kaufmann, San Francisco (1995)

22. Han, J., Kamber, M., Pei, J.: Data Mining: Concepts and Techniques. Morgan Kaufmann, San Francisco (2011)
23. Bouckaert, R.R.: Bayesian Network Classifiers in Weka, `http://weka.sourceforge.net/manuals/weka.bn.pdf`
24. Witten, I.H., Frank, E., Hall, M.A.: Data Mining: Practical Machine Learning Tools and Techniques. Morgan Kaufmann, Burlington (2011)
25. Cessie, S.L., VanHowelingen, J.C.: Ridge Estimators in Logistic Regression. Applied Statistics 41(1), 191–201 (1992)
26. Negnevitsky, M.: Artificial Intelligence: A Guide to Intelligent Systems. Addison-Wesley, Reading (2005)
27. Platt, J.: Fast Training of Support Vector Machines using Sequential Minimal Optimization. In: Schoelkopf, B., Burges, C., Smola, A. (eds.) Advances in Kernel Methods-Support Vector Learning. MIT Press, Cambridge (1998)
28. Quinlan, R.: C4.5: Programs for Machine Learning. Morgan Kaufmann, San Mateo (1993)
29. Niuniu, X., Yuxun, L.: Review of Decision Trees. In: Proc The Third IEEE International Conferrence on Computer Science and Information Technology, pp. 105–109. IEEE Press, New York (2010)
30. Breiman, L.: Random Forests. Machine Learning 45(1), 5–32 (2001)
31. Landwehr, N., Hall, M., Frank, E.: Logistic Model Trees. Machine Learning 59, 161–205 (2005)
32. Doestcsh, P., Buck, C., Golik, P., Hoppe, N.: Logistic Model Trees with AUCsplit Criterion for KDD Cup 2009 Small Challgenge. Journal of Machine Learning Research 7, 77–88 (2009)
33. Loh, W.Y.: Classification and regression trees. WIREs Data Mining and Knowledge Discovery 1, 14–23 (2011)
34. Breiman, L.: Bagging predictors. Machine Learning 4(2), 123–140 (1996)
35. Rokach, L.: Ensemble-based classifiers. Artificial Intelligence Review 33, 1–39 (2010)
36. Ting, K.M., Witten, I.H.: Stacking Bagged and Dagged Models. In: Fourteenth International Conference on Machine Learning, pp. 367–375. Morgan Kaufmann, San Francisco (1997)
37. Freund, Y., Schapire, R.: Experiments with a new boosting algorithm. In: Proc of the Thirteenth International Conference on Machine Learning, pp. 148–156. Morgan Kaufmann, San Francisco (1996)
38. Opitz, D., Maclin, R.: Popular Ensemble Methods: An Empirical Study. Journal of Artificial Intelligence Research 11, 169–198 (1999)
39. Guo, H., Viktor, H.L.: Boosting with Data Generation: Improving the Classification of Hard to Learn Examples. In: Orchard, B., Yang, C., Ali, M. (eds.) IEA/AIE 2004. LNCS (LNAI), vol. 3029, pp. 1082–1091. Springer, Heidelberg (2004)
40. Webb, G.I.: MultiBoosting: A Technique for Combining Boosting and Wagging. Machine Learning 40, 159–196 (2000)
41. Melville, P., Mooney, R.J.: Constructing Diverse Classifier Ensembles using Artificial Training Examples. In: Proceedings of the 18th IJCAI, pp. 505–510. Morgan Kaufmann, San Francisco (2003)
42. Ho, T.K.: The Random Subspace Method for Constructing Decision Forests. IEEE Transactions on Pattern Analysis and Machine Intelligence 20(8), 832–844 (1998)
43. Mangasarian, O.L., Wolberg, W.H.: Cancer diagnosis via linear programming. SIAM News 23(5), 1–18 (1990)
44. Bache, K., Lichman, M.: UCI Machine Learning Repository, `http://archieve.ics.uci.edu/ml`

Predicting Financial Distress of Banks Using Random Subspace Ensembles of Support Vector Machines

Petr Hájek[1], Vladimír Olej[1], and Renata Mysková[2]

[1] Institute of System Engineering and Informatics,
Faculty of Economics and Administration, University of Pardubice,
Pardubice, Czech Republic
{Petr.Hajek,Vladimir.Olej}@upce.cz
[2] Institute of Business Economics and Management,
Faculty of Economics and Administration, University of Pardubice,
Pardubice, Czech Republic
Renata.Myskova@upce.cz

Abstract. Models for financial distress predictions of banks are increasingly important tools used as early warning signals for the whole banking systems. In this study, a model based on random subspace method is proposed to predict investment/non-investment rating grades of U.S. banks. We show that support vector machines can be effectively used as base learners in the meta-learning model. We argue that both financial and non-financial (sentiment) information are important categories of determinants in financial distress prediction. We show that this is true for both banks and other companies.

Keywords: Banks, Financial distress, Rating grade, Random subspace, Meta-learning, Support vector machines, Sentiment analysis.

1 Introduction

Avoiding over-fitting is an important issue to address when training base learners. The random subspace (RSS) method [1] represents a parallel learning algorithm generating each base learner independently. In particular, it is favourable for parallel computing and fast learning. Thus, this computing procedure alleviates the risk of local optimum trapping. This makes RSS one of the most frequently used meta-learning method.

Meta-learning algorithms (ensembles of base learners) are increasingly important in predicting financial distress of firms [2-4]. In addition to above mentioned advantages, their use in this domain is also preferable because base learners imitate decisions of individual financial experts and the final decision corresponds to a collective decision-making process.

Various meta-learning strategies have been utilized in financial distress' prediction, including boosting, bagging and stacking [5,6]. However, considerably less attention has been paid to banks and their financial performance. More specifically, previous studies have been limited to bankruptcy prediction [7]. Moreover, previous research

R. Silhavy et al. (eds.), *Artificial Intelligence Perspectives and Applications*,
Advances in Intelligent Systems and Computing 347, DOI: 10.1007/978-3-319-18476-0_14

in predicting financial distress of banks has utilized financial performance indicators only. This study seeks to remedy these problems by employing RSS method (using support vector machines (SVMs) as base learners) to predict financial distress of selected US banks. SVMs are considered state-of-the-art method for financial distress prediction mainly owing to their generalization performance [8]. We treat the financial distress prediction as a two-class problem, where investment/non-investment rating grades provided by a rating agency are used as the indicators of distress. Additionally, we argue that both financial and non-financial (qualitative) information are important determinants of the distress. We show that the qualitative information can be effectively drawn from the textual parts of banks' annual reports using sentiment analysis.

This paper has been organized in the following way. Section 2 begins by laying out the theoretical dimensions of the research, and provides justification of using information from both financial statements and textual reports of banks as the determinants of financial distress. Section 3 describes data and their pre-processing. Section 4 presents the design of prediction models and the experimental results obtained by RSS. SVMs are used as the base learners of the RSS. The results are compared with multilayer neural networks (MLPs) which have been considered as benchmark methods in previous studies [9]. We also provide a comparison with other firms in this section. The final section concludes the paper and discusses the results.

2 Theoretical Background

The late-2000s financial crisis is considered to be the worst financial crisis since the Great Depression. It was triggered by a liquidity shortfall in the U.S. banking system in 2008 [10]. For this reason, experts examine bank concentration and its impact on the effectivity and stability of banking market [11].

Another issue addressed is the size of bank capital and the relationship between bank capital and liquidity creation (see [11]). Thus far, discussion has been concerned with the size of bank capital, risk, liquidity, as well as liquid assets and liabilities in relation to liquidity [12]. However, the issue of assessing banks' financial statements regarding the sentiment has not been addressed in previous literature. The difficulty is that the behavior of stakeholders is influenced by more than financial attributes. It has been reported that psychological factors often have substantial impact on decision-making and in many cases this may result in what is considered deviation from the normative models of action [13]. An important role of voluntarily disclosed qualitative information has been reported only recently for business companies [14-16].

Financial ratios, on the other hand, have been used in many studies on financial distress prediction of banks. These studies can be classified into two categories, bankruptcy [17-18] and credit rating prediction [19-22]. A wide range of soft computing methods have been employed in related studies, including support vector machines, neural networks, fuzzy and rough sets, evolutionary algorithms and meta-learning algorithms, see [23,24] for reviews. The Camel model is another approach used to assess the performance of banks [25]. This model assesses, in addition to financial indicators, key banking qualitative indicators in the following areas: capital adequacy, asset quality, management, earnings, and liquidity. Based on this assessment, banks can be classified into five categories, from 1 (best) to 5 (worst). The assessment is conducted by an expert from a bank regulation or supervision organization, namely the Federal Reserve, the Office of the Comptroller of the Currency, the National Credit Union Administration, and the Federal Deposit Insurance Corporation. The resulting rating is intended to serve the top managements of banks only and, thus, it is not disclosed publicly. However, banks with a deteriorating assessment are subject to increased supervision. The above consideration suggests that banks are motivated to effectively deal with the narrative parts of their annual reports. They use the sentiment in the communication with stakeholders as a tool that illustrates the overall business position of the bank.

3 Data and Their Pre-processing

Adopting the methodology used for companies in [15], the input attributes in this study cover two main categories, financial indicators and sentiment indicators. The chosen financial indicators monitor profitability (earnings per share), financial market situation (beta coefficient, high to low stock price, std. dev. of stock price, correlation with market stock index), business situation (effective tax rate), asset structure (fixed assets / total assets), leverage ratios (market debt / total capital, book debt / total capital), dividend policy (dividend yield) and ownership structure (share of insiders' and institutional holdings). Sentiment indicators, on the other hand, refer to the qualitative assessment of business position by the management of the bank. Usually, the subjects of the assessment are business performance and risks, strategic, financial and investment policy, etc. [26].

The financial indicators were collected from financial statements (Value Line database), and the sentiment indicators were drawn from annual reports (10-Ks documents) freely available at the U.S. Securities and Exchange Commission EDGAR System. Data were collected for 126 U.S. banks selected from the Standard & Poor's database in the year 2010. In the year 2011, 89 of them were classified as investment grade (IG) and 37 as non-investment grade (NG) by the Standard & Poor's rating agency. The investment/non-investment grade position is considered important to investors due to the restrictions imposed on investment instruments. Table 1 shows the list of input and output attributes, and Fig. 1 shows boxplots of input attributes separately for IG and NG class.

The sentiment indicators were processed in the following way. First, linguistic preprocessing (tokenization and lemmatization) was carried out to obtain a set of candidate terms. Second, this set was compared with the sentiment categories from the financial dictionary provided by [16]. Then, the *tf.idf* term weighting scheme was applied to obtain the importance of terms and an average weight was calculated for each sentiment category (negative, positive, uncertainty, litigious, modal strong and modal weak) [27]. For all data, we replaced missing values by median values, and all data were standardized using the Z-score to prevent problems with different scales.

Table 1. Input and output attributes

	attribute		attribute
x_1	growth in earnings per share (EPS)	x_{11}	dividend yield
x_2	expected growth in EPS	x_{12}	share of insiders' holdings
x_3	beta coefficient	x_{13}	share of institutional holdings
x_4	high to low stock price	x_{14}	frequency of negative terms
x_5	std. dev. of stock price	x_{15}	frequency of positive terms
x_6	correlation with market stock index	x_{16}	frequency of uncertainty terms
x_7	effective tax rate	x_{17}	frequency of litigious terms
x_8	fixed assets / total assets	x_{18}	frequency of strong modal terms
x_9	market debt / total capital	x_{19}	frequency of weak modal terms
x_{10}	book debt / total capital	class	{IG, NG}

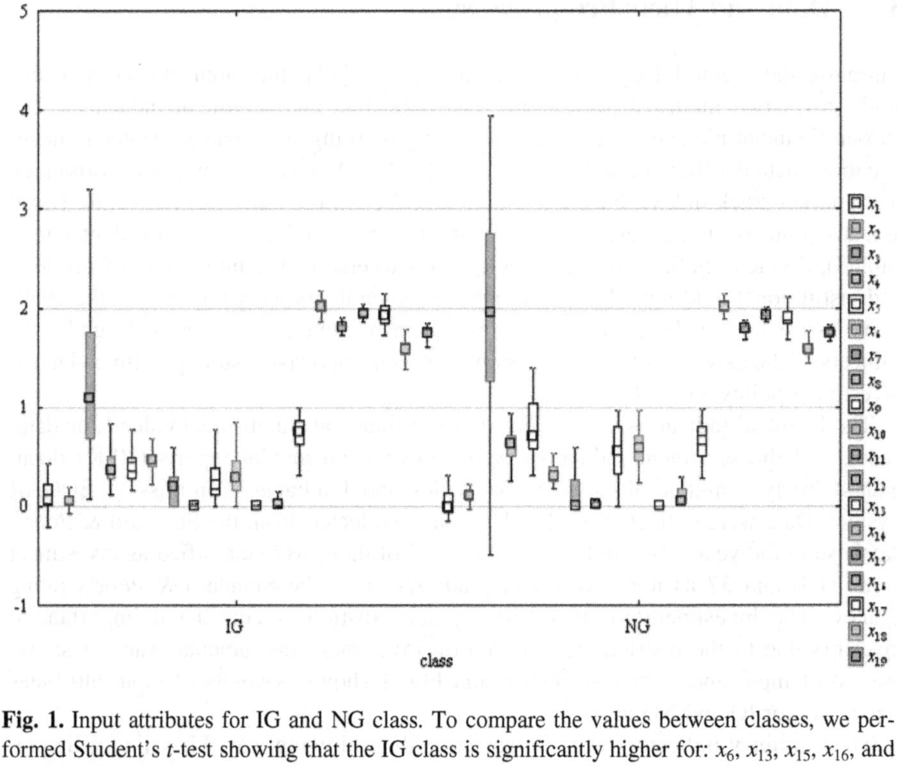

Fig. 1. Input attributes for IG and NG class. To compare the values between classes, we performed Student's *t*-test showing that the IG class is significantly higher for: x_6, x_{13}, x_{15}, x_{16}, and significantly lower for: x_3, x_4, x_5, x_8, x_9, x_{10} (both at $p<0.1$)

4 Modelling and Analysis of the Results

To predict rating grades of banks, we employed several meta-learning algorithms using SVMs as base learners. In addition, we compared the results with another base learner – MLP. The results suggest that RSS method performed significantly better when compared with the remaining meta-learning algorithms, namely multiboosting [28], adaboosting [29], bagging and dagging [30] and rotation forest [31]. Due to limited space, we report only selected results for RSS method. This method utilizes many base learners which are systematically constructed by pseudorandomly selecting subsets of components of the feature vector. Thus, it improves accuracy on testing data as it grows in complexity.

To avoid overfitting, we used 10-fold cross-validation in our experiments. The classification performance was measured by the averages of standard statistics applied in classification tasks: accuracy [%], true positives (TP rate), false positives (FP rate), and the area under the receiver operating characteristic (ROC) curve. A ROC is a graphical plot which illustrates the performance of a binary classifier system, which represents a standard technique for summarization classifier performance over a range of tradeoffs between TP and FP error rates.

Subspace size is the critical parameter in learning RSS. Therefore, we tested several settings of the subspace size to obtain the best classification performance. Fig. 2 and Fig. 3 show that the optimum subspace size was 0.4 to 0.5 for banks. The number of iterations of the RSS was set to 10. To compare the results with other dataset, we used the dataset of companies (without banks) used by [15]. This dataset describe 520 U.S. companies by 19 attributes. Similarly as in the case of the banks' dataset, 13 of the attributes are financial indicators (these are rather specific for other companies when compared with banks [32]) and 6 are sentiment indicators, see [15] for details. Although we admit that the comparison should be made with caution, it is obvious that the subspace size is more important for banks, generally requiring larger size of subspace to achieve a high accuracy. This holds true for SVMs in particular.

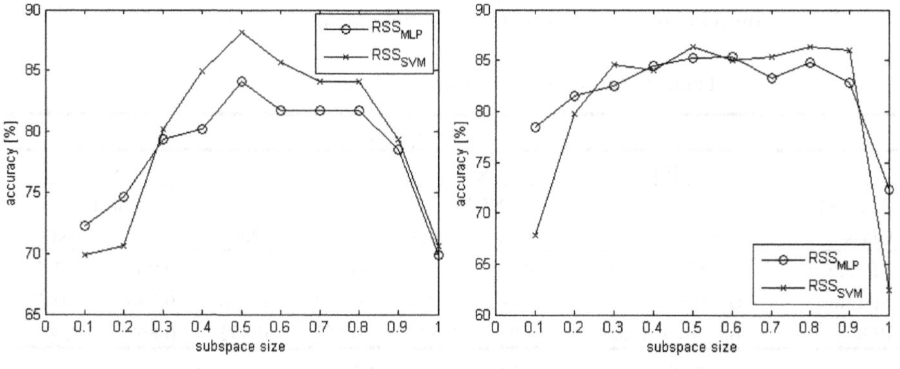

Fig. 2. Relationship between accuracy and subspace size, a) banks, b) companies. The subspace size is expressed as a ratio to the total number of attributes.

The MLP was trained using the backpropagation algorithm with momentum. The following training parameters were examined to achieve the best classification performance: the number of neurons in the hidden layer = $\{2^0, 2^1, \ldots, 2^5\}$, learning rate = {0.05, 0.1, 0.2, 0.3}, momentum = {0.1, 0.2, 0.3}, and the number of epochs = {50, 100, 300, 500, 1000}. We used grid search algorithm to find the optimum settings of these parameters.

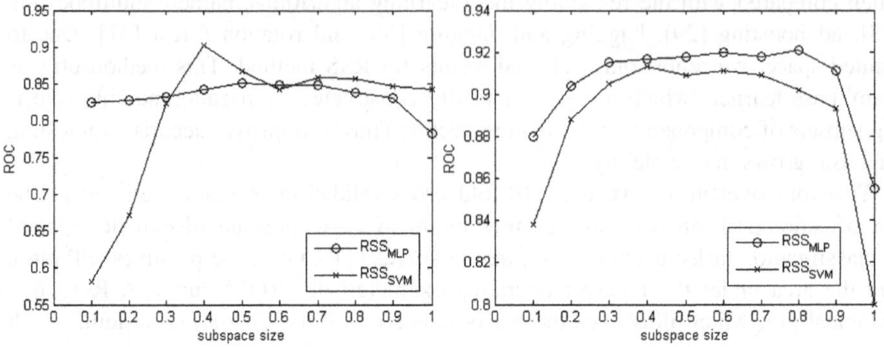

Fig. 3. Relationship between the area under ROC and subspace size, a) banks, b) companies. The subspace size is expressed as a ratio to the total number of attributes.

The SVMs was trained by the sequential minimal optimization (SMO) algorithm [33]. The best classification performance of the SVM was tested for the following user-defined parameters: kernel functions = polynomial, the level of polynomial function = 2, complexity parameter $C = \{2^0, 2^1, \ldots ,2^{10}\}$, round-off error ε = 1.0E-12, and tolerance parameter = 0.001. Again, grid search algorithm was employed to find the optimum settings.

Table 2 shows the detailed classification performance on the dataset of banks. First, SVM performed significantly better when compared with MLP (using paired Student's t-test at $p<0.1$). Second, RSS performed better than single classifiers (significantly better in the case of MLP). Third, classification accuracy was significantly higher when incorporating sentiment indicators (x_{14} to x_{19}). Fourth, all classifiers performed better on IG class, mainly due to the imbalance of the classes.

Table 2. Best results of the analyzed methods for banks

	all input attributes x_1 to x_{19}							
	MLP		SVM		RSS_{MLP}		RSS_{SVM}	
Acc [%]	81.75		**87.30**		84.13		**88.10**	
Class	IG	NG	IG	NG	IG	NG	IG	NG
TP rate	0.899	0.622	0.921	0.757	0.910	0.676	0.933	0.757
FP rate	0.378	0.101	0.243	0.079	0.324	0.090	0.243	0.067
ROC	0.824	0.824	0.839	0.839	0.852	0.852	0.868	0.868
	financial indicators only (without sentiment indicators) x_1 to x_{13}							
Acc [%]	80.16		**84.13**		83.34		**86.51**	
Class	IG	NG	IG	NG	IG	NG	IG	NG
TP rate	0.876	0.622	0.899	0.703	0.921	0.622	0.955	0.649
FP rate	0.378	0.124	0.297	0.101	0.378	0.079	0.351	0.045
ROC	0.782	0.782	0.801	0.801	0.843	0.843	0.854	0.854

In Table 3, we provide the results for companies for comparison purposes. Similarly to banks, sentiment indicators lead to significantly higher accuracy. This corroborates the findings obtained by [15]. In contrast to banks, the accuracy was higher for the NG class, which can be explained by the different frequencies of classes (also imbalanced but the NG class prevailed for companies). Altogether, the total accuracy was higher for banks, but this is mainly due to the higher accuracy on the IG class.

Table 3. Best results of the analyzed methods for companies

all input attributes x_1 to x_{19}								
	MLP		SVM		RSS_{MLP}		RSS_{SVM}	
Acc [%]	84.04		**86.15**		85.38		**86.34**	
Class	IG	NG	IG	NG	IG	NG	IG	NG
TP rate	0.728	0.908	0.785	0.908	0.759	0.911	0.790	0.908
FP rate	0.092	0.272	0.092	0.215	0.089	0.241	0.092	0.210
ROC	0.898	0.898	0.846	0.846	0.920	0.920	0.909	0.909
financial indicators only (without sentiment indicators) x_1 to x_{13}								
Acc [%]	81.92		**84.62**		83.65		**84.81**	
Class	IG	NG	IG	NG	IG	NG	IG	NG
TP rate	0.708	0.886	0.779	0.866	0.744	0.892	0.744	0.911
FP rate	0.114	0.292	0.114	0.221	0.108	0.256	0.089	0.256
ROC	0.883	0.883	0.833	0.833	0.915	0.915	0.911	0.911

5 Conclusion and Discussion

The informative value of financial statements is increasingly important for stakeholders. Although the structure of these statements is given by the accounting legislation of the corresponding country, the quality of the disclosed information is subject to expert assessment. Management communicates the financial results with stakeholders in the textual parts of annual reports. This is true for both banks and companies. In this paper, we compare these two categories of economic subjects, which differ in both business activity and legislative regulation. On the other hand, the categories have one important characteristics in common – they are established in order to make profit. Thus, they are obliged to disclose true and undistorted information about their business activity.

The extent and quality of information in the report on economic activities (annual report) depend mainly on regulation and have both quantitative and qualitative character. Therefore, we aimed at using the information to accurately predict the rating grades of banks. The results showed that sentiment information hidden in annual reports should be considered an important determinant in financial distress prediction models. This finding has serious implications for stakeholders, regulators and other authorities. More specifically, the accuracy of prediction was increased by about 1.5 % using sentiment information for both banks and companies. We also reported that RSS meta-learning algorithm significantly increases the performance of base learners

SVM and MLP in case of banks in particular. Combining the contribution of sentiment information and meta-learning approach the accuracy was increased by about 4 %. This may lead to substantial savings owing to the loss associated with potential distress. Additionally, the more accurate prediction model makes it possible to better anticipate the effects of potential financial crises.

In future research, the differences in the content of annual report between industries need to be examined, because it is the connection between sentiment and corresponding subject that provides more detailed qualitative information. Thus, a more complete picture would be extracted from annual reports. We further intend to compare the dictionary approach used in this study with machine learning algorithms such as Naïve Bayes [34].

Acknowledgments. This work was supported by the scientific research project of the Czech Sciences Foundation Grant No: 13-10331S.

References

1. Ho, T.K.: The Random Subspace Method for Constructing Decision Forests. IEEE Transactions on Pattern Analysis and Machine Intelligence 20(8), 832–844 (1998)
2. Verikas, A., Kalsyte, Z., Bacauskiene, M., Gelzinis, A.: Hybrid and Ensemble-based Soft Computing Techniques in Bankruptcy Prediction: A Survey. Soft Computing 14(9), 995–1010 (2010)
3. Hájek, P., Olej, V.: Predicting Firms' Credit Ratings Using Ensembles of Artificial Immune Systems and Machine Learning – An Over-Sampling Approach. In: Iliadis, L., Maglogiannis, I., Papadopoulos, H. (eds.) AIAI 2014. IFIP AICT, vol. 436, pp. 29–38. Springer, Heidelberg (2014)
4. Heo, J., Yang, J.Y.: AdaBoost based Bankruptcy Forecasting of Korean Construction Companies. Applied Soft Computing 24, 494–499 (2014)
5. Alfaro, E., García, N., Gámez, M., Elizondo, D.: Bankruptcy Forecasting: An Empirical Comparison of AdaBoost and Neural Networks. Decision Support Systems 45(1), 110–122 (2008)
6. Kim, M.J., Kang, D.K.: Ensemble with Neural Networks for Bankruptcy Prediction. Expert Systems with Applications 37(4), 3373–3379 (2010)
7. Shin, S.W., Lee, K.C., Kilic, S.B.: Ensemble Prediction of Commercial Bank Failure Through Diversification of Input Features. In: Sattar, A., Kang, B.-H. (eds.) AI 2006. LNCS (LNAI), vol. 4304, pp. 887–896. Springer, Heidelberg (2006)
8. Hajek, P., Olej, V.: Credit Rating Modelling by Kernel-Based Approaches with Supervised and Semi-Supervised Learning. Neural Computing and Applications 20(6), 761–773 (2011)
9. Hajek, P.: Municipal Credit Rating Modelling by Neural Networks. Decision Support Systems 51(1), 108–118 (2011)
10. Shie, F.S., Chen, M.Y., Liu, Y.S.: Prediction of Corporate Financial Distress: An Application of the America Banking Industry. Neural Computing and Applications 21(7), 1687–1696 (2012)
11. Berger, A.N., Bouwman, C.H.S.: Bank Liquidity Creation. Review of Financial Studies 22, 3779–3837 (2009)

12. de Haan, L., van den End, J.W.: Bank Liquidity, the Maturity Ladder, and Regulation. Journal of Banking & Finance 37, 3930–3950 (2013)
13. Kliger, D., Levy, O.: Mood-Induced Variation in Risk Preferences. Journal of Economic Behavior and Organization 52(4), 573–584 (2003)
14. Lu, H.M., Tsai, F.T., Chen, H., Hung, M.W., Li, S.H.: Credit Rating Change Modeling using News and Financial Ratios. ACM Transactions on Management Information Systems 3(3), 14 (2012)
15. Hájek, P., Olej, V.: Evaluating Sentiment in Annual Reports for Financial Distress Prediction Using Neural Networks and Support Vector Machines. In: Iliadis, L., Papadopoulos, H., Jayne, C. (eds.) EANN 2013, Part II. CCIS, vol. 384, pp. 1–10. Springer, Heidelberg (2013)
16. Loughran, T., McDonald, B.: When Is a Liability Not a Liability? Textual Analysis, Dictionaries, and 10-Ks. The Journal of Finance 66(1), 35–65 (2011)
17. Chauhan, N., Ravi, V., Karthik Chandra, D.: Differential Evolution Trained Wavelet Neural Networks: Application to Bankruptcy Prediction in Banks. Expert Systems with Applications 36(4), 7659–7665 (2009)
18. Ravi, V., Pramodh, C.: Threshold Accepting Trained Principal Component Neural Network and Feature Subset Selection: Application to Bankruptcy Prediction in Banks. Applied Soft Computing 8(4), 1539–1548 (2008)
19. Chen, Y.S.: Classifying Credit Ratings for Asian Banks using Integrating Feature Selection and the CPDA-based Rough Sets Approach. Knowledge-Based Systems 26, 259–270 (2012)
20. Orsenigo, C., Vercellis, C.: Linear versus Nonlinear Dimensionality Reduction for Banks' Credit Rating Prediction. Knowledge-Based Systems 47, 14–22 (2013)
21. Gogas, P., Papadimitriou, T., Agrapetidou, A.: Forecasting Bank Credit Ratings. The Journal of Risk Finance 15(2), 195–209 (2014)
22. Bellotti, T., Matousek, R., Stewart, C.: A Note Comparing Support Vector Machines and Ordered Choice Models' Predictions of International Banks' Ratings. Decision Support Systems 51(3), 682–687 (2011)
23. Demyanyk, Y., Hasan, I.: Financial Crises and Bank Failures: A Review of Prediction Methods. Omega 38(5), 315–324 (2010)
24. Ravi Kumar, P., Ravi, V.: Bankruptcy Prediction in Banks and Firms via Statistical and Intelligent Techniques – A Review. European Journal of Operational Research 180(1), 1–28 (2007)
25. DeYoung, R., Flannery, M.J., Lang, W.W., Sorescu, S.M.: The Information Content of Bank Exam Ratings and Subordinated Debt Prices. Journal of Money, Credit and Banking 33(4), 900–925 (2001)
26. Hajek, P., Olej, V.: Comparing Corporate Financial Performance and Qualitative Information from Annual Reports using Self-organizing Maps. In: 10th International Conference on Natural Computation (ICNC 2014), Xiamen, China, pp. 93–98 (2014)
27. Hajek, P., Olej, V., Myskova, R.: Forecasting Corporate Financial Performance using Sentiment in Annual Reports for Stakeholders' Decision-Making. Technological and Economic Development of Economy (2014) (in press), doi:10.3846/20294913.2014.979456
28. Webb, G.I.: MultiBoosting: A Technique for Combining Boosting and Wagging. Machine Learning 40(2) (2000)
29. Freund, Y., Schapire, R.E.: Experiments with a New Boosting Algorithm. In: 13th Int. Conf. on Machine Learning, San Francisco, CA, pp. 148–156 (1996)
30. Ting, K.M., Witten, I.H.: Stacking Bagged and Dagged Models. In: 14th Int. Conf. on Machine Learning, San Francisco, CA, pp. 367–375 (1997)

31. Rodriguez, J.J., Kuncheva, L.I., Alonso, C.J.: Rotation Forest: A New Classifier Ensemble Method. IEEE Transactions on Pattern Analysis and Machine Intelligence 28(10), 1619–1630 (2006)
32. Hajek, P., Michalak, K.: Feature Selection in Corporate Credit Rating Prediction. Knowledge-Based Systems 51, 72–84 (2013)
33. Platt, J.C.: Fast Training of Support Vector Machines using Sequential Minimal Optimization. In: Schoelkopf, B., Burges, C., Smola, A. (eds.) Advances in Kernel Methods - Support Vector Learning, pp. 185–208. MIT Press, Cambridge (1998)
34. Kearney, C., Liu, S.: Textual Sentiment in Finance: A Survey of Methods and Models. International Review of Financial Analysis 33, 171–185 (2014)

Interdependence of Text Mining Quality and the Input Data Preprocessing

František Dařena and Jan Žižka

Department of Informatics, Faculty of Business and Economics,
Mendel University, Zemědělská 1, 613 00 Brno, Czech Republic
{frantisek.darena,jan.zizka}@mendelu.cz

Abstract. The paper focuses on preprocessing techniques application to short informal textual documents created in different natural languages. The goal is to evaluate the impact on the quality of the results and computational complexity of the text mining process designed to reveal knowledge hidden in the data. Extensive number of experiments were carried out with real world text data with correction of spelling errors, stemming, stop words removal, and their combinations applied. Support vector machine, decision trees, and k-means algorithms as the commonly used methods were considered to analyze the text data. The text mining quality was generally not influenced significantly, however, the positive impact represented by the decreased computational complexity was observed.

Keywords: Text mining, text data preprocessing, stemming, spell checking, stop words, support vector machine, decision tree, k-means.

1 Introduction and Objectives

The possibilities of Internet technologies enable people to express their opinions on a variety of topics. These opinions, often in an informal textual form, hide many opportunities interesting for both commercial and non-commercial spheres. The knowledge hidden in the data is discovered in a process known as *text mining*. Text mining is in many aspects similar to *data mining* – the process of extraction of implicit, previously unknown information from data [22]. In contrast to data mining focusing on highly structured data, text mining processes collections of unstructured textual data. Thus, some additional requirements and constraints are imposed on text mining, the process of which is otherwise highly similar to the one of data mining. The main difference is in the phase of preprocessing where the unstructured textual data is transformed to a structured representation suitable for the selected data mining algorithms. The structure has typically a form of feature vectors where the features are associated to the terms extracted from the documents. Due to high dimensionality of these vectors some feature reduction techniques are often applied too [15].

Although text mining is quite a developed discipline and a lot of research has been done, there are still new challenges that must be faced. The rapid expansion

R. Silhavy et al. (eds.), *Artificial Intelligence Perspectives and Applications*,
Advances in Intelligent Systems and Computing 347, DOI: 10.1007/978-3-319-18476-0_15

of social media (such as social networks and microblogging sites) changed the way of communication. The language has changed into less formal, many non-native speakers use different than their mother tongue, or several languages are mixed in one document [3,25]. Focus on languages other than English and processing social media data belong to currently investigated topics [6].

When a new text mining algorithm or a procedure is developed, it is often tested on a traditional reference database, like the Reuters document set [23]. When text mining is applied to a specific domain in order to solve a domain dependent problem, specific techniques, including preprocessing are used to find a suitable solution. This solution sometimes lacks generality and cannot be easily adopted for a different domain.

Based on the above mentioned considerations there can be some gaps identified in the text mining research. The main focus of this paper is thus on processing rather short textual documents created informally by Internet users, in different natural languages. The goal is to evaluate the impact of text data preprocessing methods application on the quality of the achieved results and computational complexity of the text mining process. Data sets characterized by different sizes and used natural languages together with application of different preprocessing techniques and machine learning algorithms will be processed in a systematic experimental study.

2 Mining Knowledge from Text Data

Text mining is a branch of computer science that uses techniques from data mining, information retrieval, machine learning, statistics, natural language processing, and knowledge management [1]. Typical applications include, besides others, categorization of newspaper articles or web pages, e-mail filtering, organization of a library, customer complaints (or feedback) handling, marketing focus group programs, competitive intelligence, market prediction, extraction of topic trends in text streams, discovering semantic relations between events, or customer satisfaction analysis. Text mining involves tasks such as text categorization, term extraction, single- or multi-document document summarization, clustering, association rules mining, or sentiment analysis [5].

At the end of the last century, *machine learning* gained on its popularity and became a dominant approach to text mining [17]. During *supervised learning*, which is the most common type of learning problem, a classifier that generalizes the knowledge about how to assign correct labels to the data is found. Unfortunately, unavailability of the labeled data is often a major problem. As new data constantly occurs, it is nearly impossible to have the labels assigned to the data in a reasonable time and in reasonable amounts. Clustering, as the most common form of *unsupervised learning*, enables automatic grouping of unlabeled documents into subsets called clusters. The clusters are coherent internally and must be clearly different from each other to express their own distinct information. If the clusters are good and reliable with respect to the given goal, they can be successfully used as classes.

Classification algorithms used for categorization of textual data include probabilistic classifiers, decision trees, decision rules, example-based classifiers, the support vectors machine, or neural networks. It is very difficult to compare individual methods because the published results of experiments of different authors often run under different circumstances, use different sampling, preprocessing etc. Generally, support vectors machine, decision trees, instance based classifiers, and neural networks bring acceptable results [17,25].

Having a set of objects to be clustered, *partitional (flat) clustering* seeks a k-partition where k non-empty sets (clusters) are created and every object belongs to exactly one of the sets. *Hierarchical clustering* constructs a tree like, nested structure partition of the objects (a hierarchy of the clusters). Hierarchical methods include top-down (divisive) and bottom-up (agglomerative) methods and the result of their application is a dendrogram representing nesting of the clusters into a hierarchical, tree-like structure [21]. In recent years, it has been found that partitioning clustering algorithms are well suited for clustering large document data sets due to their relatively low computational requirements [24].

3 Preprocessing

Effective and efficient text mining heavily relies on application of various preprocessing techniques. Their goal is to infer or extract structured representations from unstructured plain textual data [5]. Such representations are then suitable for a particular purpose and algorithm. Preprocessing methods include, e.g., online text cleaning, white space removal, case folding, spelling errors corrections, abbreviations expanding, stemming, stop words removal, negation handling and finally feature selection [2,3,7]. Natural language processing techniques, such as tokenization (including stemming), part-of-speech tagging, syntactical or shallow parsing might, are a subset of these methods requiring the knowledge of the language to be processed [5].

Currently, there is a lack of studies that would demonstrate the impact of preprocessing techniques application on the process and outputs of text mining, especially for the short text documents created informally by Internet users. Very often the authors apply several preprocessing techniques within a certain task without evaluating their impact separately [7]. [10] experimentally determined the impact of stop words removal on the results of analysis of sequential patterns in electronic documents using sequence rule analysis. They identified an impact on the quantity and quality of extracted rules in case of paragraph sequence identification and no significant impact in case of sentence sequence identification. [12] studied the results of sentence splitting, stemming, part-of-speech tagging, and parsing application on text data collected from different sources. They, however, did not investigate the impact on machine learning algorithms outputs. In their research, [20] focused on text classification and the impact of tokenization, stop word removal, lowercase conversion, and stemming. The processed documents included e-mail and news in two different languages (Turkish and English). In the experiments, only hundreds of documents were processed. An empirical study designed to determine the value of tags in resource classification using the data generated by the users of folksonomies was carried out by [19]. The authors applied several pre-processing operations to English texts to reduce the ambiguity and noise in tags – misspelling correction to fix typing errors, synonyms to considered other words with the same meaning, and stemming.

4 Experiments

In this section we describe the experiments carried out in order to analyze the impact of different preprocessing techniques application to the text mining process. Compared to most of the existing research, large data set in multiple languages are be processed.

4.1 Data and Its Representation

The processed data collection contained opinions of several millions of hotel guests who booked accommodations in many different hotels all over the world. From the reviews written in many natural languages we focused on the ones written in English, German, Spanish, and French. Each review had two parts – negative and positive. Thus, the available document were quite carefully and reliably labeled although containing all deficiencies typical for texts written in natural languages (i.e., mistyping, transposed letters, missing letters, grammar errors, and so like).

In order to be able to apply the selected machine learning algorithms, the data needed to be transformed to a suitable representation. The commonly used vectors space model where the words in the documents were selected as meaningful units (attributes) of the texts was selected. A big advantage of such a word-based representation is its simplicity and the straightforward process of creation while still providing acceptable results [8]. Each of the documents was simply transformed into a bag-of-words, a sequence of words where the ordering was irrelevant and then into a numerical vector where individual dimensions corresponded to the words and their values were based on their occurrences in the documents.

The weights of every term are generally determined by three components – local weight representing the frequency in every single document; global weight reflecting the discriminative ability of the terms, based on the distribution of the terms over the entire document collection; and normalization factor correcting the impact of different lengths of documents [16]. The most popular methods for determining the local weights include term presence (tp) – the weights are binary (0 or 1), representing the presence or absence of the term, and term frequency (tf) – the weights correspond to numbers of times the word appeared in the text. Most commonly used global weight is inverse document frequency (idf) with the general idea that the less the word is common among all texts, the more specific and thus more important it is. The local weight tf is often combined with the global weight *idf* into a weight called *term frequency–inverse document frequency, tf-idf* [4]. Because the tf-idf representation lead to better performance of machine learning algorithms, we present only the results with this document representation.

The detailed description of the data sets, including the numbers of documents and dictionary sizes, might be found in [25]. The shortest reviews contained just one word, the longest more than 300 (308 for German, 346 for French, 352 for Spanish, and 389 for English). Average review length was 18 for German, 20 for French and Spanish, and 24 for English. Sparsity of the vectors representing the data was almost 100%.

4.2 Applied Preprocessing Techniques

Correction of spelling errors was performed using a tool based on the functionality of MS Word 2010. This software has a large built-in collection of dictionaries in many languages. For the words that are not found in the dictionary several alternatives are suggested. These suggested words are handled according to the probability with which they appear in the entire document collection. The misspelled words are replaced by the alternative with the highest probability.

Stemming for different languages was based on Snowball – a language used to express the rules of stemming algorithms. There exist definitions of such rules for several languages. From these rules fast stemmer programs in C, Java, or Perl can be generated and subsequently applied to the data [13].

The words that were considered to be *stop words* were based on general stop words lists created under stop-words project [26], where lists for 29 languages were available.

4.3 Applied Machine Learning Algorithms

This study focused on application of the support vector machine (SVM), decision trees, and k-means algorithms to the data. These algorithms belong to commonly used ones, are computationally efficient (SVM, k-means), or provide the output in a form that can be easily interpreted by humans (decision trees).

Support vector machine (SVM) classifiers try to partition the data by finding a linear boundary (hyperplane) between the classes. The margin widths between the class boundary and training patterns are maximized during the training process. When it is not possible to separate the data in a given n-dimensional space linearly, a kernel function that projects the data to a space of higher dimension is used [11].

A *decision tree* is a classifier that is used in order to give an answer to the given problem (here the answer is the category of the object to be classified) performing a sequence of tests. These tests are based on the values of attributes characterizing the object. A decision tree might be represented by a directed rooted tree. Vertices (also the nodes) of the tree, except the leaves, represent the questions (tests) that must be answered (for example, whether the document contains a certain word or not), and the leaves contain the answer, i.e., the category into which the object should be placed. One of the most popular tree generators is the algorithm known as C4.5 that builds a tree using minimization of entropy [14].

A commonly used measure that is often used in classification is known as the F-measure. It combines values of precision (the fraction of instances that are from a desired, i.e., relevant class) and recall (the fraction of relevant instances that are retrieved). The value of F-measure is typically calculated as the harmonic mean of precision and recall [18].

K-means is the most widely used flat clustering algorithm. In the first step, k randomly selected cluster centers are selected. Then, all objects are assigned to a cluster which is the closest in order to minimize the residual sum of squares. In the following step, the cluster centroids are re-computed according to the positions of the objects in the clusters. Both steps are repeated until a stopping criterion has been met [9].

Having the labels of the data to be clustered the quality of the created clusters might be measured by Purity and Entropy. The Purity measures how pure individual clusters are, i.e., to which extent the clusters contain instances from one class. The Purity of the entire clustering solution is an average of purities for all clusters weighted according to their sizes [24]. The lowest entropy and thus the best quality of a cluster is achieved when the cluster contains instances solely from one class. The highest entropy and thus the worst quality of a cluster is achieved when there are instances of more classes in the cluster and the classes are equally distributed.

5 Results

The following graphs summarize the results achieved by application of the above mentioned algorithms to the data prepared using the mentioned preprocessing techniques and their combinations. The main emphasis was on the impact of text data preprocessing on the quality of the results and computational complexity. The presented values are often normalized in order to reasonably compare results achieved for different data set sizes.

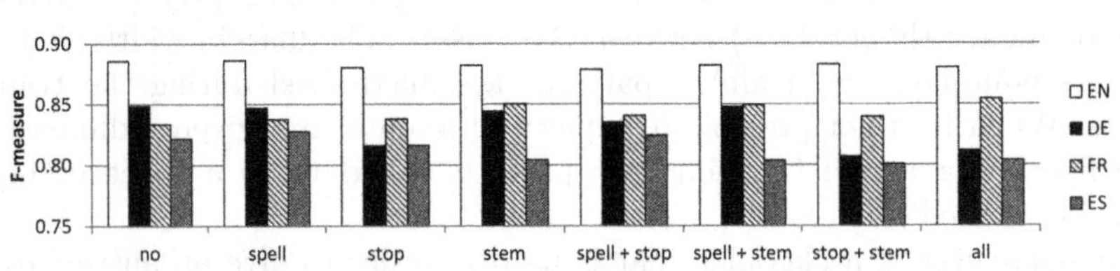

Fig. 1. Comparison of different preprocessing methods impact on quality of classification using decision trees for document collections in different languages (training data set size is 25,000 documents)

Selected results for different data set sizes and languages are presented. For comparison of the results achieved for different languages, the data sets with the biggest sizes were chosen because of the best validity of the results (given by higher volumes of the data available for training).

In the presented graphs, several abbreviations are used to distinguish the results influenced by different preprocessing techniques: *no* – no preprocessing is applied, *spell* – spell checking is performed, *stop* – stop words are removed, *stem* – stemming is performed, and *all* – all three preprocessing methods are applied.

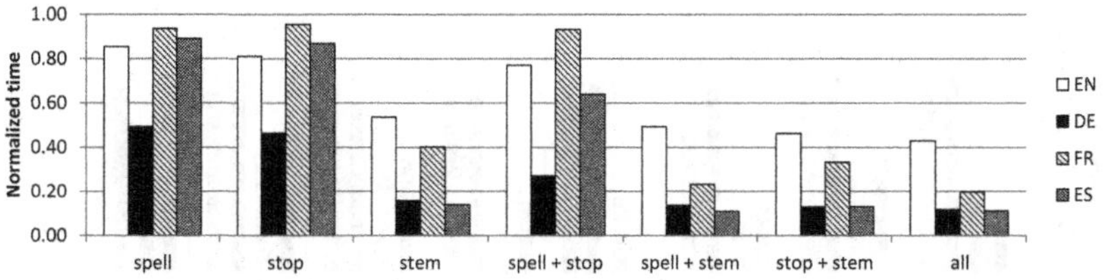

Fig. 2. Comparison of different preprocessing methods impact on time needed for training decision trees for document collections in different languages (training data set size was 25,000 documents)

In supervised learning problems the classifiers were trained on data sets of the given sizes. Their performance was then tested on a test set with one-third size of the training set.

Fig. 1 demonstrates the impact of different preprocessing methods on quality of classification using decision trees. It is obvious that classification of documents in different languages lead to different values of classification performance metrics (the best quality for English, the worst for Spanish). Stop words removal had a negative impact on classification of German documents; stop words removal and stemming negatively influenced classification of Spanish documents. For the remaining languages and preprocessing methods and their combinations was the change of classification quality insignificant.

Fig. 2 demonstrates the dependence between preprocessing techniques application and time needed for building a classifier. Preprocessing techniques generally decrease the number of unique words (dictionary size) and the time of building the classifier is generally lower. This is obvious especially for algorithms with time complexity extensively dependent on the number of attributes (like the decision trees).

Stop words removal had a negative impact on classification of documents using SVM in all of the investigated languages; stemming positively influenced classification of Spanish documents. The remaining languages and preprocessing methods and their combinations did not change the values of classifier performance metrics quality too much, see Fig. 3. The speed of training support vector machine classifiers for all preprocessing methods compared to no preprocessing remained almost the same. The reason is that SVM is very efficient even for

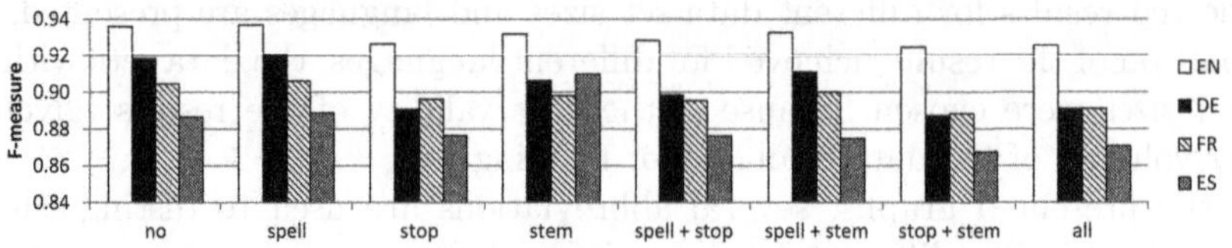

Fig. 3. Comparison of different preprocessing methods impact on quality of classification using support vector machine for document collections in different languages (training data set size was 50,000 documents)

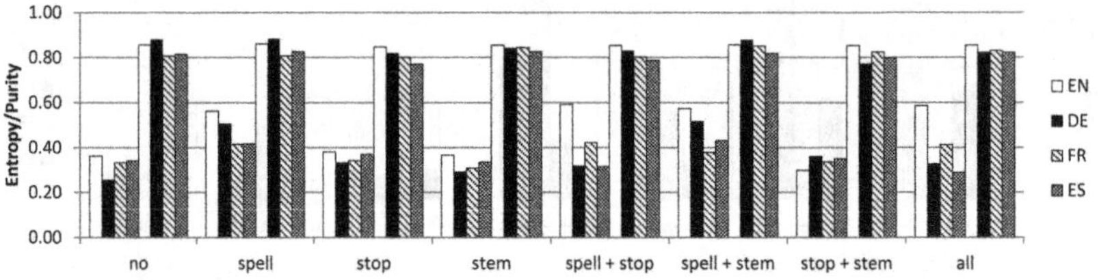

Fig. 4. Comparison of different preprocessing methods impact on quality of clustering using the k-means algorithm for document collections in different languages (three clusters, data set size was 200,000 documents)

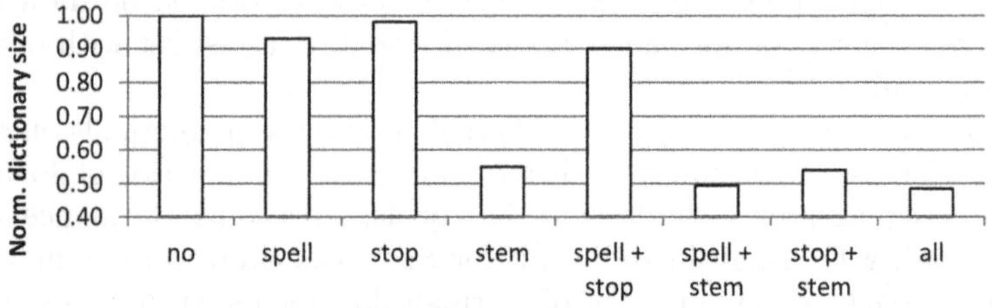

Fig. 5. Impact on dictionary size after application of different preprocessing methods for the data set containing documents in Spanish

large feature spaces. A small increase can be seen for application of stop words removal in Spanish, German, and English documents (ca. 10%, 12%, and 33%).

For evaluating the impact of preprocessing methods application to the data for clustering the quality of clustering solutions, expressed by Entropy and Purity, with 2, 3, 5, 10, and 50 clusters was analyzed, see Fig. 4. The best quality was achieved for three created clusters; thus, the following figures show the results achieved for three-clusters solutions. When measuring the quality by Entropy, a strong negative impact was identified after correcting spelling errors (separately or with combination with other methods). The biggest decline in quality is obvious for English and German. Using Purity as the measure of cluster quality, almost no impact was identified for English. German was negatively influenced by application of stop words removal and stemming. Application of stemming had a positive and stop words removal a negative effect when applied to French and Spanish texts.

Because of high efficiency of the clustering algorithm the change in the duration of the clustering process was rather small. It slightly decreased with application of preprocessing methods and their combinations.

6 Conclusion

In the results of the executed experiments, several patterns might be found. Especially smaller data sets bring unstable results that are influenced by the nature of the processed data. Higher data volumes more significantly reflect the distribution of terms in the collection and the observed findings are more reliable. More data available for training also generally bring better results.

The experiments included also working with data representations not employing the global idf weight. With using it better results were generally achieved. On the other hand, the process of data preparation is slightly more demanding and the size of the data in the vector representation is significantly bigger (real numbers are stored instead of integer numbers). Working with term frequency (TF) and term presence (TP) weights had almost no impact on the experiments since most of the terms (except very common words, like the articles) appeared only once in each document.

It was not possible to clearly and generally determine and quantify the impact of individual preprocessing techniques on text mining results. The results were different according to the applied data mining algorithm and also according to the language of the documents. This is given by the nature of each of the investigated language (size of the dictionary, using inflections, conjugations, diacritics etc.) and by the performance of preprocessing tools for the languages. Each of the preprocessing techniques changed the original data differently, see for example Fig. 5 for the impact on data in Spanish. The text mining quality, as measured by F-measure score or Entropy/Purity, was generally not influenced significantly. Because preprocessing techniques application generally reduces the dictionary size of the data collection the positive impact represented by decreased computational complexity might be observed, especially for algorithms highly sensitive to number of dimensions.

Acknowledgments. This work was supported by the research grant IGA of Mendel University in Brno No. 16/2014. We thank Michaela Kotíková for her assistance and effort in the experimental work.

References

1. Berry, M.W., Kogan, J.: Text Mining: Applications and Theory. Wiley, Chichester (2010)
2. Carvalho, G., Matos, D.M., Rocio, V.: Document Retrieval for Question Answering: A Quantitative Evaluation of Text Preprocessing. In: PIKM 2007, pp. 125–130. ACM (2007)

3. Clark, E., Araki, K.: Text normalization in social media: progress, problems and applications for a pre-processing system of casual English. Procedia – Social and Behavioral Sciences 27, 2–11 (2011)
4. Cummins, R., O'Riordan, C.: Evolving local and global weighting schemes in information retrieval. Information Retrieval 9, 311–330 (2006)
5. Feldman, R., Sanger, J.: The Text Mining Handbook: Advanced Approaches in Analyzing Unstructured Data. Cambridge University Press (2006)
6. Habernal, I., Ptáček, T., Steinberger, J.: Supervised sentiment analysis in Czech social media. Information Processing and Management 50, 693–707 (2014)
7. Haddi, E., Liu, X., Shi, Y.: The Role of Text Pre-processing in Sentiment Analysis. Procedia Computer Science 17, 26–32 (2013)
8. Joachims, T.: Learning to classify text using support vector machines. Kluwer Academic Publishers, Norwell (2002)
9. Manning, C.D., Raghavan, P., Schütze, H.: Introduction to Information Retrieval. Cambridge University Press (2008)
10. Munková, D., Munk, M., Vozár, M.: Data Pre-Processing Evaluation for Text Mining: Transaction/Sequence Model. Procedia Computer Science 18, 1198–1207 (2013)
11. Noble, W.S.: What is a support vector machine? Nature Biotechnology 24, 1564–1567 (2006)
12. Petz, G., et al.: Computational approaches for mining user's opinions on the Web 2.0. Information Processing & Management 50, 899–908 (2014)
13. Porter, M.F.: Snowball: A language for stemming algorithms (2001)
14. Quinlan, J.R.: C4.5: Programs for Machine Learning. Morgan Kaufmann, San Francisco (1993)
15. Rajman, M., Vesely, M.: From Text to Knowledge: Document Processing and Visualization: A Text Mining Approach. In: Sirmakessis, S. (ed.) Text Mining and Its Applications: Results of the NEMIS Launch Conference, pp. 7–24. Springer (2004)
16. Salton, G., McGill, M.J.: Introduction to Modern Information Retrieval. McGraw-Hill, New York (1983)
17. Sebastiani, F.: Machine Learning in Automated Text Categorization. ACM Computing Surveys 34, 1–47 (2002)
18. Sokolova, M., Japkowicz, N., Szpakowicz, S.: Beyond Accuracy, F-Score and ROC: A Family of Discriminant Measures for Performance Evaluation. In: Sattar, A., Kang, B.-H. (eds.) AI 2006. LNCS (LNAI), vol. 4304, pp. 1015–1021. Springer, Heidelberg (2006)
19. Tourné, N., Godoy, D.: Evaluating tag filtering techniques for web resource classification in folksonomies. Expert Systems with Applications 39, 9723–9729 (2012)
20. Uysal, A.K., Gunal, S.: The impact of preprocessing on text classification. Information Processing & Management 50, 104–112 (2014)
21. Xu, R., Wunsch, D.C.: Clustering. Wiley, Hoboken (2009)
22. Witten, I., Frank, E., Hall, M.: Data Mining: Practical Machine Learning Tools and Techniques. Morgan Kaufmann Publishers (2011)
23. Zhang, W., Yoshida, T., Tang, X.: Text classification based on multi-word with support vector machine. Knowledge-Based Systems, 879–886 (2008)
24. Zhao, Y., Karypis, G.: Criterion Functions for Document Clustering: Experiments and Analysis. Technical Report, University of Minnesota (2001)
25. Žižka, J., Dařena, F.: Mining Significant Words from Customer Opinions Written in Different Natural Languages. In: Habernal, I., Matoušek, V. (eds.) TSD 2011. LNCS, vol. 6836, pp. 211–218. Springer, Heidelberg (2011)
26. `https://code.google.com/p/stop-words`

Using Fuzzy Logic Controller in Ant Colony Optimization

Victor M. Kureichik and Asker Kazharov

Chekhova St.,22, Taganrog, Russian Federation, 347928
kur@tgn.sfedu.ru,
aakazharov@sfedu.ru

Abstract. The new modification of ant colony optimization has been proposed to solve travelling salesman problem. This modification is based on using fuzzy rules and fuzzy terms like «a little», «much», «almost» etc. Fuzzy logic controller was developed to define fuzzy rules. This controller allows to regulate values of heuristic coefficients of ant colony optimization dynamically. Experimental research was carried out. The results received show high effectiveness of fuzzy logic controller using in ant colony optimization. The modified ant colony optimization algorithm finds shorter routes on 1-3%. This modification can be used to solve other problems.

Keywords: ant colony optimization, fuzzy logic, travelling salesman problem.

1 Introduction

This paper is dedicated to development fuzzy ant colony optimization (ACO) to solve travelling salesman problem. Classical ACO was based on idea of ant colony behavior modeling. AA belongs to the category of "swarm intelligence". Its author is Marko Dorigo, who made researches in this area in the middle 90-s of the XX century [1]. ACO got spread currency in graphic problems solving. One of the popular applications is transport and logistics problems solving. To improve the quality of solutions and its compliment, new operators were used, and the agent's behavior was changing [2]. There existing various modifications of AA, hybrid algorithms, based on AA, bee, genetic algorithms, which are among them [3]. The main idea is in ant behavior modeling [4]. Each ant is revealed as a simple agent, conforming to the elementary rules, but in general, it is revealed as an intelligent multi-agent system.

One of the main drawbacks of AA is a complicated and lengthy process of the algorithmic parameter setting. Some parameters, such as the amount of ants in the colony is nonlinearly depends on the size of a problem, for example, on the amount of nodes in a graph. In case of taking the natural analogy it is similar the necessity of a larger colony in searching of a larger square to cover it completely. For a small site territory the colony of moderate size is enough. It is important to find an optimal size of the ant colony, because its oversize in the algorithm causes not only the growth of the computing exercises, but also the redundancy in pheromone secreting, which has a negative impact on ACO.

R. Silhavy et al. (eds.), *Artificial Intelligence Perspectives and Applications*,
Advances in Intelligent Systems and Computing 347, DOI: 10.1007/978-3-319-18476-0_16

A new method of ACO (heuristic coefficients) parameters regulation is proposed in this paper. This method is based on using of fuzzy logic controller (FLC) that uses fuzzy terms like «a little», «much», «almost» etc. FLC using allows find optimal solutions more fast.

2 Ant Colony Optimization

The main idea of the ant algorithm is the movement of ants modeling. The choice of the direction of ants' movement is performed on food sources proximity (for graph problems they are the verges to adjacent vertices vertexes) and pheromones level. Pheromones are special chemical compounds, naturally shed by ants. In arithmetic model pheromones are stored in some weight matrix $\tau_{i,j}$ (i, j – vertex numbers), and are represented as real numbers. In actual model the ants shed pheromones as markers on their route. When the ants find the food source, the level of pheromones shed by them increases. The negative feedback is an exhaling of pheromones. The exhaling of pheromones increases on long routes, that is why with the passage of time the pheromones level increases on the shortest route.

Modeling of food searching by ant colony is displayed on fig. 1-3. This modeling was realized by program "AntSim v1.1".

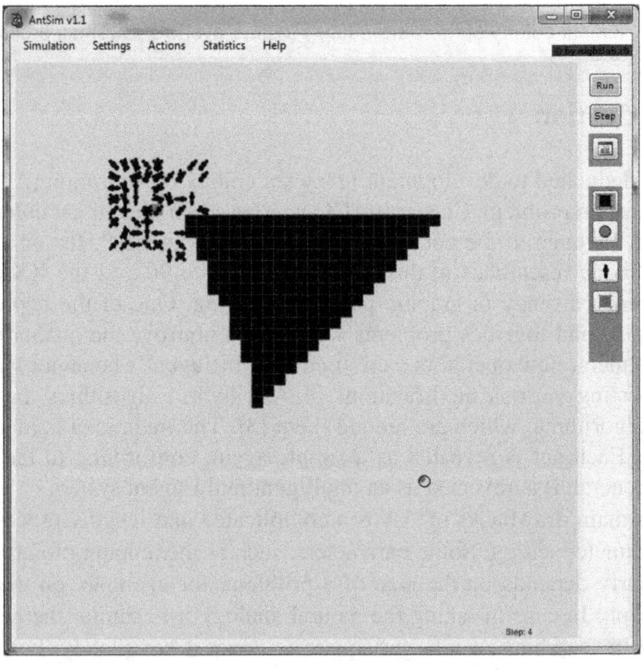

Fig. 1. 10-th iteration

According to fig. 1 obstacle is marked by solid black color. Food source is marked right bottom as small circle. Ant colony is marked left top. Ants become search for food. Food search process is displayed in fig. 2.

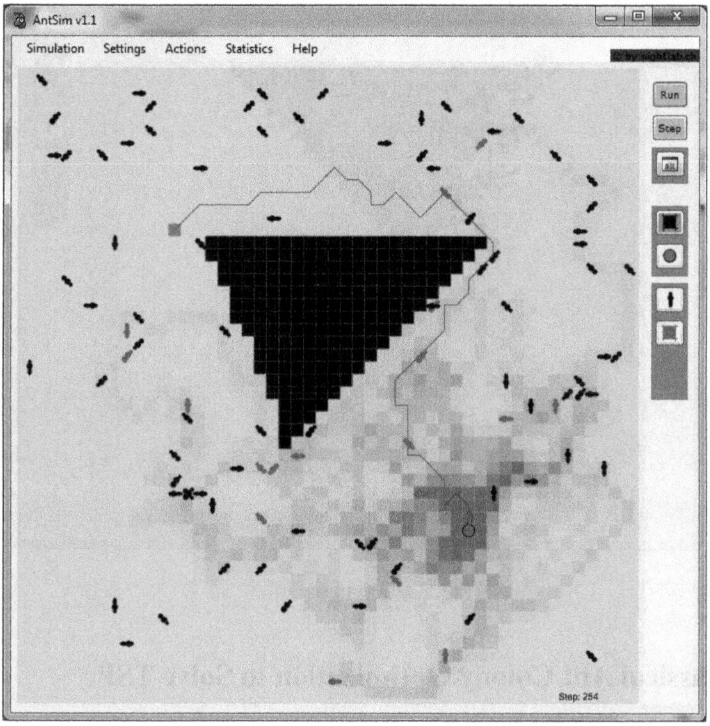

Fig. 2. 1500-th iteration

Pheromones of ants are displayed in fig. 2. Ants mark routes by pheromones. The best ant route of 1500-th iteration marked by thin broken line. Pheromone level on "bad" routes is decreased according to evaporation. Pheromone level is fixed and increased only on more short routes. The best ant route is displayed in fig. 3. As we can see on fig. 3 pheromones concentrate and accumulate on the best route.

The work [6] of Kureichik V.M. and Kazharov A.A. introduces a survey, research and modification of ant algorithms for solving of transportation problems, including the traveling salesman problem. In these problems the results quality, obtaining by means of ACO, excels the results quality of other algorithms. Experimental researches were made for the traveling salesman problem.

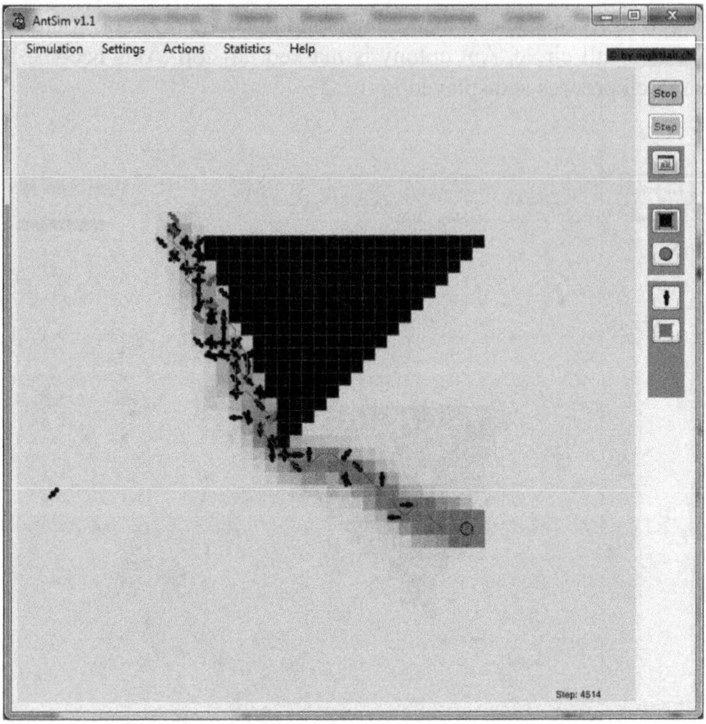

Fig. 3. 3000-th iteration

3 Classical Ant Colony Optimization to Solve TSP

To solve this problem we have got: a graph G=(X,U), where $|X|$ = n is a vertex set (cities), $|U|$ = m is a branches set (possible ways between the cities). Values matrix $D(i, j)$, where $i, j \in 1, 2, \ldots, n$, represent a cost of moving from vertex x_i to x_j. It requires finding the permutation φ of the elements of the set X, such as that the value of the objective function will be:

$$F(\varphi) = D(\varphi_1, \varphi_n) + \sum_{i=1}^{n-1} D(\varphi_i, \varphi_{i+1}) \rightarrow min.$$

Hence, the criterion is the way length and the criterion of minimization is the aim of the problem. If the graph is not fully connected, in the matrix D, in the cells, corresponding the absent branches in the graph, the perpetuity is assigned with the result that the underpass with respect to the given branch is excluded [7].

Behavior of ants is determined by the algorithm selection of the following route point. A migration probability of an ant from the point i to the point j is determined by the following formula [4]:

$$\begin{cases} P_{ij,k}(t) = \dfrac{[\tau_{ij}(t)]^{\alpha} \cdot [\eta_{ij}(t)]^{\beta}}{\sum\limits_{l \in J_{i,k}} [\tau_{il}(t)]^{\alpha} \cdot [\eta_{il}(t)]^{\beta}} & , j \in J_{i,k} \\ P_{ij,k}(t) = 0, j \notin J_{i,k} & \\ \eta_{ij} = \dfrac{1}{D_{ij}} & \end{cases} \tag{1}$$

where α, β are parameters, which assume a track weight coefficient of the pheromone, are the coefficients of heuristics. The pheromone level in the time point t at the branch D_{ij} corresponds $\tau_{ij}(t)$. Parameters α and β determine the contribution of two parameters and their influence on the equation.

They determine ants "greed". As α=0 an ant longs to chose the shortest branch, as β=0 – the branch with most pheromone [8]. Easy to notice that given formula has the phenomenon of a "roulette wheel". Except listed parameters m, the quantity of ants in the colony is of the prime importance. Upon this parameter depends $\tau_{ij}(t)$, a cumulative size of emitted pheromones at the iteration t at the verge from the vertex i to the vertex j. The work is devoted to the defining of optimal values α, β. This affords us to find out quasioptimal solutions for smaller period.

4 Fuzzy Logic Controller

The main idea of this work is creation of ACO using fuzzy logic, i.e. FACO [5]. FACO uses fuzzy terms like "a little", "much", "almost" etc. The total of this work is FLC creation that process as input data the changes of average objective function (OF) and best OF. The output data of FLC is changes of ACO parameters. So ACO with dynamic parameters was developed to improve solutions quality and decreasing search time.

As mentioned above the main idea of this algorithm is FLC using that process as input data the changes of average OF and best OF [5]. The proposed FLC have two parameters e_1 и e_2:

$$e_1(t) = \frac{f_{ave}(t) - f_{best}(t)}{f_{ave}(t)} \tag{2}$$

$$e_2(t) = \frac{f_{ave}(t) - f_{ave}(t-1)}{f_{best}(t)} \tag{3}$$

where

t is iteration number,

$f_{best}(t)$ is the best OF for t-th iteration,
$f_{ave}(t)$ is average OF for t-th iteration,
$f_{ave}(t - 1)$ is average OF for $(t-1)$-th iteration.

Fuzzy values definitions are displayed graphically on fig. 4-6. NL – negative large (much decreasing of OF), NS – negative small (a little decreasing of OF), ZE – zero (almost unchanged), PS – positive small (a little increasing of OF), PL – positive large (much increasing of OF).

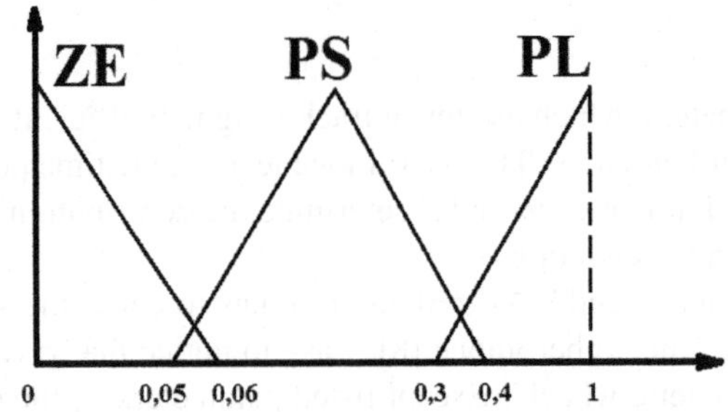

Fig. 4. Borders of fuzzy values of parameter e_1

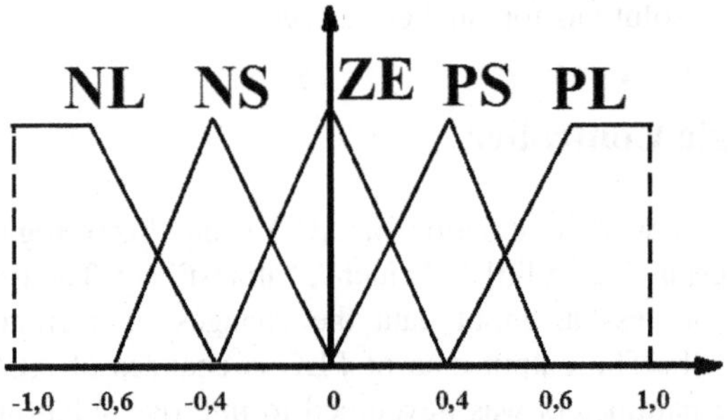

Fig. 5. Borders of fuzzy values of parameter e_2

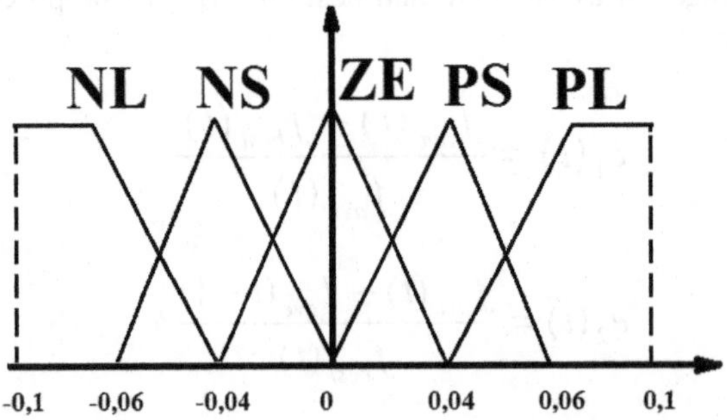

Fig. 6. Borders of fuzzy values of parameters $\Delta\alpha(t)$ and $\Delta\beta(t)$

Output parameters of $\Delta\alpha(t)$ and $\Delta\beta(t)$ are same. According to fig. 6 values of α and β can be changed on $\Delta\in[-0,1;0,1]$. So α and β evaluated as follows:

$$\alpha(t) = \alpha(t-1) + \Delta\alpha(t) \tag{4}$$

$$\beta(t) = \beta(t-1) + \Delta\beta(t) \tag{5}$$

Thus, FLC can be represents as two-dimensional set of fuzzy rules:

Table 1. Fuzzy rules for parameter α ($\Delta\alpha(t)$)

				e_2		
		NL	***NS***	***ZE***	***PS***	***PL***
e_1	***PL***	PS	ZE	PS	ZE	PL
	PS	NL	ZE	ZE	ZE	NS
	ZE	ZE	ZE	NS	NS	NS

Table 2. Fuzzy rules for parameter β ($\Delta\beta(t)$)

				e_2		
		NL	***NS***	***ZE***	***PS***	***PL***
e_1	***PL***	PL	PS	PS	ZE	NS
	PS	PS	ZE	PL	ZE	NS
	ZE	ZE	ZE	NS	NL	NL

For example, if the input data of FLC is e_1 = PS и e_2 = ZE then

$\Delta\alpha(t)$ = ZE, $\Delta\beta(t)$ = PL
=> $\alpha(t) = \alpha(t-1)$ + ZE, $\beta(t) = \beta(t-1)$ + PL.

It means, if e_1 was increased "a little" according to (2) and e_2 was not changed almost according to (3), so parameter α was not changed «almost», parameter β was changed much.

5 Conclusion

Developed FACO allows to find shorter routes for TSP in comparison with the classical ACO. The routes found by FLC are shorter on 1-3% in comparison with the standard ACO. The offered modification can be used for determination of various parameters of ACO, including evaporation coefficient, ant colony size etc.

Acknowledgement. The work is supported by the Russian science foundation (project №14-11-00242) in Southern Federal University.

References

1. Dorigo, M., Di Caro, G., Gambardella, L.M.: Ant Algorithms for discrete Optimizations. Artificial Life 2(5), 137–172 (1999)
2. Kureichik, V.M., Kazharov, A.A., Lyapunova, I.A.: Parameter analysis of ant algorithm. Life Science Journal 2014 11(10s), 402–406 (2014)
3. Cristofides, N., Eilon, S.: An algorithm for the vehicle dispatching problem. Opl. Res. Quart. (1969)
4. Kazharov, A.A., Kureichik, V.M.: The Development of the Ant Algorithm for Solving the Vehicle Routing Problems. World Applied Sciences Journal 26(1), 114–121 (2013)
5. Jones, T.: AI Application Programming. Cengage Learning Press (2005)
6. Cormen, T.H., Leiserson, C.E., Rivest, R.L., Stein, C.: Introduction to Algorithms, 3rd edn. The MIT Press, Massachusetts (2009)
7. Stovba, S.D.: Ant algorithms. Exponents Pro. Mathematics in Application 4, 70–75 (2003)
8. Kazharov, A.A., Kureichik, V.M.: Ant colony optimization algorithms for solving transportation problems. Journal of Computer and Systems Sciences International 49(1), 30–43 (2010)
9. Liu, H., Xu, Z., Abraham, A.: Hybrid fuzzy-genetic algorithm approach for crew grouping. In: 5th International Conference on Intelligent Systems Design and Applications (ISDA 2005), pp. 332–337 (September 2005)
10. Cordon, O., Gomide, F., Herrera, F., Hoffmann, F., Magdalena, L.: Ten years of genetic fuzzy systems: current framework and new trends. Fuzzy Sets and Systems 141, 5–31 (2004)

WSM Tuning in Autonomous Search via Gravitational Search Algorithms

Ricardo Soto[1,2,3], Broderick Crawford[1,4,5], Rodrigo Herrera[1], Rodrigo Olivares[6], Franklin Johnson[7], and Fernando Paredes[8]

[1] Pontificia Universidad Católica de Valparaíso, Chile
[2] Universidad Autónoma de Chile, Chile
[3] Universidad Central de Chile, Chile
[4] Universidad Finis Terrae, Chile
[5] Universidad San Sebastián, Chile
[6] Universidad de Valparaíso, Chile
[7] Universidad de Playa Ancha, Chile
[8] Escuela de Ingeniería Industrial, Universidad Diego Portales, Chile
{ricardo.soto,broderick.crawford}@ucv.cl, rodrigo.herrera.laferte@gmail.com, rodrigo.olivares@uv.cl, franklin.johnson@upla.cl, fernando.paredes@udp.cl

Abstract. Autonomous search is a recent approach that allows the solver to adapt their search so as to be more efficient without the manual configuration of an expert user. The goal is to provide more capabilities to the solver in order to improve the search process based on some performance indicators and self-tuning. This approach has effectively been applied to different optimization and satisfaction techniques such as constraint programming, SAT, and various metaheuristics. This paper focuses on automated self-tuning of constraint programming solvers. We employ a classic decision making method called weighted sum model (WSM) to evaluate the search process performance. This evaluation is used by the solver to re-configure its parameters in benefit of reaching a better performance. However, reaching good configurations straightly depends on the correct tuning of the WSM. This is known to be hard as the WSM is problem-dependent and good settings are not commonly stable along the search. To this end, we introduce a gravitational search algorithm (GSA), which is able to find good WSM configurations when solving constraint satisfaction problems. We illustrate experimental results where the GSA-based approach directly competes against previously reported autonomous search methods for constraint programming.

Keywords: Adaptive Systems, Constraint Satisfaction, Gravitational Search, Optimization.

1 Introduction

Autonomous search (AS) is a modern approach used to allow solvers to automatically re-configure its solving parameters in order to improve the process when poor performances are detected. The solving performance is evaluated via

R. Silhavy et al. (eds.), *Artificial Intelligence Perspectives and Applications*,
Advances in Intelligent Systems and Computing 347, DOI: 10.1007/978-3-319-18476-0_17

indicators that gather relevant information during search. Then, the searching parameters are updated according the results yielded by the performance evaluation. This approach has effectively been applied to different optimization and satisfaction techniques such as constraint programming [5], SAT [8], mixed integer programming [10,9] and various metaheuristics [11,13,12].

In this paper, we focus on automated self-tuning of constraint programming (CP) solvers. The goal is to provide a mechanism for CP solvers in order to let them control and adapt their search parameters in benefit of reaching efficient solving processes. This is possible by integrating a component able to evaluate the solving performance. In particular, we employ a classic decision making method called weighted sum model (WSM), which allows one to evaluate a number of alternatives in terms of a number of decision criteria in a simple manner. This evaluation is then used by the solver which re-configure its parameters in order to improve the solving performance. However, selecting the correct solving parameters implies to correctly tuning the WSM. This is known to be a hard task as the WSM is problem-dependent and an efficient configuration is not commonly stable along the search. A common way to tune parameters is to incorporate an optimizer so as to maximize the quality of the solving process of a given optimization or constraint satisfaction problem. To this end, we employ a relatively modern but efficient metaheuristic, based on the Newtonian gravity and the laws of motion, called gravitational search. This incorporation results in an efficient tuning approach that directly competes against previously reported autonomous search methods for solving constraint satisfaction and optimization problems.

The remainder of this paper is organized as follows. The related work is given in Section 2. Section 2 presents an overview of constraint programming. The WSM tuning is described in Section 4 followed by the associated GSA optimizer. Finally, we provide experiments, conclusions, and future work.

2 Related Work

A pioneer work in AS for CP is the one presented in [2]. This framework introduced a four-component architecture, allowing the dynamic replacement of enumeration strategies. The strategies are evaluated via performance indicators of the search process, and better evaluated strategies replace worse ones during solving time. Such a pioneer framework was used as basis of different related works. For instance, a more modern approach approach based on this idea is reported in [4]. This approach employs a two-layered framework where an hyper-heuristic placed on the top-layer controls the dynamic selection of enumeration strategies of the solver placed on the lower-layer. An hyper-heuristic can be regarded as a method to choose heuristics [7]. In this approach, two different top-layers have been proposed, one using a genetic algorithm [15,3] and another using a particle swarm optimizer [5]. Similar approaches have also been implemented for solving optimization problems instead of pure CSPs [14]. In Section 5 we provide a comparison of our approach with the best AS optimizers reported in the literature.

3 Constraint Programming

In this Section we provide an overview of constraint programming and associated concepts.

3.1 Constraint Satisfaction Problems

A constraint satisfaction problem (CSP) $\mathcal{P}$ is defined by a triple $P = \langle \mathcal{X}, D, C \rangle$ where $\mathcal{X}$ is an n-tuple of variables $\mathcal{X} = \langle x_1, x_2, \ldots, x_n \rangle$. $\mathcal{D}$ is a corresponding n-tuple of domains $\mathcal{D} = \langle d_1, d_2, \ldots, d_n \rangle$ such that $x_i \in d_i$, and d_i is a set of values, for $i = 1, \ldots, n$; and $\mathcal{C}$ is an m-tuple of constraints $\mathcal{C} = \langle c_1, c_2, \ldots, c_m \rangle$, and a constraint c_j is defined as a subset of the Cartesian product of domains $d_{j_1} \times \cdots \times d_{j_{n_j}}$, for $j = 1, \ldots, m$. A solution to a CSP is an assignment $\{x_1 \rightarrow a_1, \ldots, x_n \rightarrow a_n\}$ such that $a_i \in d_i$ for $i = 1, \ldots, n$ and $(a_{j_1}, \ldots, a_{j_{n_j}}) \in c_j$, for $j = 1, \ldots, m$.

Constraint Solving. CSPs are usually solved by combining enumeration and propagation phases. The enumeration strategy decides the order in which variables and values are selected by means of the variable and value ordering heuristics, respectively. The propagation phase tries to delete from domain the values that do not drive the search process to a feasible solution.

4 WSM Tuning

In this work, we aim at online controlling a set of enumeration strategies which are dynamically interleaved during solving time. Our purpose is to select the most promising one for each part of the search tree. To this end we need to evaluate the strategies by penalizing the ones exhibiting poor performances and giving more credits to better ones. Based on the work done on [5], we can evaluate the performance of strategies via a set of indicators that are able to measure the quality of the search. The performance is evaluated via a weighted sum model (WSM), which is a well-known decision making method from multi-criteria decision analysis for evaluating alternatives in terms of decision criteria. Formally, we define a weighted sum model $A_t(S_j)^{WSM-score}$ that evaluates a strategy S_j in time t as follows:

$$A_t(S_j)^{WSM-score} = \sum_{i=1}^{IN} w_i a_{it}(S_j) \tag{1}$$

Where IN corresponds to the indicator set, w_i is a weight that controls the importance of the ith-indicator within the WSM and $a_{it}(S_j)$ is the score of the ith-indicator for the strategy S_j in time t. A main component of this model are the weights, which must be finely tuned by an optimizer. This is done by carrying out a sampling phase where the CSP is partially solved to a given cutoff. The performance information gathered in this phase via the indicators is used as input data of the optimizer, which attempt to determine the most successful weight set for the WSM. This tuning process is very important as the correct configuration of the WSM may have essential effects on the ability of the solver to properly solve specific CSPs. Parameter (weights) tuning is hard to achieve as parameters are problem-dependent and their best configuration is not stable along the search [13]. As previously mentioned, we are in the presence of an hyper-optimization problem, which in practice is the optimization of the process (the optimal configuration of the WSM) for solving an optimization (satisfaction) problem. In Section 4.1, we present the approach used to optimize the WSM.

4.1 Gravitational Search Algorithm

As previously mentioned, to determine the most successful weight set for performance indicators, the WSM must be finely tuned by an optimizer. To this end we propose the use of a GSA algorithm, which is a recent metaheuristic [16] based by the Newtonian laws of gravity and motion. In GSA, the agents are considered as objects and their performance is measured by their masses. All agents interact with each other based on the Newton laws on gravity and motion[6,17]. All object attract each other by the gravity force, and this force causes a global movement of all objects towards the objects with heavier masses. The heavy masses correspond to good solutions and move more slowly than lighter ones, this guarantees the exploitation step of the algorithm.

In GSA, each agent (mass) has four specifications: its position, its inertial mass, its active gravitational mass and its passive gravitational mass. The position of the mass corresponds to a solution of the problem, and its gravitational and inertial masses are determined using a fitness function. Each agent (mass) presents a solution, and the algorithm is navigated by the properly adjusting the gravitational and inertia masses. By lapse of time, we expect that masses be attracted by the heaviest mass. This mass will present an optimum solution in the search space. In GSA, the position of the ith agent may be defined as:

$$X_i = (x_i^1, x_i^2, ..., x_i^d, ..., x_i^n) \qquad for \quad i = 1, 2, ..., N, \tag{2}$$

where x_i^d is the position of the ith agent in the dth dimension, n is the dimension of the search space, and N is the number of agents. At a specific time 't', we define the force acting on mass 'i' from mass 'j' as following:

$$F_{ij}^d(t) = G(t)\frac{M_{pi}(t)M_{aj}(t)}{R_{ij}(t) + \varepsilon}(x_j^d(t) - x_i^d(t)) \tag{3}$$

where M_{aj} is the active gravitational mass related to agent j, M_{pi} is the passive gravitational mass related to agent i, $G(t)$ is gravitational constant at time t, ε is a small constant, and $R_{ij}(t)$ is the Euclidian distance between the ith and the jth agents defined as:

$$R_{ij}(t) = \|X_i(t), X_j(t)\|_2 \tag{4}$$

$$G(t) = G_0(1 - \frac{t}{T}) \tag{5}$$

In GSA, G is considered as a linear decreasing function 5,where G_0 is a constant and set to 100 and T is the total number of iterations (the total age of system). The total force acting on the ith agent is given as 6.

$$F_i^d(t) = \sum_{j=1, j \neq i}^{N} rand_j F_{ij}^d(t), \quad (6) \qquad a_i^d(t) = \frac{F_i^d(t)}{M_{ii}(t)} \quad (7)$$

We suppose that the total force that acts on agent i in a dimension d be a randomly weighted sum of dth components of the forces exerted from other agents. $rand_j$ is a random number in the interval [0, 1]. According to Newtons law of motion, the acceleration of the ith agent at time t, in dth dimension, is given as 7, where M_{ii} is the inertial mass of ith agent. The next velocity of an agent is considered a fraction of its current velocity added to its acceleration. Therefore, the position and the velocity of the ith agent at time t, in dth dimension may be formulated as follows:

$$v_i^d(t+1) = rand_i \times v_i^d(t) + a_i^d(t) \quad (8) \qquad x_i^d(t+1) = x_i^d(t) + v_i^d(t+1) \quad (9)$$

where $rand_i$ is a uniform random variable in the interval [0, 1]. We use this random number to give a randomized characteristic to the search. In GSA, gravitational and inertial masses are considered the same. However, they may use different values. A large inertial mass performs a search operation more precise because the movement of agents is slower. Moreover, a larger gravitational mass causes a greater attraction of agents allows faster convergence rate. Accordingly, we update the gravitational and inertial masses by the following equations:

$$M_{ai} = M_{pi} = M_{ii} = M_i, \quad i = 1, 2, ..., N, \quad (10)$$

$$m_i(t) = \frac{fit_i(t) - worst(t)}{best(t) - worst(t)}, \quad (11) \qquad M_i(t) = \frac{m_i(t)}{\sum_{j=1}^{N} m_j(t)} \quad (12)$$

where $fit_i(t)$ represent the fitness value of the ith agent at time t, and, $worst(t)$ and $best(t)$ are defined as follows:

$$best(t) = \min_{j \in \{1,...,N\}} fit_j(t) \quad (13) \qquad worst(t) = \max_{j \in \{1,...,N\}} fit_j(t) \quad (14)$$

One way to make a good compromise between exploration and exploitation is to reduce the number of agents as time passes in the Eq.(6). Thus, we propose that only a set of agents, those with the greatest mass, the force applied to the other. However, be careful with the use of this policy, as this may reduce the power of exploration and increase the capacity of exploitation.

To avoid falling into a local minimum, the algorithm must use the exploration at the beginning. For periods of time, exploration should disappear and exploitation should appear. To improve the performance of GSA, *Kbest* agents only attract other agents. *Kbest* is a function of time, with K_0 at the beginning and decreasing over time. Thus, initially, all agents apply their force on the other, and as time passes, Kbest will reduce linearly causing the end only an agent apply its strength to all the rest. Therefore, the Eq.(6) could be replaced by:

$$F_i^d(t) = \sum_{j \in Kbest, j \neq i} rand_j F_{ij}^d(t), \tag{15}$$

where *Kbest* is the set of first K agents with the best fitness value and biggest mass. The GSA algorithm would be as follows (Algorithm 1):

Algorithm 1. GSA

Load initial solution to each agent x_i;
Calculate fitness for each agent x_i;
while $it < it\_max$ **do**
  Update Kbest and G (5), *asign best and worst*;
  for all x_i **do**
    Calculate m (11) *and Mass* (10);
  end for
  for all x_i **do**
    for all *Kbest* **do**
      if $x_i \neq Kbest$ **then**
        Calculate the force between the x_i *agent and the Kbest agent* (15);
      end if
    end for
    For each x_i *agent calculate aceleration* (7) *and the new velocity* (8);
    For each x_i *agent calculate their new position* (9);
  end for
end while
return *The best solution*;

5 Experimental Evaluation

We have performed an experimental evaluation of the proposed approach on different instances of classic constraint satisfaction problems:

- The n-queens (nQ) problem with $n = \{8,10,12,15,20,50,75\}$
- The magic squares (nMS) with $size = \{3,4,5,6,7\}$
- The Sudoku puzzle {1,2,5,7,9}
- The Latin Square (nLS) with $size = \{4,5,6,7,8\}$

The adaptive enumeration component has been implemented on the Ecl^{i}ps^e Constraint logic Programming Solver v5.10, and the GSA-optimizer has been developed in Java. The experiments have been launched on a 3.3GHz Intel Core i3 with 4Gb RAM running Windows 7 Professional 32 bits. The instances are solved to a maximum number of 65535 steps as equally done in previous work [3]. If no solution is found at this point the problem is set to t.o. (time-out). Let us recall that a step refers to a request of the solver to instantiate a variable by enumeration. The adaptive enumeration uses a portfolio of 24 enumeration strategies, which is detailed in table 1.

Table 1. Portfolio used

Id	Variable ordering	Value ordering
S_1	First variable of the list	min. value in domain
S_2	The variable with the smallest domain	min. value in domain
S_3	The variable with the largest domain	min. value in domain
S_4	The variable with the smallest value of the domain	min. value in domain
S_5	The variable with the largest value of the domain	min. value in domain
S_6	The variable with the largest number of attached constraints	min. value in domain
S_7	The variable with the smallest domain. If are more than one, choose the variable with the bigger number of attached constraints.	min. value in domain
S_8	The variable with the biggest difference between the smallest value and the second more smallest of the domain	min. value in domain
S_9	First variable of the list	mid. value in domain
S_{10}	The variable with the smallest domain	mid. value in domain
S_{11}	The variable with the largest domain	mid. value in domain
S_{12}	The variable with the smallest value of the domain	mid. value in domain
S_{13}	The variable with the largest value of the domain	mid. value in domain
S_{14}	The variable with the largest number of attached constraints	mid. value in domain
S_{15}	The variable with the smallest domain. If are more than one, choose the variable with the bigger number of attached constraints.	mid. value in domain
S_{16}	The variable with the biggest difference between the smallest value and the second more smallest of the domain	mid. value in domain
S_{17}	First variable of the list	max. value in domain
S_{18}	The variable with the smallest domain	max. value in domain
S_{19}	The variable with the largest domain	max. value in domain
S_{20}	The variable with the smallest value of the domain	max. value in domain
S_{21}	The variable with the largest value of the domain	max. value in domain
S_{22}	The variable with the largest number of attached constraints	max. value in domain
S_{23}	The variable with the smallest domain. If are more than one, choose the variable with the bigger number of attached constraints	max. value in domain
S_{24}	The variable with the biggest difference between the smallest value and the second more smallest of the domain.	max. value in domain

We compare the proposed approach based on GSA-optimization (GSA) with the two previously reported optimized online control systems, one based on genetic algorithm (GA) [3] and the other one based on particle swarm optimization (PSO) [5]. For the evaluation, we consider number of backtracks and runtime needed to reach a solution, both being widely employed indicators of search performance.

We employ the following WSM for the experiments: $w_1SB + w_2In1 + w_3In2$, where SB is the number of shallow backtracks [1] (SB), In1=$CurrentMaximumDepth - PreviousMaximumDepth$, and In2=$CurrentDepth - PreviousDepth$, where *Depth* refers to the depth reached within the search tree. This WSM was the best performing one after the corresponding training phase of the algorithm. For the GSA we use 100 iterations and 10 agents, which suffices to solve all instances for the aforementioned problems.

Table 2. Backtracks

S_j	Problems													
	N-Queens							$\bar{X}_Q$	Sudoku					$\bar{X}_S$
	8Q	**10Q**	**12Q**	**15Q**	**20Q**	**50Q**	**75Q**		**1S**	**2S**	**5S**	**7S**	**9S**	
PSO	3	1	1	1	11	0	818	119,29	0	2	13	256	0	54,2
GA	1	4	40	68	38	15	17	26,14	732	6541	4229	10786	7	4459
GSA	1	0	0	1	2	0	2	0,85	0	2	37	292	0	66,2

S_j	Problems											
	Magic Square					$\bar{X}_{MS}$	Latin Square					$\bar{X}_{LS}$
	3MS	4MS	5MS	6MS	7MS		4LS	5LS	6LS	7LS	8LS	
PSO	0	0	14	>47209	>56342	>56342	0	0	0	0	0	0
GA	0	42	198	>176518	>213299	>213299	0	0	0	0	0	0
GSA	0	0	12	734	1874	524	0	0	0	0	0	0

Table 2 depicts the backtracks required to reach a solution. Considering N-Queens problems the minimum average is obtained by the GSA, while for Sudokus GSA takes the second place. Taking into account Magic Squares instances GSA performs better again, and for Latin Squares all techniques reach perfect enumerations (no backtracks needed). In Table 3, solving times are provided. Here, for N-Queens GSA takes the second place, but for Sudokus, Magic and Latin Squares, GSA is notably faster. Let us also remark that GSA is the only approach able to solve all instances before time-out.

Table 3. Solving Times

S_j	Problemas													
	N-Queens							$\bar{X}_Q$	Sudoku					$\bar{X}_S$
	8Q	**10Q**	**12Q**	**15Q**	**20Q**	**50Q**	**75Q**		**1S**	**2S**	**5S**	**7S**	**9S**	
PSO	4982	7735	24369	10483	52827	980195	997452	296863	12043	10967	979975	967014	10351	200465
GA	645	735	875	972	7520	6530	16069	4763,71	2270	15638	8202	25748	740	12964,5
GSA	625	779	972	1322	2046	10760	28254	6394	879	760	836	847	841	832,6

S_j	Problemas											
	Magic Square					$\bar{X}_{MS}$	Latin Square					$\bar{X}_{LS}$
	3MS	4MS	5MS	6MS	7MS		4LS	5LS	6LS	7LS	8LS	
PSO	2745	15986	565155	t.o.	t.o.	t.o.	3323	6647	12716	20519	31500	14941
GA	735	1162	1087	t.o.	t.o.	t.o.	695	692	725	777	752	728,2
GSA	731	1053	1004	1526	2654	1393,6	425	482	556	610	733	561,2

6 Conclusions

Autonomous search is an interesting approach to provide more capabilities to the solver in order to improve the search process based on some performance indicators and self-tuning. In this paper we have focused on the automated self-tuning of constraint programming solvers. To this end, we have presented a gravitational search algorithm able to find good WSM configurations when solving constraint satisfaction problems. GSA is a relatively modern but efficient metaheuristic, based on the Newtonian gravity and the laws of motion. We have illustrated experimental results where the GSA-based approach directly competes against previously reported autonomous search methods for constraint programming. Indeed it was the only one able to solve all instances of tested problems. As future work, we plan to test new modern metaheuristics for supporting WSMs for AS. The incorporation of additional strategies to the portfolio would be an interesting research direction to follow as well.

Acknowledgements. Ricardo Soto is supported by Grant CONICYT/ FONDECYT/INICIACION/ 11130459, Broderick Crawford is supported by Grant CONICYT/FONDECYT/ 1140897, and Fernando Paredes is supported by Grant CONICYT/FONDECYT/ 1130455.

References

1. Barták, R., Rudová, H.: Limited assignments: A new cutoff strategy for incomplete depth-first search. In: Proceedings of the 20th ACM Symposium on Applied Computing (SAC), pp. 388–392 (2005)
2. Castro, C., Monfroy, E., Figueroa, C., Meneses, R.: An Approach for Dynamic Split Strategies in Constraint Solving. In: Gelbukh, A., de Albornoz, Á., Terashima-Marín, H. (eds.) MICAI 2005. LNCS (LNAI), vol. 3789, pp. 162–174. Springer, Heidelberg (2005)
3. Crawford, B., Soto, R., Castro, C., Monfroy, E.: A Hyperheuristic Approach for Dynamic Enumeration Strategy Selection in Constraint Satisfaction. In: Ferrández, J.M., Álvarez Sánchez, J.R., de la Paz, F., Toledo, F.J. (eds.) IWINAC 2011, Part II. LNCS, vol. 6687, pp. 295–304. Springer, Heidelberg (2011)
4. Crawford, B., Soto, R., Castro, C., Monfroy, E., Paredes, F.: An Extensible Autonomous Search Framework for Constraint Programming. Int. J. Phys. Sci. 6(14), 3369–3376 (2011)
5. Crawford, B., Soto, R., Monfroy, E., Palma, W., Castro, C., Paredes, F.: Parameter tuning of a choice-function based hyperheuristic using Particle Swarm Optimization. Expert Systems with Applications 40(5), 1690–1695 (2013)
6. Halliday, D., Resnick, R., Walker, J.: Fundamentals of Physics. Halliday & Resnick Fundamentals of Physics. John Wiley & Sons Canada, Limited (2010)
7. Hamadi, Y., Monfroy, E., Saubion, F.: Autonomous Search. Springer (2012)
8. Hutter, F., Hamadi, Y., Hoos, H.H., Leyton-Brown, K.: Performance Prediction and Automated Tuning of Randomized and Parametric Algorithms. In: Benhamou, F. (ed.) CP 2006. LNCS, vol. 4204, pp. 213–228. Springer, Heidelberg (2006)
9. Hutter, F., Hoos, H.H., Leyton-Brown, K.: Automated configuration of mixed integer programming solvers. In: Lodi, A., Milano, M., Toth, P. (eds.) CPAIOR 2010. LNCS, vol. 6140, pp. 186–202. Springer, Heidelberg (2010)
10. Hutter, F., Hoos, H.H., Leyton-Brown, K.: Sequential model-based optimization for general algorithm configuration. In: Coello Coello, C.A. (ed.) LION 2011. LNCS, vol. 6683, pp. 507–523. Springer, Heidelberg (2011)
11. Maturana, J., Lardeux, F., Saubion, F.: Autonomous operator management for evolutionary algorithms. J. Heuristics 16(6), 881–909 (2010)

12. Maturana, J., Saubion, F.: On the design of adaptive control strategies for evolutionary algorithms. In: Monmarché, N., Talbi, E.-G., Collet, P., Schoenauer, M., Lutton, E. (eds.) EA 2007. LNCS, vol. 4926, pp. 303–315. Springer, Heidelberg (2008)
13. Maturana, J., Saubion, F.: A compass to guide genetic algorithms. In: Rudolph, G., Jansen, T., Lucas, S., Poloni, C., Beume, N. (eds.) PPSN 2008. LNCS, vol. 5199, pp. 256–265. Springer, Heidelberg (2008)
14. Monfroy, E., Castro, C., Crawford, B., Soto, R., Paredes, F., Figueroa, C.: A Reactive and Hybrid Constraint Solver. J. Exp. Theor. Artif. Intell. (2012) (in press), doi:10.1080/0952813X.2012.656328
15. Soto, R., Crawford, B., Monfroy, E., Bustos, V.: Using Autonomous Search for Generating Good Enumeration Strategy Blends in Constraint Programming. In: Murgante, B., Gervasi, O., Misra, S., Nedjah, N., Rocha, A.M.A.C., Taniar, D., Apduhan, B.O. (eds.) ICCSA 2012, Part III. LNCS, vol. 7335, pp. 607–617. Springer, Heidelberg (2012)
16. Rashedi, E., Nezamabadi-pour, H., Saryazdi, S.: GSA: A gravitational search algorithm. Inf. Sci. 179(13), 2232–2248 (2009)
17. Schutz, B.: Gravity from the ground up. Cambridge University Press (2003)

Enumeration Strategies for Solving Constraint Satisfaction Problems: A Performance Evaluation

Ricardo Soto[1,2,3], Broderick Crawford[1,4,5], Rodrigo Olivares[6], Rodrigo Herrera[1], Franklin Johnson[7], and Fernando Paredes[8]

[1] Pontificia Universidad Católica de Valparaíso, Chile
[2] Universidad Autónoma de Chile, Chile
[3] Universidad Central de Chile, Chile
[4] Universidad Finis Terrae, Chile
[5] Universidad San Sebastián, Chile
[6] Universidad de Valparaíso, Chile
[7] Universidad de Playa Ancha, Chile
[8] Escuela de Ingeniería Industrial, Universidad Diego Portales, Chile
{ricardo.soto,broderick.crawford}@ucv.cl,
rodrigo.olivares@uv.cl, rodrigo.herrera.1@mail.pucv.cl,
franklin.johnson@upla.cl, fernando.paredes@udp.cl

Abstract. Constraint programming allows to solve constraint satisfaction and optimization problems by building and then exploring a search tree of potential solutions. Potential solutions are generated by firstly selecting a variable and then a value from the given problem. The enumeration strategy is responsible for selecting the order in which those variables and values are selected to produce a potential solution. There exist different ways to perform this selection, and depending on the quality of this decision, the efficiency of the solving process may dramatically vary. A main concern in this context is that the behavior of the strategy is notably hard to predict. In this paper, we present a performance evaluation of 24 enumeration strategies for solving constraint satisfaction problems. Our goal is to provide new and interesting knowledge about the behavior of such strategies. To this end, we employ a set of well-known benchmarks that collect general features that may be present on most constraint satisfaction and optimization problems. We believe this information will be useful to help users making better solving decisions when facing new problems.

Keywords: Constraint programming, constraint satisfaction problems, enumeration strategies, heuristics.

1 Introduction

Constraint Programming (CP) can be described as a modern software technology devoted to efficient resolution of combinatorial problems [6]. Under this framework, problems are formulated as a set of variables and constraints. The variables

R. Silhavy et al. (eds.), *Artificial Intelligence Perspectives and Applications*,
Advances in Intelligent Systems and Computing 347, DOI: 10.1007/978-3-319-18476-0_18

represent the unknowns of the problem and are linked to a non-empty domain of possible values, while the constraints define relations among those variables. This formal representation is known as Constraint Satisfaction Problem (CSP), which consist of a sequence of variables $X = x_1, x_2, ..., x_n$, with its respective domains $D = D_{x_1}, D_{x_2}, ..., D_{x_n}$, and a finite set C of constraints restricting the values that the variables can take simultaneously [6,2]. A solution to the problem is defined as an assignment of values to variables that satisfy all the constraints. The resolution process is carried out by a search engine, commonly called solver, which attempt to reach a result by building and exploring a search tree of potential solutions. This process considers enumeration and propagation phases. The enumeration phase is responsible for creating tree branches by assigning permitted values to the variables. The propagation phase aims at removing from domains, unfeasible values by employing consistency techniques.

In CP, the selection of an enumeration strategy is essential for the performance of the resolution process, where a correct selection can dramatically reduce the computational cost of finding a solution. However, it is well-known that deciding a priori the correct heuristic is quite difficult, as the effects of the strategy can be unpredictable. A CP framework can be seen in figure 1.

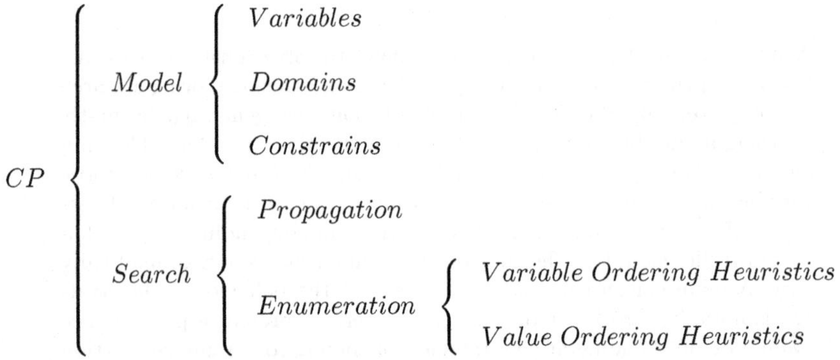

Fig. 1. CP framework

In this paper, we focus on the enumeration phase of constraint programming, where heuristics for selecting variables and values are critical. The correct use of enumeration strategy can improve the performance of the resolution process. Our goal is to provide new, interesting, and relevant information about the behavior of enumeration strategies. To this end, we employ 24 strategies to solve a set of well-known benchmarks that collect general features that may be present on most constraint satisfaction and optimization problems. Particularly, we use the N-Queens problem, the Sudoku puzzle, the Magic & Latin Square, the Knight Tour as well as the Langford and Quasigroup problem. We believe that this information will be useful in order to help users to perform a better strategy selection when facing new problems. As far as we know, the knowledge provided in this work as not been reported yet in the literature.

This work is organized as follows: In Section 2 we briefly describe the problems. Section 3 provides an overview of the resolution technique with the 24 enumeration strategies used. Section 4 presents the experimental results followed by conclusions and future work.

2 Problems Description

In this Section, we provide a briefly description of tested problems. For space reasons, we do not include additional information such as the mathematical model and examples, which can be found in the associated reference. As previously mentioned, all tested problems collect general features that may be present on most constraint satisfaction and optimization problems such as NP-hardness or NP-completeness, global constraints as well as a considerable amount of variables and constraints. Such features make the problem in general to demand huge computational resources in terms of memory and time to be solved.

2.1 N-Queens

The n-queens puzzle is the problem of placing n chess queens on an $n \times n$ chessboard so that no two queens threaten each other. Thus, a solution requires that no two queens share the same row, column, or diagonal. There exist solutions for all natural numbers n with the exception of $n = 2$ or $n = 3$ [7].

2.2 Sudoku

Sudoku is a logic-based, combinatorial number-placement puzzle. The objective is to fill a 9 grid with digits so that each column, each row, and each of the nine 3 sub-grids that compose the grid (also called "boxes", "blocks", "regions", or "sub-squares") contains all of the digits from 1 to 9. The puzzle setter provides a partially completed grid, which for a well-posed puzzle has a unique solution [9,11].

2.3 Magic Square

A magic square is an arrangement of distinct numbers, usually integers, in a square grid, where the numbers in each row, and in each column, and the numbers in the main and secondary diagonals, all add up to the same number. A magic square has the same number of rows as it has columns, "n" stands for the number of rows (and columns) it has [12].

2.4 Latin Square

A Latin Square puzzle of order n is defined as an $n \times n$ matrix, where all its elements are numbers between 1 and n, with the property that each one of the n numbers appear exactly once in each column of the matrix [3].

2.5 Knight's Tour

A knight's tour is a sequence of moves of a knight on a chessboard such that the knight visits every square only once. If the knight ends on a square that is one knight's move from the beginning square (so that it could tour the board again immediately, following the same path), the tour is closed, otherwise it is open [10].

2.6 Langford

Each number occurs two times (hence the 2 in names of the sequences). Between equal numbers k there are exactly k other numbers. The sequences are named $L(m,n)$ with m the multiplicity, and n the order of the sequence. Not all sequences $L(m,n)$ have solutions: there are no instances of $L(2,5)$ and $L(2,6)$. Both $L(2,3)$ and $L(2,4)$ have exactly one instance (apart from the obvious symmetry by reversing the sequence). Some sequences have a large number of solutions. For instance $L(2,7)$ has 26 solutions, and $L(2,8)$ has 150 solutions [5].

2.7 Quasigroup

A quasigroup corresponding to an ordered pair (Q, ·) where Q is a set and $\cdot$ is a binary operation on Q, so that the equations $x \cdot a = b$ and $y \cdot a = b$ only can be resolved by each pair of elements a, b belonging to Q. The order of a quasigroup n is the cardinality of the set Q. The best way to understand the structure of a quasigroup is to consider $n \times n$ as a multiplication table, defined by its binary operation [8].

3 CP Resolution Technique Overview

Different techniques can be used in the resolution process of Constraints Satisfaction Problem, currently they are solved using complete techniques, incomplete techniques and hybrids of both techniques [6]. In this work, we employ the classic constraint programming approach where enumeration and propagation phases are interleaved in a backtracking-based algorithm. The propagation phase prunes the search tree by eliminating values that can not participate in a solution, while the enumeration divides the original problem into two smaller ones, creating one branch by instantiating a variable with a variables form its domain. Additional information about this approach can be seen in [2].

The goal here is to employ 24 enumeration strategies which are constituted by the combination of 8 variable ordering heuristics and 3 value ordering heuristics, which are described in the following. The resulting strategies are depicted in Table 1.

3.1 Variable Ordering Heuristics

Input Order (First): the first entry in the list is selected.
First Fail (MRV): the entry with the smallest domain size is selected.
Anti First Fail (AMRV): the entry with the largest domain size is selected.
Smallest (S): the entry with the smallest value in the domain is selected.
Largest (L): the entry with the largest value in the domain is selected.
Occurrence (O): the entry with the largest number of attached constraints is selected.
Most Constrained (MC): the entry with the smallest domain size is selected. If several entries have the same domain size, the entry with the largest number of attached constraints is selected.
Max Regret (MR): the entry with the largest difference between the smallest and second smallest value in the domain is selected. This method is typically used if the variable represents a cost, and we are interested in the choice which could increase overall cost the most if the best possibility is not taken. Unfortunately, the implementation does not always work: If two decision variables incur the same minimal cost, the regret is not calculated as zero, but as the difference from this minimal value to the next greater value.

3.2 Value Ordering Heuristics

Indomain Min (Min): Values are tried in increasing order. On failure, the previously tested value is removed. The values are tested in the same order as for indomain, but backtracking may occur earlier.
Indomain Middle (Mid): Values are tried beginning from the middle of the domain. On failure, the previously tested value is removed.
Indomain Max (Max): Values are tried in decreasing order. On failure, the previously tested value is removed.

Table 1. Enumeration Strategies

Ordering Heuristics								
Id	Variable	Value	Id	Variable	Value	Id	Variable	Value
S_1	**First**	Min	S_9	**First**	Mid	S_{17}	**First**	Max
S_2	**MRV**		S_{10}	**MRV**		S_{18}	**MRV**	
S_3	**AMRV**		S_{11}	**AMRV**		S_{19}	**AMRV**	
S_4	**O**		S_{12}	**O**		S_{20}	**O**	
S_5	**S**		S_{13}	**S**		S_{21}	**S**	
S_6	**L**		S_{14}	**L**		S_{22}	**L**	
S_7	**MC**		S_{15}	**MC**		S_{23}	**MC**	
S_8	**MR**		S_{16}	**MR**		S_{24}	**MR**	

4 Experimental Results

We have performed an experimental evaluation of the proposed approach on different instances of the aforementioned problems, detailed in the following:

- N-Queens with N=$\{8, 10, 12, 15, 20, 50, 75\}$.
- Sudokus numbers $1, 2, 5, 7$, and 9 taken from [1].
- Magic Square with N=$\{3, 4, 5\}$.
- Latin Square with N=$\{4, 5, 6, 7, 8\}$.
- Knight's Tour with N=$\{5, 6\}$.
- Quasigroup with N=$\{5, 6, 7\}$.
- Langford with $M = 2$ and N=$\{16, 20, 23\}$.

Each instance was solved with the 24 strategies depicted in Table 1. The instances are solved to a maximum number of 65535 steps as equally done in previous work [4]. Each run has a time bound of 5 minutes. Instances with no solution before time bound are tagged with the symbol "t.o." (time-out). The solver has been implemented on the Ecl^{i}ps^e Constraint Logic Programming Solver v6.10. The experiments have been launched on a 3.30GHz Intel Core i3-2120 with 4Gb RAM running Windows 7 Professional 32 bits. We have employed two well-known solving indicators in order to evaluate performance:

Backtracking (B): it shows the amount of bad decisions made during the search process of the solutions, that is, calculations or decisions executed without leading to a solution.
Runtime (t): it measures the required time to solve the problem.

Tables 2, 3, and 4 present the results for each enumeration strategies (S_1, ..., S_{24}) in terms of backtracks. Tables 5, 6, and 7 depicts the results in terms of solving time. In general, a huge variation in terms of performance can be seen in results depending on the strategy employed. For instance, S_3 is dramatically faster for 20-Queens compared to S_1, the same occurs comparing S_1 with S_2 for Magic Squares (N=4), and S_10 with S_11 for Quasigroup (N=7), among several other examples. More critical situations can be seen for harder problems such as the Knight's tour where several strategies are unable to solve the problem. This clearly validates the fact the correct selection of the strategy may dramatically impact on the resolution. Additionally, we can observe that regardless of the value ordering heuristic, the participation of Anti First Fail (AMRV) or Most Constrained (MC) within the strategy may lead to better performance in general. This may be explained by the fact that ARMV and MC bet by the variable going quickly towards an insolvent space avoiding a priori unnecessary calculations.

Table 2. Backtracks N-Queens & Sudoku

S_j	Problems													
	N-Queens							$\bar{X}_Q$	Sudoku					$\bar{X}_S$
	8	**10**	**12**	**15**	**20**	**50**	**75**		**1**	**2**	**5**	**7**	**9**	
S_1	10	6	15	73	10026	>121277	>118127	>121277	0	18	4229	10786	0	3006.6
S_2	11	12	11	808	2539	>160845	>152812	>160845	1523	10439	>89125	>59828	114	>89125
S_3	10	4	16	1	11	177	818	**148.1**	0	4	871	773	0	**329.6**
S_4	10	6	15	73	10026	>121277	>118127	>121277	0	18	4229	10786	0	3006.6
S_5	3	6	17	40	862	>173869	>186617	>186617	7	155	>112170	>81994	0	>112170
S_6	9	6	16	82	15808	>143472	>137450	>143472	0	764	>83735	>80786	0	>83735
S_7	10	4	16	1	11	177	818	**148.1**	0	4	871	773	0	**329.6**
S_8	3	5	12	28	63	>117616	>133184	>133184	0	2	308	10379	0	2137.8
S_9	10	6	15	73	10026	>121277	>118127	>121277	0	18	4229	10786	0	3006.6
S_{10}	11	12	11	808	2539	>160845	>152812	>160845	1523	10439	>89125	>59828	114	>89125
S_{11}	10	12	16	1	11	177	818	**148.1**	0	4	871	773	0	**329.6**
S_{12}	10	6	15	73	10026	>121277	>118127	>121277	0	18	4229	10786	0	3006.6
S_{13}	3	6	17	40	862	>173869	>186617	>186617	7	155	>112174	>81994	0	>112174
S_{14}	9	6	16	82	15808	>143472	>137450	>143472	0	764	>83735	>80786	0	>83735
S_{15}	10	4	16	1	11	177	818	**148.1**	0	4	871	773	0	**329.6**
S_{16}	3	5	12	28	63	>117616	>133184	>133184	0	2	308	10379	0	2137.8
S_{17}	10	6	15	73	10026	>121277	>118127	>121277	3	2	>104148	1865	1	>104148
S_{18}	11	12	11	808	2539	>160845	>152812	>160845	8482	6541	>80203	>80295	7	>80295
S_{19}	10	4	16	1	11	177	818	**148.1**	0	9	963	187	0	**231.8**
S_{20}	10	6	15	73	10026	>121277	>118127	>121277	3	2	>104148	1865	1	>104148
S_{21}	9	6	16	82	15808	>173869	>186617	>186617	49	89	>78774	>93675	0	>93675
S_{22}	3	6	17	40	862	>143472	>137450	>143472	1039	887	>101058	>91514	0	>101058
S_{23}	10	4	16	1	11	177	818	**148.1**	4	9	963	187	0	**232.6**
S_{24}	2	37	13	127	1129	>117616	>133184	>133184	72	12	>92557	2626	0	>92557

Table 3. Backtracks Magic & Latin Square

S_j	Problems									
	Magic Square			$\bar{X}_{MS}$	Latin Square					$\bar{X}_{LS}$
	3	**4**	**5**		**4**	**5**	**6**	**7**	**8**	
S_1	0	12	910	**307.3**	0	0	0	12	0	**2.4**
S_2	4	1191	>191240	>191240	0	9	163	>99332	0	>99332
S_3	0	3	185	**62.7**	0	0	0	0	0	**0**
S_4	0	10	5231	**1747**	0	0	0	14	0	**2.8**
S_5	0	22	>153410	>153410	0	7	61	>99403	0	>99403
S_6	4	992	>204361	>204361	0	0	0	71	0	**14.2**
S_7	0	3	193	**65.3**	0	0	0	0	0	**0**
S_8	0	13	854	**289**	0	0	0	0	0	**0**
S_9	0	12	910	**307.3**	0	0	0	9	0	**1.8**
S_{10}	4	1191	>191240	>191240	0	9	163	>99332	0	>99332
S_{11}	0	3	185	**62.7**	0	0	0	0	0	**0**
S_{12}	0	10	5231	**1747**	0	0	0	9	0	**1.8**
S_{13}	0	22	>153410	>153410	0	7	61	>99539	0	>99539
S_{14}	4	992	>204361	>204361	0	0	0	71	0	**14.2**
S_{15}	0	3	193	**65.3**	0	0	0	0	0	**0**
S_{16}	0	13	854	**289**	0	0	0	0	0	**0**
S_{17}	1	51	>204089	>204089	0	0	0	9	0	**1.8**
S_{18}	0	42	>176414	>176414	0	9	163	>99481	0	>99481
S_{19}	1	3	>197512	>197512	0	0	0	0	0	**0**
S_{20}	1	29	54063	**18031**	0	0	0	9	0	**1.8**
S_{21}	1	95	>201698	>201698	0	0	0	71	0	**14.2**
S_{22}	0	46	47011	**15685.7**	0	7	61	>99539	0	>99539
S_{23}	1	96	>190692	>190692	0	0	0	0	0	**0**
S_{24}	1	47	>183580	>183580	0	0	0	0	1	**0.2**

Table 4. Backtracks Knight's Tour. Langford & Quasigroup

S_j	Problems										
	Knight's Tour		$\bar{X}_{KT}$	**Quasigroup**			$\bar{X}_{QG}$	**Langford**			$\bar{X}_L$
	5	**6**		**5**	**6**	**7**		**16**	**20**	**23**	
S_1	767	37695	**19231**	>145662	30	349	>145662	39	77	26	**47.3**
S_2	>179097	>177103	>179097	>103603	>176613	3475	>176613	24310	>98157	>97621	>98157
S_3	767	37695	**19231**	8343	0	1	**2781.3**	97	172	64	**111**
S_4	>97176	35059	>97176	>145656	30	349	>145656	39	77	26	**47.3**
S_5	>228316	>239427	>239427	>92253	>83087	4417	>92253	599	26314	29805	**18906**
S_6	>178970	>176668	>178970	>114550	965	4417	>114550	210	1	3	**71.3**
S_7	>73253	14988	>73253	8343	0	1	**2781.3**	97	172	64	**111**
S_8	>190116	>194116	>194116	>93315	>96367	4	>96367	0	64	7	**26.3**
S_9	767	37695	**19231**	>145835	30	349	>145835	39	77	26	**47.3**
S_{10}	>179126	>177129	>179126	>103663	>176613	3475	>176613	24310	>98157	>97621	>98157
S_{11}	767	37695	**19231**	8343	0	1	**2781.3**	97	172	64	**111**
S_{12}	>97176	35059	>97176	>145830	30	349	>145830	39	77	26	**47.3**
S_{13}	>228316	>239427	>239427	>92355	>83087	583	>92355	599	26314	29805	**18906**
S_{14}	>178970	>176668	>178970	>114550	965	4417	>114550	210	1	3	**71.3**
S_{15}	>73253	14998	>73253	8343	0	1	**2781.3**	97	172	64	**111**
S_{16}	>190116	>194116	>194116	>93315	>93820	4	>93820	0	64	7	**26.3**
S_{17}	767	37695	**19231**	7743	2009	3	**3258.3**	39	77	26	**47.3**
S_{18}	>179126	>177129	>179126	>130635	>75475	845	>130635	24592	>98028	>97649	>98028
S_{19}	767	37695	**19231**	0	89	1	**30**	98	172	64	**111**
S_{20}	>97178	35059	>97178	7763	2009	3	**3258.3**	39	77	26	**47.3**
S_{21}	>178970	>176668	>178970	>96083	>108987	773	>108987	210	1	3	**71.3**
S_{22}	>228316	>239427	>239427	>94426	>124523	1	>124523	599	26314	29805	**18906**
S_{23}	767	14998	**7882.5**	0	89	1	**30**	98	172	64	**111.3**
S_{24}	>190116	>160789	>190116	>95406	>89888	1	>95406	239	4521	0	**1586.7**

Table 5. Runtime N-Queens & Sudoku in ms

S_j	Problems													
	N-Queens							$\bar{X}_Q$	**Sudoku**					$\bar{X}_S$
	8	**10**	**12**	**15**	**20**	**50**	**75**		**1**	**2**	**5**	**7**	**9**	
S_1	5	5	12	57	20405	t.o.	t.o.	t.o.	5	35	7453	26882	4	**8593.7**
S_2	5	8	11	903	4867	t.o.	t.o.	t.o.	2247	30515	t.o.	t.o.	209	t.o.
S_3	5	3	11	3	16	532	4280	**692.86**	6	10	2181	2135	8	**1083**
S_4	4	4	11	59	20529	t.o.	t.o.	t.o.	6	50	8274	25486	5	**8454**
S_5	2	4	13	28	1294	t.o.	t.o.	t.o.	18	225	t.o.	t.o.	5	t.o.
S_6	4	5	14	79	26972	t.o.	t.o.	t.o.	6	1607	t.o.	t.o.	3	t.o.
S_7	4	3	11	3	15	524	4217	**682.43**	6	10	2247	2187	8	**1112.5**
S_8	2	4	10	24	93	t.o.	t.o.	t.o.	8	10	897	31732	8	**8161.7**
S_9	5	5	11	58	20349	t.o.	t.o.	t.o.	4	35	7521	26621	4	**8545.2**
S_{10}	5	8	11	971	4780	t.o.	t.o.	t.o.	4754	29797	t.o.	t.o.	215	t.o.
S_{11}	4	7	11	3	18	532	4336	**701.57**	6	10	2394	2069	8	**1119.7**
S_{12}	5	5	11	61	23860	t.o.	t.o.	t.o.	6	50	9015	26573	5	**8911**
S_{13}	2	5	13	29	1250	t.o.	t.o.	t.o.	18	225	t.o.	t.o.	4	t.o.
S_{14}	4	5	14	83	36034	t.o.	t.o.	t.o.	7	1732	t.o.	t.o.	3	t.o.
S_{15}	4	3	11	3	17	533	4195	**680.86**	6	10	2310	2094	8	**1105**
S_{16}	2	4	10	25	87	t.o.	t.o.	t.o.	8	10	972	30767	8	7939.2
S_{17}	5	4	11	58	22286	t.o.	t.o.	t.o.	8	5	t.o.	3725	6	t.o.
S_{18}	5	7	10	953	4547	t.o.	t.o.	t.o.	2497	18836	t.o.	t.o.	20	t.o.
S_{19}	4	2	11	3	16	520	4334	**698.57**	15	30	2590	338	7	**743.2**
S_{20}	4	4	11	59	13135	t.o.	t.o.	t.o.	15	5	t.o.	5350	8	t.o.
S_{21}	4	5	14	79	26515	t.o.	t.o.	t.o.	74	100	t.o.	t.o.	4	t.o.
S_{22}	2	4	13	28	1249	t.o.	t.o.	t.o.	3615	1710	t.o.	t.o.	4	t.o.
S_{23}	4	3	11	3	16	521	4187	**677.86**	15	30	2670	378	8	**773.2**
S_{24}	2	5	8	102	1528	t.o.	t.o.	t.o.	125	40	t.o.	9168	6	t.o.

Table 6. Runtime Magic & Latin Square in ms

S_j	Problems									
	Magic Square			$\bar{X}_{MS}$	Latin Square					$\bar{X}_{LS}$
	3	4	5		4	5	6	7	8	
S_1	1	14	1544	**519.7**	2	3	5	9	11	**6**
S_2	5	2340	t.o.	t.o.	2	11	102	t.o.	14	t.o.
S_3	1	6	296	**101**	1	3	5	7	12	**5.6**
S_4	1	21	6490	**2170.7**	1	3	5	9	12	**6**
S_5	1	21	t.o.	t.o.	2	8	60	t.o.	14	t.o.
S_6	4	1500	t.o.	t.o.	2	4	7	88	12	**22.6**
S_7	1	6	203	**70**	1	2	4	7	12	**5.2**
S_8	1	11	1669	**560.3**	1	2	5	7	12	**5.4**
S_9	1	13	1498	**504**	2	3	5	13	11	**6.8**
S_{10}	4	2366	t.o.	t.o.	1	11	103	t.o.	14	t.o.
S_{11}	1	6	297	**101.3**	2	3	6	8	12	**6.2**
S_{12}	1	21	6053	**2025**	2	3	6	14	13	**7.6**
S_{13}	1	21	t.o.	t.o.	2	8	62	t.o.	15	t.o.
S_{14}	4	1495	t.o.	t.o.	2	4	7	92	14	**23.8**
S_{15}	1	6	216	**74.3**	2	3	5	9	14	**6.6**
S_{16}	1	11	1690	**567.3**	2	3	5	9	14	**6.6**
S_{17}	1	88	t.o.	t.o.	2	3	6	14	12	**7.4**
S_{18}	1	37	t.o.	t.o.	2	12	107	t.o.	14	t.o.
S_{19}	1	99	t.o.	t.o.	2	3	5	8	12	**6**
S_{20}	1	42	165878	**55307**	2	4	5	16	13	**8**
S_{21}	1	147	t.o.	t.o.	2	4	7	93	13	**23.8**
S_{22}	1	37	153679	**51239**	2	8	64	t.o.	15	t.o.
S_{23}	1	102	t.o.	t.o.	2	3	6	8	13	**6.4**
S_{24}	1	79	t.o.	t.o.	2	3	5	10	17	**7.4**

Table 7. Runtime Knight's Tour. Quasigroup & Langford in ms

S_j	Problems										
	Knight's Tour		$\bar{X}_{KT}$	Quasigroup			$\bar{X}_{QG}$	Langford			$\bar{X}_L$
	5	6		5	6	7		16	20	23	
S_1	1825	90755	**46290**	t.o.	45	256	t.o.	70	191	79	**113.3**
S_2	t.o	t.o	t.o	t.o.	t.o.	8020	t.o.	70526	t.o.	t.o.	t.o.
S_3	2499	111200	**56849.5**	7510	15	10	**2511.7**	231	546	286	**354.3**
S_4	t.o	89854	t.o	t.o.	45	307	t.o.	115	318	140	**191**
S_5	t.o	t.o	t.o	t.o.	t.o.	943	t.o.	1217	61944	68254	**43805**
S_6	t.o	t.o	t.o	t.o.	3605	16896	t.o.	489	11	19	**173**
S_7	t.o	39728	t.o	9465	15	10	**3163.3**	237	553	285	**358.3**
S_8	t.o	t.o	t.o	t.o.	t.o.	16	t.o.	7	240	19	**88.7**
S_9	1908	93762	**47835**	t.o.	40	240	t.o.	69	185	79	**111**
S_{10}	t.o	t.o	t.o	t.o.	t.o.	13481	t.o.	55291	t.o.	t.o.	t.o.
S_{11}	2625	102387	**52506**	9219	15	10	**3081.3**	250	538	285	**357.7**
S_{12}	t.o	109157	t.o	t.o.	45	348	t.o.	118	312	140	**190**
S_{13}	t.o	t.o	t.o	t.o.	t.o.	1097	t.o.	1273	61345	71209	**44609**
S_{14}	t.o	t.o	t.o	t.o.	3565	18205	t.o.	530	11	19	**186.7**
S_{15}	t.o	46673	t.o	10010	15	11	**3345.3**	235	541	278	**351.3**
S_{16}	t.o	t.o	t.o	t.o.	t.o.	15	t.o.	8	237	19	**88**
S_{17}	1827	96666	**49246.5**	9743	7075	9	**5609**	66	170	75	**103.7**
S_{18}	t.o	t.o	t.o	t.o.	t.o.	1878	t.o.	55687	t.o.	t.o.	t.o.
S_{19}	2620	97388	**50004**	20	125	12	**52.3**	245	562	272	**359.7**
S_{20}	t.o	90938	t.o	10507	6945	9	**5820.3**	107	294	126	**175.7**
S_{21}	t.o	t.o	t.o	t.o.	t.o.	1705	t.o.	510	11	20	**180.3**
S_{22}	t.o	t.o	t.o	t.o.	t.o.	9	t.o.	1297	58732	73168	**44399**
S_{23}	2975	40997	**21986**	21	130	12	**54.3**	240	569	276	**361.7**
S_{24}	t.o	t.o	t.o	t.o.	t.o.	14	t.o.	584	15437	10	**5343.7**

5 Conclusions

In this paper, we have presented a performance evaluation of 24 enumeration strategies for solving CSPs. Our main goal was to provide new and interesting knowledge about the behavior of such strategies. We have tested those strategies on a set of well-known benchmarks that collect general features that may be present on most constraint satisfaction and optimization problems. The resolution performance was evaluated on the basis of performance indicators. The work included the modeling and resolution of classic combinatorial problems that collect general features that may be present on most constraint satisfaction and optimization problems. The results have demonstrated that variable and value selection heuristics notably influence the efficiency of the resolution process of combinatorial problems. Additionally, regardless of the value selection, selecting ARMV and MC as variable ordering heuristics exhibited better results in general compared to other strategies. This may be explained by the fact that ARMV and MC bet by the variable going quickly towards an insolvent space avoiding a priori unnecessary calculations. As future work, we plan to extend the benchmark set as well as the enumeration strategies employed, and to incorporate propagation strategies to the study.

Acknowledgements. Ricardo Soto is supported by Grant CONICYT/FONDECYT/INICIACION/ 11130459, Broderick Crawford is supported by Grant CONICYT/FONDECYT/ 1140897, and Fernando Paredes is supported by Grant CONICYT/FONDECYT/ 1130455.

References

1. The ECLiPSe Constraint Programming System (2008), `http://www.eclipse-clp.org/` (visited January 2015)
2. Apt, K.R.: Principles of Constraint Programming. Cambridge University Press (2003)
3. Borkowski, J.: Network inclusion probabilities and horvitz-thompson estimation for adaptive simple latin square sampling. Environmental and Ecological Statistics 6(3), 291–311 (1999)
4. Crawford, B., Castro, C., Monfroy, E., Soto, R., Palma, W., Paredes, F.: Dynamic Selection of Enumeration Strategies for Solving Constraint Satisfaction Problems. Rom. J. Inf. Sci. Tech. 15(2), 421–430 (2012)
5. Davies, R.: On langford's problem (II). Math. Gaz. 43, 253–255 (1959)
6. Rossi, F.: Handbook of Constraint Programming. Elsevier (2006)
7. Hoffman, E., Loessi, J., Moore, R.: Construction for the solutions of the m queens problem. Mathematics Magazine 42(2), 62–72 (1969)
8. Kjellerstrand, H.: Hakank's home page, `http://www.hakank.org/eclipse/quasigroup_completion.ecl` (visited January 2015)
9. Lambert, T., Monfroy, E., Saubion, F.: A generic framework for local search: Application to the sudoku problem. In: Alexandrov, V.N., van Albada, G.D., Sloot, P.M.A., Dongarra, J. (eds.) ICCS 2006. LNCS, vol. 3991, pp. 641–648. Springer, Heidelberg (2006)

10. Lin, S.-S., Wei, C.-L.: Optimal algorithms for constructing knight's tours on arbitrary chessboards. Discrete Applied Mathematics 146(3), 219–232 (2005)
11. Soto, R., Crawford, B., Galleguillos, C., Monfroy, E., Paredes, F.: A hybrid ac3-tabu search algorithm for solving sudoku puzzles. Expert Syst. Appl. 40(15), 5817–5821 (2013)
12. Westbury, B.W.: Sextonions and the magic square. Journal of the London Mathematical Society 73(2), 455–474 (2006)

Evaluation of the Accuracy of Numerical Weather Prediction Models

David Šaur

Tomas Bata University in Zlin,
Faculty of Applied Informatics,
Nad Stranemi 4511
saur@fai.utb.cz

Abstract. This article is focused on numerical weather prediction models, which are publicly available on the Internet and their evaluation of the accuracy of predictions. The first part of the article deals with the basic principles of creating a weather forecast by numerical models, including an overview of selected numerical models. The various methods of evaluation of the accuracy of forecasts are described in the following part of the article; the results of which are shown in the last chapter. This article aims to bring major information on the numerical models with the greatest accuracy convective precipitation forecasts based on an analysis of 30 situations for the year 2014. These findings can be useful, especially for Crisis Management of the Zlin Region in extraordinary natural events (flash floods).

Keywords: Numerical weather prediction models, flash floods, crisis management, convective precipitation.

1 Introduction

A central issue in current meteorology is forecasting of convective precipitation. Five summer floods with torrential rainfall occurred in the Czech Republic in the years 2009-2014. The fundamental problem in forecasting of convective precipitation lies in its specific spatial and temporal evolution. Convective precipitation is formed over the territory of a small size (from 1x1km up to 10x10 km), takes approximately 30-60 minutes and is accompanied by extreme weather events (heavy rainfall, hail, strong wind gusts, tornadoes and downbursts). In order to obtain an accurate and a quality prediction of convective precipitation is necessary to analyze the available numerical weather prediction model by appropriate evaluation methods. There have been several investigations into the causes of accuracy evaluation of numerical weather prediction models (Keil et al. 2014, Singh et al. 2014, Liu et al. 2013 and Comellas et al. 2011). These studies several researchers examined the accuracy of predictions of convective precipitation using numerical weather prediction models for selected areas. [7], [8], [9], [10] Analyzed results of numerical weather prediction models were investigated

R. Silhavy et al. (eds.), *Artificial Intelligence Perspectives and Applications*,
Advances in Intelligent Systems and Computing 347, DOI: 10.1007/978-3-319-18476-0_19

in the summary reports of floods in the Czech Republic (Sandev et al. 2010, Kubát 2009). The causes of verification of the convective precipitation forecast have been investigated (Dorninger and Gorgas 2013, Sindosi et al. 2012, Amodei et al. 2009, Zacharov 2004 and Rezacova 2005, Atger 2001 and McBride et al. 2000). [11], [12], [13], [14], [15]

None of these studies managed to provide results of weather forecast evaluation from a larger number of numerical models. The main contribution of this paper is to create an overview of the most accurate numerical weather prediction models. The outputs from numerical weather prediction models, that have achieved the best ratings, will be used in the prediction system of convective precipitation for crisis management Zlín Region. Forecasting system will generate a summary report about the future development of convective precipitation, which will be distributed to other crisis management bodies of the region. Forecasting system of convective precipitation will be designed as a software application that will be part of the solution dissertation and regional project "Information, notifying and warning system of the Zlín Region".

2 Numerical Weather Prediction Models

Numerical weather prediction models are information resources that are designated to collect data from meteorological and aerological stations, processing, evaluation and creation of forecasts. These models are based on the initial state of the Earth's atmosphere (measured station data), which are part of the differential equations describing the laws of physics, especially thermodynamic and kinetic laws. The principle of the creation of weather forecasts can be expressed in these primitive equations: [1], [2]

$$(A)=\Delta A/\Delta t \quad (1)$$

Where:

ΔA - weather change at a specific location
Δt - period during which the change occurred and
$F(A)$ - functions describing the change of weather changes meteorological situation A. [1], [2]

Equation 1 can be written as:

$$F(A) = \frac{A_{forecast} - A_{now}}{\Delta t} \quad (2)$$

$$A_{forecast} = A_{now} + F(A)\Delta t \quad (3)$$

The most important parameter in the selection of the numerical model is the resolution of the grid model indicated by the size of the area, characterized by convective cell size. Each grid cell contains meteorological data in the horizontal and vertical directions. Numerical weather prediction models are divided according to the size of the resolution:

- Global models with lower resolution (>0.10°-10x10km) and
- Regional models with higher resolution (<0.10°-10x10km). [1], [2]

Based on years of experience and work with numerical weather prediction models were selected the following models:

1. Global numerical models – model COAMPS, EURO4, GEM, GFS, NAVGEM, MM5, RHMC and UKMET (numerical model ECMWF was not included here due to lack of data on precipitation for selected weather situations).
2. Regional numerical models – model ALADIN CR and ALADIN SR.

Table 1. The parametres of numerical models (forecast meteorological parametres - legend: T - temperature, s – rainfall, v - wind, t - pressure of 500 hPa, o - clouds, d - visibility, RH - relative humidity CAPE - Convective Available Potential Energy , LI - Lifted Index, EHI - Energy Helicity Index) [3]

Models	COAMPS	EURO4	GEM	GFS
Country of origin	USA	GB	France, USA, Canada	USA
Type of model	global	global	global	global
Resolution (km)	20x20	11x11	11x11	25x25
Area prediction	Europe	Europe	Europe	The whole world
The number of predicted days	4 days	2 days	10 days	16 days
Predicted meteorological parametres	T, s, v, t	T, s, v, t, o, d, RH	T, s, v, t, RH, CAPE, vorticity, LI, EHI	T, s, v, t, o, d, RH, all indexes instability

Models	NAVGEM	MM5	RHMC	UKMET	ALADIN CR a SR
Country of origin	USA	USA	Russia	GB	Czech and Slovakia Republic
Type of model	global	regional	global	global	regional
Resolution (km)	100x100	9x9	250x250	11x11	4x4
Area prediction	The whole world	Czech Republic	The whole world	The whole world	Czech Republic
The number of predicted days	6 days	3 days	3 days	3 days	2,5 (ČR) 3 (SR) days
Predicted meteorological parametres	T, s, v, t, RH	T, s, v, t, o	T, s, v, t, RH	T, s, v, t, vorticity	T, s, v, t, o, RH, VI

Table 1 shows individual numerical weather prediction models, which will be part of evaluation of accuracy predicted convective precipitation.

3 Methods of Evaluation of the Accuracy of Weather Forecasts

Evaluation of the accuracy and quality of weather forecast of numerical weather prediction models is realized by these methods:

– Percentage evaluating of the accuracy of numerical weather prediction models
– Verification of convective precipitation forecast

3.1 The Percentage Evaluating of the Accuracy of Numerical Weather Prediction Models

This method compares the precipitation total predicted by numerical weather prediction models with the precipitation measured on the ground meteorological stations. The maximal predicted precipitation total is compared with the measured precipitation total. The accuracy of the numerical model is given by the following formula:

$$X = \frac{S_{predicted}}{S_{measured}} \times 100\ (\%) \quad (4)$$

Where Spredicted is the maximal predicted precipitation totals in milimetres and Smeasured is the maximal precipitation totals measured on a ground meteorological station. The maximal precipitation totals are evaluated for convective precipitation clouds with characteristic occurrence of an extreme local precipitation. The aim of this method is to determine which numerical models predict with greater accuracy these local extreme phenomena. We can also evaluate the accuracy of forecasting of precipitation in given territory, time of the occurrence of a precipitation and more options. This method is commonly used in the evaluation of the weather forecast in the summary report of flood events in the Czech Republic in 2009 and 2010. [4]

The percentage evaluating of the accuracy of numerical weather prediction models contains a table with the date and the location, the measured and the predicted values of precipitation totals, including the evaluation of the percentage of the numerical model.

3.2 Verification of Convective Precipitation Forecast

The convective precipitation forecast can be verified by various techniques. Methods of verification predictions of convective precipitation are:

– Standard methods with verification criteria Skill Scores (SS),
– Non-standard methods using radar precipitation estimates.

Standard methods are most commonly used for verification of convective precipitation to achieve more accurate results than non-standard evaluation methods because radar precipitation estimates are very imprecise, especially convective precipitation clouds.

The standard method is based on the pivot table which consists of four fields. This table lists the frequency of cases where the phenomenon was predicted and where it actually occurred, and in all possible combinations. [5], [6]

		Forecast	
		+	-
Measurement	+	a Intervention	b Error
	-	c False Alarm	d Correct preclusion

Fig. 1. The pivot table in standard method [6]

Where:

- **a - Intervention** is the number of cases when the phenomenon was predicted and actually occurred – good forecast of phenomenon.
- **b - Error** is the number of cases when the phenomenon was not predicted and occurred – wrong forecast of phenomenon.
- **c - False** alarm is the number of cases when the phenomenon was predicted and did not occur – wrong forecast of phenomenon.
- **d - Correct** preclusion is the number of cases when the phenomenon was not predicted and did not occur – good forecast of phenomenon. [6]

The pivot table describes that the categories a and d are successful while b and c are unsuccessful. The value of d very often exceeds the value a in the case of extreme events.

Standard methods analyze forecasts with verification criteria, which are divided into two categories:

- category **d** – criteria True Skill Statistic (TSS), Probability Skill Score, Fraction Correct (PSS, FRC) and Heidke Skill Score (HSS),
- category **a,b,c** – criteria Probability of Detection (POD), False Alarm Ratio (FAR) and Critical Success Index, "Threat Score" (CSI).

Verifikační kriterium, Skill Score	Kód	Reference	Rovnice	Hranice
Probability Of Detection	POD	Wilks (1995), Huntrieser (1997), Metelka (2001), Marzban (1998)	$POD = \frac{a}{a+b}$	$0 \le POD \le 1$
False Alarm Ratio	FAR	Wilks (1995), Huntrieser (1997), Metelka (2001), Marzban (1998)	$FAR = \frac{c}{a+c}$	$0 \le FAR \le 1$
Critical success index, "Threat score"	CSI	Wilks (1995), Huntrieser (1997), Marzban (1998)	$CSI = \frac{a}{a+b+c}$	$0 \le CSI \le 1$
True Skill Statistics	TSS	Huntrieser (1997), Marzban (1998)	$TSS = \frac{a}{a+b} - \frac{c}{c+d}$	$-1 \le TSS \le 1$
Heidke Skill Score	HSS	Wilks (1995), Huntrieser (1997), Marzban (1998)	$HSS = \frac{2(ad-bc)}{(a+b)(b+d)+(a+c)(c+d)}$	$-1 \le HSS \le 1$
Probability Skill Score, FRaction Correct	PSS, FRC	Metelka (2001), Marzban (1998)	$PSS = FRC = \frac{a+d}{a+b+c+d}$	$0 \le PSS \le 1$

Fig. 2. Verification criteria Skill Scores [6]

For purposes of this article, this information is defined as a basic knowledge to evaluate the accuracy of numerical weather prediction models.

4 Comparison of Methods to Evaluation of the Accuracy of Numerical Models

The percentage evaluating of the accuracy and verification of prediction of numerical models was performed by extensive analysis, which is part of the IGA / FAI / 2014/003. The first method Percentage evaluating of the accuracy of numerical weather prediction models was performed only in this project, but second method Verification of convective precipitation will be also included in this part of this article. The objective comparison of these methods will create a rank of the most accurate numerical weather prediction models.

4.1 The Percentage Evaluating of Accuracy of Numerical Weather Prediction Models

The results of the first method "The percentage evaluating of the accuracy of numerical weather prediction models" are presented based on the analysis of numerical weather prediction models in the IGA project for the year 2014. The accuracy of predictions X is calculated as the ratio of the maximal predicted precipitation by numerical models and maximum measured precipitation.

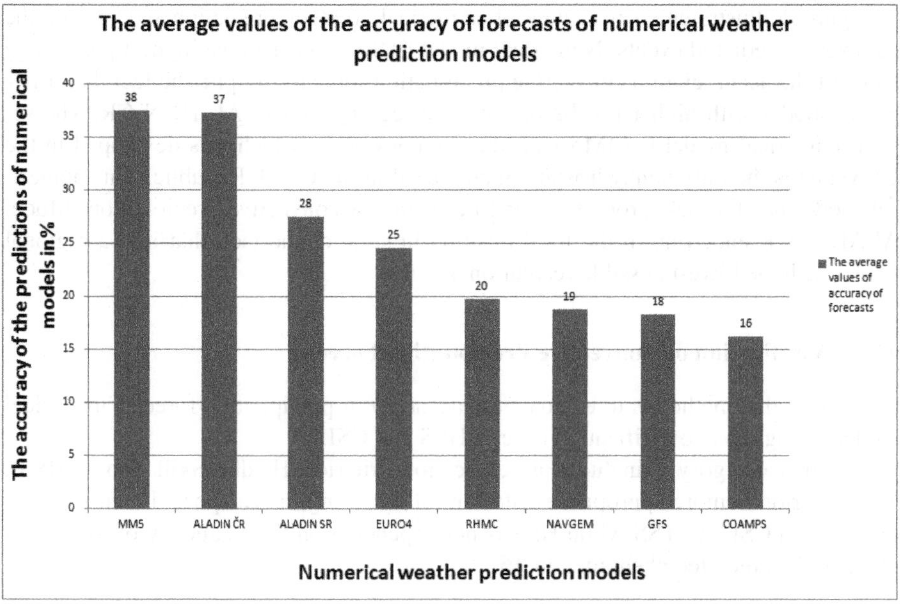

Fig. 3. The average values of the accuracy of forecasts of numerical models

Figure 3 shows the average values of the accuracy of convective precipitation forecasts by numerical models. These values were counted from the analysis of 30 situations that happened in the Zlin Region in 2014. Highest average values of prediction accuracy were achieved by MM5 and ALADIN numerical models due to a better resolution compared to other numerical models.

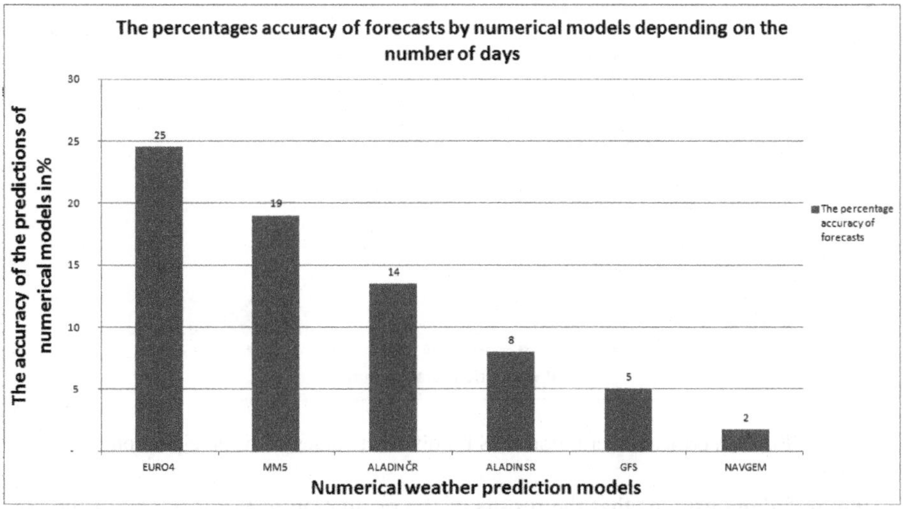

Fig. 4. The percentage of the accuracy of forecasts of numerical models

Figure 4 illustrates an overview of numerical weather prediction models for the number of predicted events. Numerical model EURO4 reached the highest percentage value of the accuracy of a convective precipitation forecast, despite the fact that it is a global model with higher resolution than the regional model ALADIN CR. The second numerical model is MM5 with the accuracy 19%, which was developed in the USA, and is the only nonhydrostatic model used in the Czech Republic. This numerical model has the best properties for forecasting of convective precipitation. Model ALADIN CR ended up in the third position in spite of the fact, that it is a regional model with the lowest possible resolution 4x4 km.

4.2 Verification of Convective Precipitation Forecast

The aim of this method is to evaluate the accuracy of precipitation forecast numerical models using the two verification criteria HSS and CSI.

The first category d includes a verification criterion Heidke Skill Score (HSS). This criterion is more appropriate criterion for assessing the category d than verification criteria CSI and TSS while HSS is not dependent on the frequency of the occurrence of the predicted phenomenon. [6]

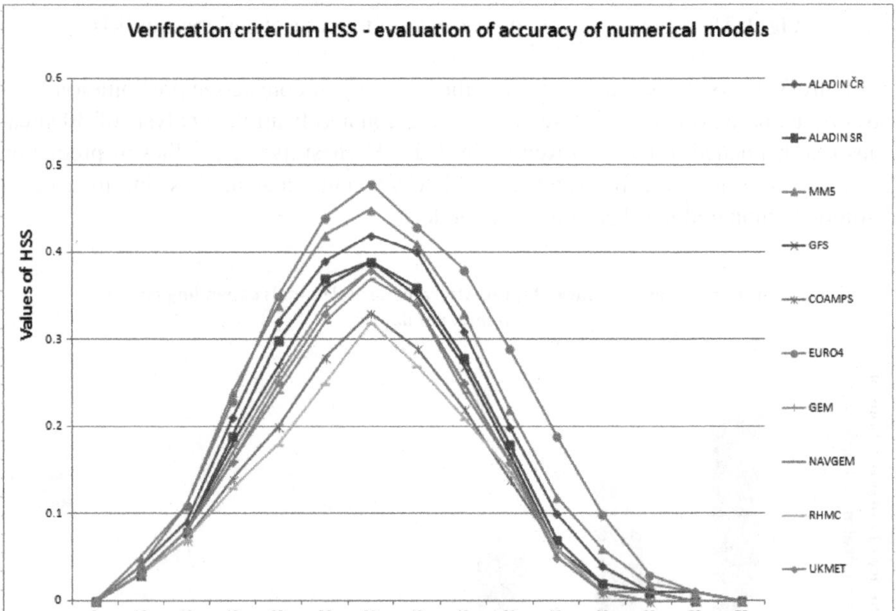

Fig. 5. Verification criterion HSS for different values of the precipitation

Figure 5 demonstrates that the results of the evaluation of the accuracy of the precipitation forecast for the individual numerical models using the verification criterion

HSS. The highest values were achieved at precipitation totals 35-40 mm by numerical models EURO4, MM5 and ALADIN CR.

The second category of without a value d (contains the values of a, b, c) includes a verification criterion Critical Success Index, "Threat Score" (CSI). CSI criterion is often used criterion for the prediction of extreme events. CSI represents the ratio of correct predictions of the number of phenomena and false alarms, and depend on the frequency of the occurrence of the predicted precipitation totals. [6]

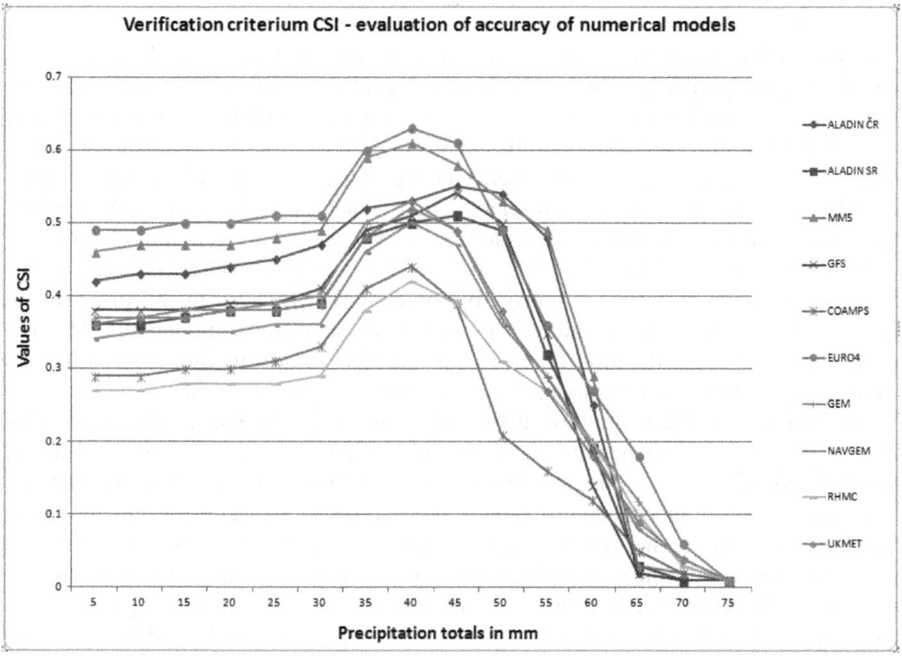

Fig. 6. Verification criterion CSI for different values of the precipitation

Figure 6 illustrates that the results of the evaluation of the accuracy of precipitation forecast for the individual numerical models using the verification criterion CSI. Generally, the less the phenomenon occurs, the CSI is higher and vice versa. The CSI highest values were achieved for the same numerical models as the previous criteria (model EURO4, MM5 and ALADIN CR). [6]

4.3 Summary Evaluation of Numerical Weather Prediction Models

For the best results, evaluation of the accuracy of the convective precipitation forecast achieved these numerical models by methods:

- The percentage evaluating of the accuracy of numerical weather prediction models:
  - The average value of the accuracy of forecast – numerical model MM5, ALADIN CR and ALADIN SR.

  - The order of numerical models with the greatest accuracy predicted by the number of situations – numerical models EURO4, MM5 and ALADIN CR.

- Verification of convective precipitation forecast – for both verification criteria HSS and CSI, the best results were obtained by numerical models EURO4, MM5 and ALADIN CR.

5 Conclusion

The aim of this paper is to analyze selected numerical weather prediction models based on a comparison of the two methods of evaluation of the forecasts accuracy. Evaluation of numerical models related only to situations with the occurrence of convective precipitation (intense rainfall and storms). This type of precipitation clouds occurs very frequently in the summer, causing extensive material damage. Consequently, these situations have been selected for evaluation for the year 2014, in which there were heavy (over 30 mm / hour) to very heavy rainfall (over 50 mm / hour).

Numerical models EURO4, MM5, ALADIN CR and ALADIN SR attained the best results of the accuracy of convective precipitation forecast in the first method Percentage evaluation the accuracy of numerical weather prediction models. The outputs of the graphs verification criteria HSS and CSI inform us that the best results were achieved by numerical models EURO4, MM5 and ALADIN CR.

For the purpose of providing for sufficient foreknowledge to crisis management of the Zlin Region has been found that the best results obtained the global numerical model EURO4 and MM5, and the regional model ALADIN CR. The overall results indicate that these numerical models can be used in practice for a short-term forecast for one to two days in advance. The limitations of this study are clear: Forecasting of convective precipitation is a very difficult problem due to insufficient resolution of the numerical models and the occurrence of a large number of factors affecting the development of this type of precipitation. The results of the percentage evaluation also showed that the accuracy of convective precipitation forecast has not reached by 50%. The lack of resolution models mean that we cannot be certain the accuracy of the numerical models outputs, and therefore it is necessary to complement with warning information from the Integrated Warning Service of the Czech Hydrometeorological Institute (CHMI SIVS) and predictive information from the Portal of the CHMI.

Acknowledgments. This article was supported by the Department of Security Engineering under internal grant IGA/FAI/2015/025 "Forecast system of convective precipitation" (sponsor and financial support acknowledgment goes here).

References

1. Saur, D., Lukas, L.: Computational models of weather forecasts as a support tool for Crisis Management. In: Proceedings of the Contributory 7th International Scientific Conference Safe Slovakia and the European Union, p. 10. College Security Management in Košice, Kosice, `http://conference.vsbm.sk`, ISBN 978-80-89282-88-3

2. Jaros, V.: Computational modeling of weather. Brno, The diploma thesis. Masaryk University, Faculty of Informatics (2010), http://is.muni.cz/th/140509/fi_m/dp140509.pdf
3. WeatherOnline, http://www.weatheronline.cz/cgi-bin/expertcharts?LANG=cz&CONT=czcz&MODELL=gfs&VAR=prec
4. Ministry of the Environment of the Czech Republic. Evaluation of spring floods in June and July 2009, the Czech Republic, http://voda.chmi.cz/pov09/doc/01.pdf (cit. December 08, 2014)
5. Zacharov, P., Rezacova, D.: Comparison of efficiency of diagnostic and prognostic characteristic the convection environment. Meteorological Bulletin (2005), http://www.mzp.cz/ris/ekodisk-new.nsf/3c715bb7027b1c65c1256bb3007b7af2/272beff860536f1dc12573fc00429823/$FILE/MZ%202005_3.pdf#page=3 (cit. December 08, 2014), ISSN: 0026 – 1173
6. Zacharov, P.: Diagnostic and prognostic precursors of convection. [Diploma thesis] Prague: Faculty of Mathematics and Physics UK, KMOP, 61 p. (2004), https://is.cuni.cz/webapps/zzp/detail/44489/
7. Keil, C., Heinlein, F., Craig, G.C.: The convective adjustment time-scale as indicator of predictability of convective precipitation. Quarterly Journal of the Royal Meteorological Society 140(679), 480–490 (2014), Dostupné z: http://doi.wiley.com/10.1002/qj.2143, doi:10.1002/qj.2143 (cit. February 04, 2015)
8. Singh, D., Bhutiany, M.R., Ram, T., Ye, A., Tao, Y., Miao, C., Mu, X., Schaake, J.C.: Station-based verification of qualitative and quantitative MM5 precipitation forecasts over Northwest Himalaya (NWH). Meteorology and Atmospheric Physics 125(3-4), 107–118 (2014), Dostupné z: http://link.springer.com/10.1007/s00703-014-0321-9, doi:10.1007/s00703-014-0321-9 (cit. February 04, 2015)
9. Liu, Y., Duan, Q., Zhao, L., Ye, A., Tao, Y., Miao, C., Mu, X., Schaake, J.C.: Evaluating the predictive skill of post-processed NCEP GFS ensemble precipitation forecasts in China's Huai river basin. Hydrological Processes 27(1), 57–74 (2013), Dostupné z: http://doi.wiley.com/10.1002/hyp.9496, doi:10.1002/hyp.9496 (cit. February 04, 2015)
10. Comellas, A., Molini, L., Parodi, A., Sairouni, A., Llasat, M.C., Siccardi, F.: Predictive ability of severe rainfall events over Catalonia for the year 2008. Natural Hazards and Earth System Science 11(7), 1813–1827 (2011)
11. Dorninger, M., Gorgas, T.: Comparison of NWP-model chains by using novel verification methods. Meteorologische Zeitschrift 22(4), 373–393 (2013)
12. Sindosi, O.A., Bartzokas, A., Kotroni, V., Lagouvardos, K.: Verification of precipitation forecasts of MM5 model over Epirus, NW Greece, for various convective parameterization schemes. Natural Hazards and Earth System Science 12(5), 1393–1405 (2012)
13. Amodei, M., Stein, J.: Deterministic and fuzzy verification methods for a hierarchy of numerical models. Meteorological Applications 16(2), 191–203 (2009), Dostupné z: http://doi.wiley.com/10.1002/met.101, doi:10.1002/met.101 (cit. February 04, 2015)
14. Atger, F.: Verification of intense precipitation forecasts from single models and ensemble prediction systems. Nonlinear Processes in Geophysics 8(6), 401–417 (2001)
15. Mcbride, J.L., Ebert, E.E.: Verification of quantitative precipitation forecasts from operational numerical weather prediction models over Australia. Weather and Forecasting 15(1), 103–121 (2000)

Design of Fuzzy Controller for Hexacopter Position Control

Jan Bacik, Daniela Perdukova, and Pavol Fedor

Technical University of Kosice
Letna 9, Kosice, Slovakia
jan.bacik.2@tuke.sk

Abstract. The paper deals with the design of a fuzzy controller for controlling the position of a hexacopter represented by a simulation model in Gazebo robot simulation environment, which in terms of control presents a highly nonlinear system with 6 degrees of freedom. The fuzzy controller design was based on the pilot´s experience and on analysis of experimental data collected during a controlled hexacopter flight, without the knowledge of its structure and parameters. The fuzzy controller properties were verified by real time experimental measurements, with sampling time 10 milliseconds. The obtained results have confirmed good dynamic properties of the PI fuzzy controller, which can in future be also applied in a real physical hexacopter model.

Keywords: Fuzzy, Control, Hexacopter, Gazebo.

1 Introduction

The research and development of unmanned aerial vehicles offers many opportunities for effective application of intelligent control methods [1, 2, 3, 4, 5, 6]. Issues related to unmanned aerial vehicles are numerous and can in general be divided into smaller areas of interest such as sensory system development, 3D modelling, mathematic modelling and simulation and control [7]. It is the area of modelling and control that is suitable for the application of neural networks and fuzzy logic, as in general vehicles with a rotating wing are systems with 6 DOF and they are characterised by a high rate of nonlinearity [9, 10, 11, 12].

This paper describes the design of a PI fuzzy controller for controlling the position of a hexacopter. Membership functions and fuzzy controller rules were based on the experience of a pilot (expert) and on analysis of experimental data collected during a pilot-controlled flight, where the pilot´s control commands were sent via joystick into the simulation environment and together with the data from the sensors were saved in the relevant file.

The simulations were carried out in Gazebo robot simulation environment running on a computer with Linux operating system. All algorithms and programmes were written in C++ language and were running in real time.

The objective of the paper has been the application of fuzzy logic principles in the design of a controller for a hexacopter as a nonlinear system. It was our aim

R. Silhavy et al. (eds.), *Artificial Intelligence Perspectives and Applications*,
Advances in Intelligent Systems and Computing 347, DOI: 10.1007/978-3-319-18476-0_20

todemonstrate that even without the knowledge of the structure and the parameters of the system we are able to design a fuzzy controller that has the ability to control the altitude and the position of the hexacopter.

2 Hexacopter Simulation Model

We considered the hexacopter simulation model to be a system with unknown parameters, subsystems and relations between them. The only information available was that about the system´s inputs (Table 1) and outputs (Table 2). Fig. 1 shows the block diagram of the hexacopter system with inputs, outputs and added control structure.

Table 1. System inputs

Input	Description	Units
Throttle	Thrust of the motors	[N]
Roll	Longitudinal tilt of hexacopter	[rad]
Pitch	Lateral tilt of hexacopter	[rad]
Heading	Angular velocity along normal axis	[rad/s]

Table 2. System outputs

Output	Description	Units
X	Position of hexacopter in Earth frame	[m]
Y	Position of hexacopter in Earth frame	[m]
Z	Position of hexacopter in Earth frame	[m]

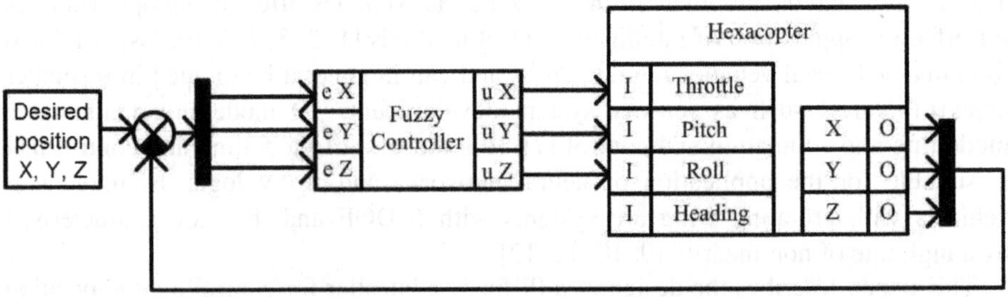

Fig. 1. Block diagram of hexacopter with control structure

The hexacopter 3D model visualization was developed in Gazebo simulation environment (Fig. 2) which is provided with a robust physics engine, with quality graphics and conventional programming and graphic interface. The model runs in real time which enables its direct control by pilot via joystick connected to a PC that enables control of all four inputs. The control data, together with data on position and time were saved in the relevant file and they presented the database needed for the fuzzy controller design.

Fig. 2. Model of hexacopter in Gazebo

In order to obtain experimental data several pilot-controlled flights were carried out during which we tried to maintain the hexacopter at a certain predefined altitude. The data collected served for a better understanding of and searching for relations between information on the hexacopter altitude, input throttle value and its dynamic change. The following figures (Fig. 3 to Fig. 5) illustrate experimental data collection results for Z-axis for one pilot-controlled flight.

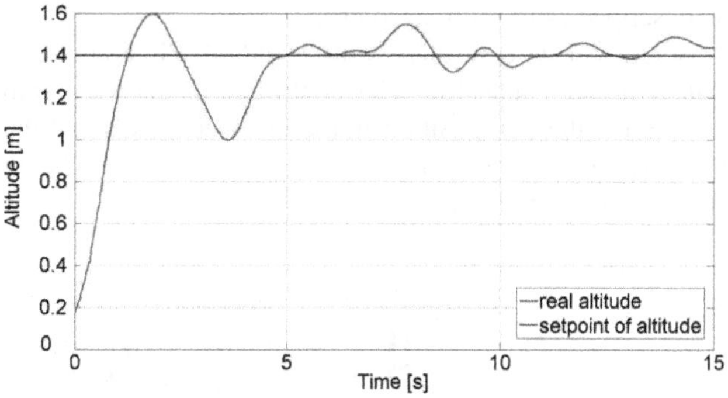

Fig. 3. Real altitude and set point of altitude

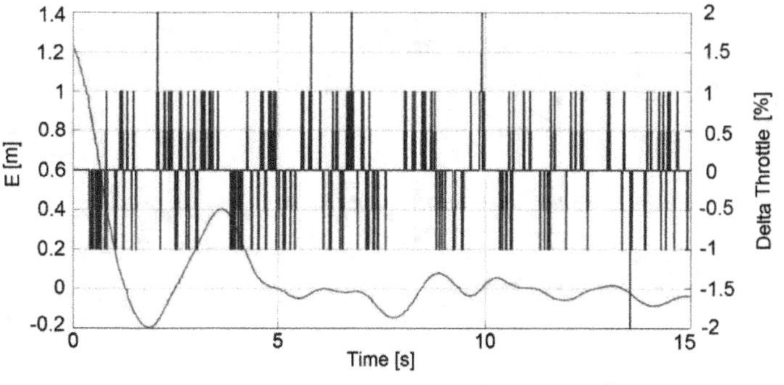

Fig. 4. Real altitude and throttle input

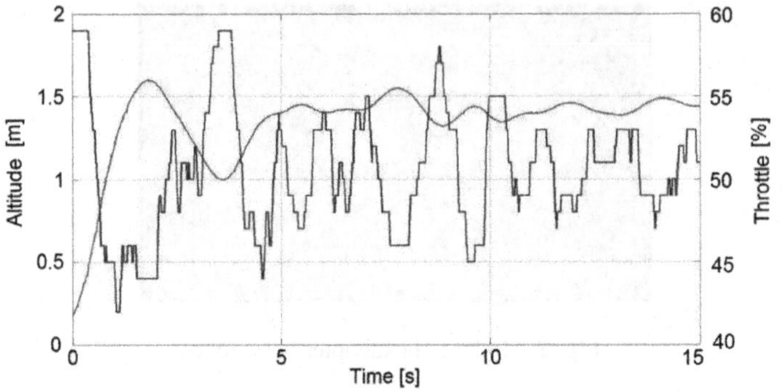

Fig. 5. Altitude deviation and delta throttle input

3 Design of Fuzzy Controller

The objective of the fuzzy controller design was the control of hexacopter position in space, i.e. position control in all three axes X, Y and Z. Hexacopter position control in Z-axis is also described as its altitude control.

A separate discrete fuzzy controller with standard PI structure was designed for position control in each axis [5]. The controller inputs are control deviations between the desired and the real position in the individual axes and their difference, and the output is the gain of the corresponding control action. The resulting fuzzy controller diagram is shown in Fig. 6.

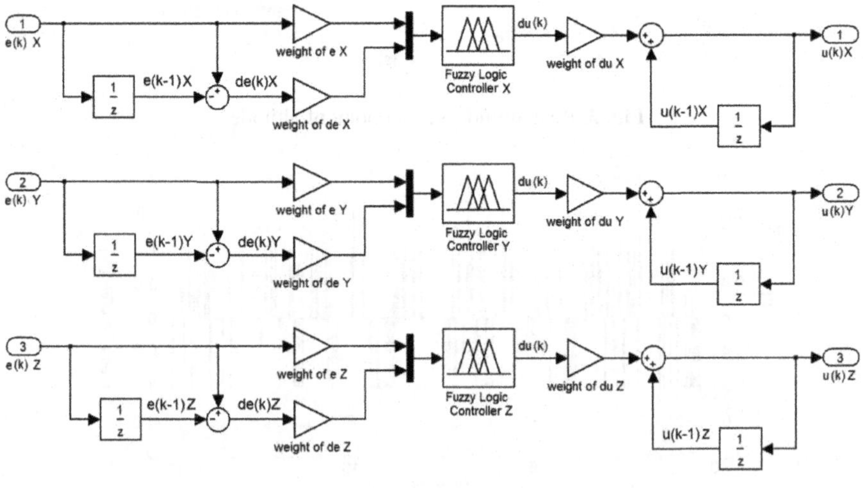

Fig. 6. Internal structure of fuzzy controller

3.1 Design of Fuzzy Controller for Hexacopter Position Control in Z-axis

In the design of hexacopter altitude fuzzy controller (position in Z-axis) we used a standard Mamdani type controller [8], and the fuzzification of variables and proposal of rules were based on the analysis of experimentally measured data and experience of a pilot – expert.

For the sake of simplification, we chose triangular membership function universes of discourse, i.e. each real value of a variable is tuned via scaling universes by means of relevant weight coefficients. On basis of experience from pilot control of flights and analysis of experimentally measured data we proposed a „smoother" distribution of individual membership functions about zero, and a „more rough" distribution for marginal values which represent large deviations and large control actions. The universe of discourse for 2 input and 1 output variable was divided into five levels. (Table 3). The fuzzification of input and output variables is presented in Fig. 7 and Fig. 8.

Table 3. System outputs

Value	Description
Z	**Z**ero
NZL	**N**ear from **Z**ero to the **L**eft
NZR	**N**ear from **Z**ero to the **R**ight
FZL	**F**ar from **Z**ero to the **L**eft
FZR	**F**ar from **Z**ero to the **R**ight

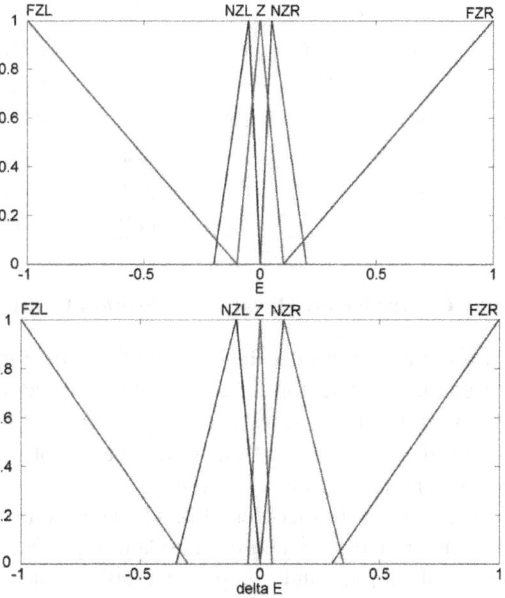

Fig. 7. Membership function of input variables

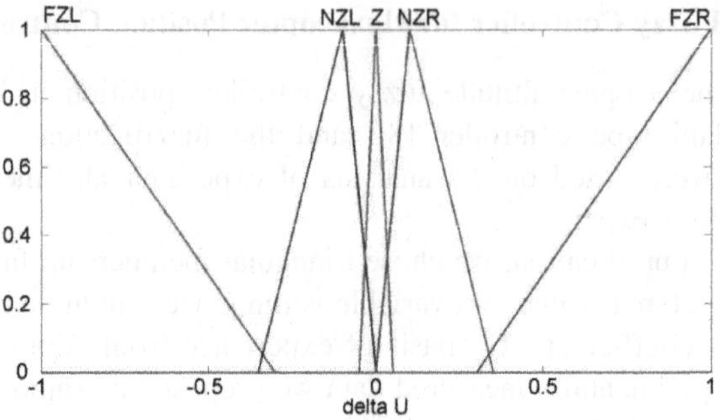

Fig. 8. Membership function of output variable

The main problem in hexacopter altitude control (motor thrust control) is the fact that its propellers are not capable of generating negative thrust. If we want to stop the hexacopter at a certain altitude, we have to apply the brake before reaching this altitude, while the braking is possible only through reducing motor thrust, or stopping the motors. Altitude control as such therefore means balancing between motor thrust force and gravitational force.

The above facts were taken into consideration in the fuzzy controller rules table design (Table 4) which was set up mainly on basis of the pilot-expert´s experience and partially on basis of experimental data analysis.

The resulting hexacopter altitude fuzzy controller is of the Mamdani type and has 25 rules.

Table 4. Rules of relationship among e, delta e and delta u

de/e	FZL	NZL	Z	NZR	FZR
FZL	FZL	FZL	FZL	FZL	NZL
NZL	FZL	FZL	FZL	NZL	Z
Z	FZL	NZL	Z	NZR	NZR
NZL	NZL	Z	Z	NZR	NZR
FZR	Z	Z	NZR	FZR	FZR

3.2 Design of Fuzzy Controller for Hexacopter Position Control in X- and Y-axis

Contrary to hexacopter altitude control is the control of its position in X- and Y-axis. Position control is based on thrust vector deflection into the direction we want to head the hexacopter. Thrust vector deflection is realised by means of tilting the hexacopter body. This means that if we want to control the hexacopter position we have to control its longitudinal and transverse deflection.

Same as in the design of the hexacopter altitude fuzzy controller, in the design of its position in X- and Y-axes we chose Mamdani type fuzzy controllers with identical fuzzification of input and output variables and with maintaining the triangular form of membership functions. There was a change in the fuzzy controller rules design (Table 5), where the fact that with tilting the hexacopter we can bring about the same deflection into the positive and the negative side was accounted for. For this reason the fuzzy controller rules table shows symmetricity.

The developed fuzzy controllers for hexacopter position control in X- and Y-axes are Mamdani type controllers and they have 25 rules.

Table 5. Rules of relationship among e, delta e and delta u

de/e	FZL	NZL	Z	NZR	FZR
FZL	FZL	FZL	FZL	Z	Z
NZL	FZL	FZL	NZL	Z	NZR
Z	FZL	FZL	Z	NZR	FZR
NZL	NZL	Z	NZR	NZR	FZR
FZR	Z	Z	FZR	FZR	FZR

4 Experimental Results

The properties of the designed fuzzy controller for control of hexacopter position in space were verified by experimental measurements in real time. The fuzzy controller structure was realised in C++ language using the FuzzyLite library. This fuzzy controller communicated with Gazebo robot simulation environment via a predefined communication interface. The Gazebo simulation environment provided information on the hexacopter position (X, Y, and Z) obtained on basis of simulation from visual odometry of the position sensor and sent this information to the fuzzy controller input. From the inputs the fuzzy controller calculated the corresponding control action and sent it back to the simulation environment with 100Hz frequency.

The experimentally measured responses of fuzzy control of hexacopter position in space to step changes of the set-point for the individual axes are shown in Fig. 9 to Fig. 11.

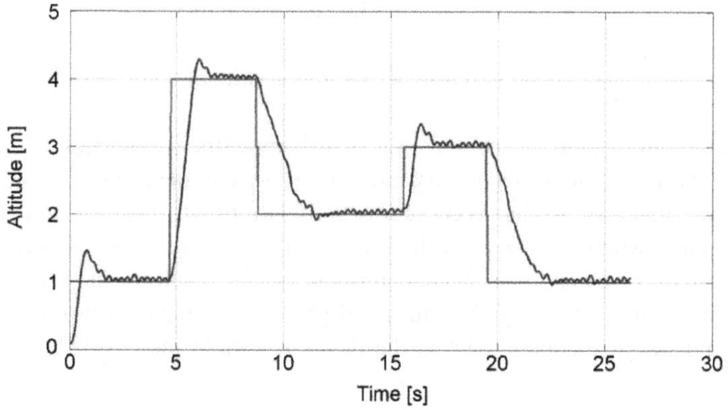

Fig. 9. Altitude (Z) fuzzy controller

In position control in X- and Y-axes floating oscillations about the desired value can be observed. These are caused by the hexacopter moment of inertia during rotation about the relevant axes and by the limited tilt during its rotation.

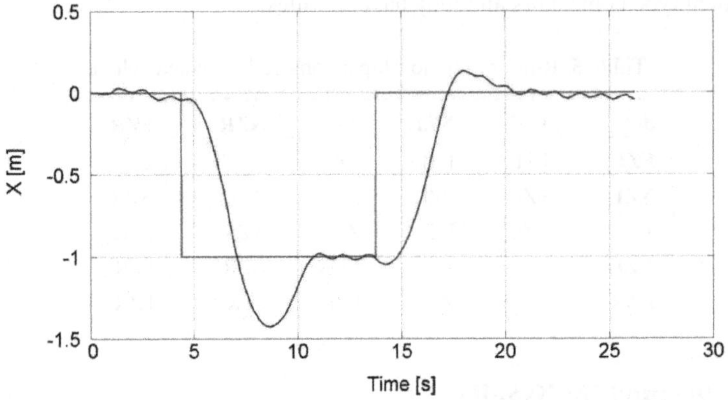

Fig. 10. Position (X) fuzzy controller

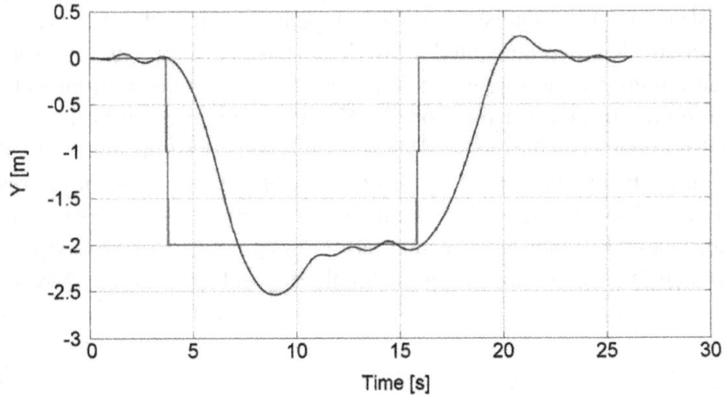

Fig. 11. Position (Y) fuzzy controller

Rules have an important role as they influence the properties of the fuzzy controller. Therefore, in order to try and improve the designed fuzzy controller properties, we also tested other acceptable combinations of rules. As an example we present the case where two Zero values in the last line of Table 4 were replaced by Near Zero Left values. This change influenced the resulting hexacopter altitude fuzzy controller in such a way that during flight of hexacopter upwards there is no overshoot and the hexacopter reaches the desired altitude more slowly and steadily (Fig. 12).

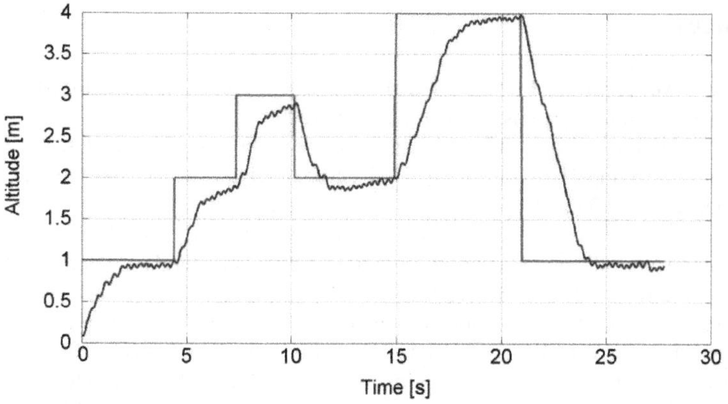

Fig. 12. Behaviour of altitude fuzzy controller with changed rules

5 Conclusion

The paper deals with the design of a fuzzy controller for controlling the position of a hexacopter in space, with absence of knowledge about the structure and parameters of this considerably nonlinear system with 6 degrees of freedom. For the fuzzification of variables and proposal of rules for a Mamdani type controller we used experimental data collected in Gazebo robot simulation environment and also the experience of a pilot – expert. The properties of the controller designed in this way were verified by real time experimental measurements.

The results of experimental measurements have confirmed the rightness of the designed fuzzy controller and have demonstrated its good dynamic properties. The fuzzy controller is capable of holding the hexacopter at the desired altitude and position even at step changes of the desired values. During the hexacopter upward flight the originally designed controller showed a small overshoot which, as it was demonstrated, can be removed, for example, by a minor changing of the rules. Our objective will be to apply the designed controller in the future in a real physical hexacopter model.

The results presented have confirmed the rightness of fuzzy systems deployment in the control of unmanned aerial vehicles. However, it is not always possible to obtain the expert knowledge required for fuzzy controller design, or to use universal „meta rules", especially in the case of nonlinear high order systems. For this reason, research efforts concerning fuzzy logic control in unmanned aerial vehicles in the following years should be devoted to systematic analysis and design of fuzzy control systems that do not demand heuristic searching for linguistic control rules. This most often involves methods based on fuzzy models of the controlled systems.

Acknowledgements. The work has been supported by project KEGA 011TUKE-4/2013.

References

1. Norgaard, M., Ravn, O., Pulsen, N.K., Hansen, L.K.: Neural Networks for Modelling and Control of Dynamic Systems. Springer, London (2000)
2. Stojcsics, D.: Fuzzy Controller for Small Size Unmanned Aerial Vehicles. In: IEEE 10th International Symposium on Applied Machine Intelligence and Informatics (SAMI), Slovakia, pp. 91–95 (2012)
3. Bickraj, K., Pamphile, T., Yenilmez, A., Li, M., Tansel, I.N.: Fuzzy Logic Based Integrated Controller for Unmanned Aerial Vehicles. In: Florida Conference on Recent Advances in Robotics, FCRAR (2006)
4. Kurnaz, S., Çetin, O.: Autonomous Navigation and Landing Tasks for Fixed Wing Small Unmanned Aerial Vehicles. Acta Polytechnica Hungarica 7(1), 87–102 (2010)
5. Shengyi, Y., Kunqin, L., Jiao, S.: Optimal tuning method of PID controller based on gain margin and phase margin. In: International Conference on Computational Intelligence and Security, pp. 634–638 (2009)
6. Beard, R.: Autonomous Vehicle Technologies for Small Fixed-Wing UAVs. Journal of Aerospace Computing Information and Communication 2(1), 92–108 (2005)
7. Korba, P., Pila, J.: Aplikácia CAx systémov pri projektovaní konštrukčných uzlov vrtuľníka. Zaklad Poligraficzny, Wisla (2007)
8. Zilkova, J., Timko, J.: A Fuzzy vector control of asynchronous motor. Acta Technica 55(3), 259–274 (2010)
9. Sastry, S.: Nonlinear Systems: Analysis, Stability, and Control. Springer, New York (2010)
10. Modrlak, O.: Fuzzy řízení a regulace. Technická univerzia v Liberci, Liberec (2004)
11. Wicaksono, H., Christophorus, B.: T1-Fuzzy vs T2-Fuzzy Stabilize Quadrotor Hover with Payload Position Disturbance. International Journal of Applied Engineering Research 9(22), 15251–15262 (2014)
12. Chao, H., Luo, Y., Di, L., Chen, Y.: Fractional order flight control of a small fixed-wing UAV: Controller design and simulation study. In: Proceedings of the ASME 2009 International Design Engineering, Technical Conferences & Computers and Information in Engineering Conference, pp. 621–628 (2009)

Implementation of Two Stages *k*-Means Algorithm to Apply a Payment System Provider Framework in Banking Systems

Omid Mahdi Ebadati E.[1] and Sara Sadat Babaie[2]

[1] #242, Somayeh Street, Between Qarani & Vila, Department of Mathematics and Computer Science, University of Economic Sciences, Tehran, Iran
omidit@gmail.com

[2] #242, Somayeh Street, Between Qarani & Vila, Department of Knowledge Engineering and Decision Science, University of Economic Sciences, Tehran, Iran
sara.babaiee@gmail.com

Abstract. Payment Systems Providers (PSPs) are companies, which provide services of payment for their customers. Recently, according to some changes in Iran central bank rules, providing services of payment are not monitored by banks anymore. This duty is assigned to some organizations called PSPs and becomes one of the most challenging topics for them. Clustering the datasets, assessment and the way of expressing customers' demands and the provinces of requests should be recognized for improving services to the customers, banks, financial and credit institutes. The proposed framework consists of two stages using *k*-means algorithm and Euclidean square distances. The *k*-means algorithm is applied in the first stage for five provinces, which have the highest demands. In the second stage, the mean of centroids obtained from *k*-means are calculated and repeat clustering according to the minimum Euclidean square distances to the new centroids then comparing the information gained by two stages.

Keywords: Payment Systems Providers, Data mining, *k*-means algorithm, Artificial Intelligence in Bank Systems, Euclidean distances.

1 Introduction

Providing services for bank customers and ensuring customers' satisfaction is the best way for each bank to win the competition among several competitors. In previous years each bank work on the services independently, but recently some changes in the Central Banks of Iran (CBI) rules abdicate these duties to PSPs (Payment System Provider). In banking systems these agencies are companies, which provide services of payment for their customers and they are not monitored by banks anymore. The customers of PSPs are usually banks, financial and credit institutes. According to these alterations in the rules, it puts them in the trouble to provide services for the whole of the country. Providing favorite and optimized services and also developing

R. Silhavy et al. (eds.), *Artificial Intelligence Perspectives and Applications*,
Advances in Intelligent Systems and Computing 347, DOI: 10.1007/978-3-319-18476-0_21

that are so efficient for increasing the customer's satisfaction. In this paper, we try to recognize the customer's requests and cluster base on their locations (provinces). In the first stage we use *k*-means algorithm as a data driven techniques for clustering our datasets. Then, in the second stage, the mean of the centroids obtained from the previous stage is calculated, and each request cluster by Euclidean square distances to new centroids. The result of two stages is compared and detects the similarity, dissimilarity and the most efficient factors in the consequences. The changes in the central bank have happened since last year, however, there is not enough research on this subject. On the other side, many problems have been caused by these changes. Recognizing the number of each request and the importance of each one in different provinces and also in various banks would be very useful for improving our services. As an example, according to our dataset, if in an agency the number of preparing POS (Point Of Sail) was significant, then we should decide to choose one of the ways below to reduce the troubles like: 1. The quality of the resource 2. The quality of transferring POS 3. Sampling for admitting POS. The framework for extracting the cluster of customer's requests helps us to improve our services payment base on breakdown provinces. A data clustering algorithm is a popular data mining technique like SOM [1] fuzzy clustering [2,3], and vector quantization [4], ANNs [5,6,7,8,9,10,11,12] are performed for cluster data in their best position. Also *k*-means algorithm, which is used in this paper is one of the best computational algorithm in clustering [13,14,15,16]. The rest of the paper is arranged as follows. A literature of the clustering methods, the application of clustering in the bank and use of *k*-means algorithm is laid at the rest of this section. The research methods, data description, the way of clustering and introducing the applied software are presented in Section 2. In Section 3, the results of the experimental work and gained information are introduced. The discussions, conclusion and future research are presented in Section 4.

This rest of this section introduces some general background of data mining. Introducing and reviewing of papers that used cluster analysis as one of the most famous and popular techniques in data mining, is considered in this part. It is also included *k*-means method as a useful algorithm in clustering and the fields of the application of that and hybrid of *k*-means with metaheuristic optimization algorithms to increase the efficiency of *k*-means is reviewed. In addition, other techniques such as ANN, LDA and LRA in artificial intelligence that is used in the banking systems are also presented.

1.1 Data Clustering

Cluster analysis is an unsupervised learning algorithm, which partitioning datasets by considering the similarity of data according to the patterns into some groups [17,18]. Data clustering has been used in various subjects, such as computer science, including: Image segmentation [19,20,21], pattern recognition [22], information retrieval [23,24,25], documents [26], marketing [27], vector quantization[28], biology, psychiatry, psychology, archaeology, geology, etc. There are several clustering methods [18], DBSCAN [29], Probabilistic Latent Semantic Analysis (PLSA) [30] and

Latent Dirichlet Allocation (LDA) [31]. CLIQUE [32], *k*-means [13] that reduce dimensionality from high to a low dimension using the Laplacian Eigen map [33].

1.2 The Application of *k*-Means Method in Clustering

K-means clustering is one of the most popular and simplest data driven algorithm [34,35], which was demonstrated by MacQueen [13] in 1967 though the idea goes back to Steinhaus [36]. Considering the local heuristic k and standard *k*-means clustering algorithm was first proposed by Lloyd [37]. After considering k, each element is assigned to the closest cluster according to the local minimum of the objective function, which is the sum of the squared distance of each data point to the centroids of its clusters. This procedure repeat until the cluster associate with the points does not change [13]. Simplicity and efficiency are the best reasons that the *k*-means is still one of the most popular algorithms [38]. In [39] by Fuzzy Genetic and *k*-means clustering try to recognize customer Buying Patterns. There are several ways for initialization for Cluster-Based Population and *k*-means is very efficient in this subject [40]. Also the algorithm can be applied in intrusion detection [41]. Hybrid of *k*-means algorithm and meta-heuristic algorithm decreased the runtime and also help to escape from a local optimum [42,43].

1.3 The Application of Data Mining in Banks

Durand [44] first actualized credit scoring by LDA, by searching the differences between good and bad credit groups. Since then, statistical techniques, primarily LDA and LRA, have been used in financial prediction studies [45,46,47,48], [29]. Since the 1990s, ANN has been also used in modeling credit scoring. Desai et al. [7] developed credit scoring models with ANN on a data set of 1962 credit consumers, obtained from three different credit unions. Among the subjected models ANN, LDA and LRA [10] got similar results on a set of 1078 data, obtained from twelve different credit unions. Clustering e-Banking Customer by discussion about data driven algorithm such as *k*-means, SOM and Marketing technique-RFM analysis is used in [49]. Segmenting the Banking Market by *k*-means algorithm is applied in [50] and Life Time Value of the customers for generating a profile of them are used. The efficiency of branches of banks can be ranked by *k*-means and Grey relational method [51].

2 Methods

This research from the point of view of doing is placed in applicable-developmental research categories.

Step One - Data Collection: This step is associated with the production process and data collection, the data used in this paper collected from the unit of the PSP in Tehran. It has 80,532 records from 25 various provinces with 5 characters, includes: name of the bank, type of requests, port of payment, type of customers and status requests that each of them contains the following sections: Eight different Banks, two types of customers includes holder and the recipient, paid ports in three ways: Telephone ports, terminals and telephone cards, request status in two situations: in doing and done. Customers` request, including twenty-one diverse request such as: organizer to declare the transaction to a recipient, requests to fix terminal, fix block cards, fix inconsistent, request roll for POS, etc. A small portion of the dataset is shown in Table 1.

Table 1. A small portion of the datasets

	Provinces	**Bank**	**Type of request**	**Customer Type**	**Status Request**	**Port of Payment**
Request1	Tehran	A	declare transactions to recipient	recipient	done	telephone ports
Request2	Khorasan	B	requests to fix terminal	recipient	done	telephone cards
Request3	Mazandaran	A	declare transactions to recipient	recipient	done	telephone ports
Request4	Mazandaran	B	requests to fix terminal	recipient	done	telephone ports
Request5	Tehran	A	requests to fix terminal	recipient	In doing	telephone ports
Request6	Mazandaran	A	declare transactions to recipient	holder	done	telephone ports
Request7	Alborz	A	Resolve discrepancies less than 72 hours	recipient	done	telephone ports
.	.	.	.	.	.	.

Step two - Pre-processing of data, after data collection, errors contained in them should be removed and cleaned. These Probable errors include out-of-treatment values, missing values, and duplicate attributes for modeling data that are not in proper form.

Data preprocessing includes the following sections:

2A: Preparing the raw data in the form of a data set that is used in other data mining processes.

2B: Selecting cases and variables, which are necessary for analysis.

2C: If needed, transform data.

According to the previous steps, these data were prepared in the Excel software then the selected variables were extracted and the other variables were removed. After this step the data decreased to 78977 records.

Step Three - Model Selection and proper implementation of data mining is the main task at this stage.

Different parts of this step are:

3A: Select and use appropriate modeling techniques.

3B: Manipulation and adjustment of the model and algorithms to achieve optimal results.

3C: If necessary, return to the pre-processing step.

In this step, *k*-means algorithm is chosen. The data transferred to the MATLAB and then mining of the data and clustering done in the software. In this paper the *k*-means algorithm is applied in two stages, in the first stage the algorithm used for each of populated provinces, then after clustering all provinces, the mean of centroids of all clusters, which is obtained in the previous stage calculated and again this algorithm is applied this centroids and all of the provinces all together are clustered with the new centroids. The whole process is shown in Pseudo-code in Fig. 1.

Step Four - Evaluation and derivation of the model: In this step evaluate the performed clustering to answer basic research questions.

The framework presented in this paper is shown in Fig. 2 and Fig. 3.

```
READ data records
   SET the row data in the form which can be used in the process
      FOR <X=1: all number of records from the PSP>
         FOR< Y=1: all number of features>
            IF <values are out-of-treatment or missing>
               THEN transform data
            ELSE
               Do nothing
            END IF
         END FOR
      END FOR
   SET k-means algorithm FOR CLUSTERING
      IF < A fixed number of iterations has been completed>
         ELSE IF < Assignment of documents does not change>
            ELSE IF< Centroids do not change between iterations>
               THEN Terminate k-means
            ELSE
               Return next iteration
      END IF
   SET mean of centroids from previous steps as new centroids
      FOR< X=1: all number of records from the PSP>
         FOR< Y=1: all number of features>
               Calculate Euclidean square distances for each records(X, Y) to
               new centroids
         END FOR
      END FOR
END
```

Fig. 1. Pseudo-code of the proposed model

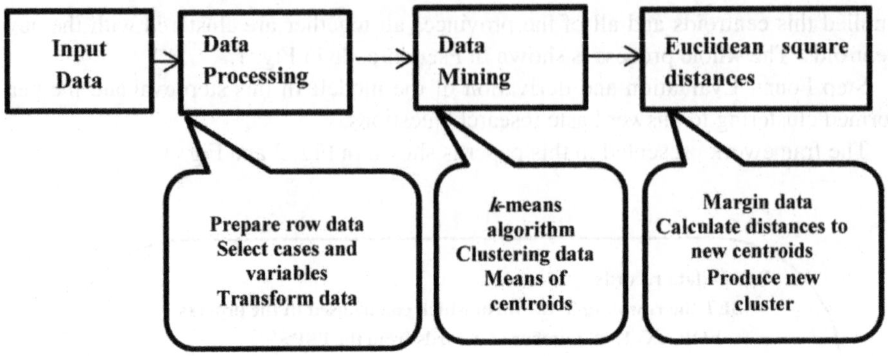

Fig. 2. The proposed model of two-stage framework for PSPs agencies

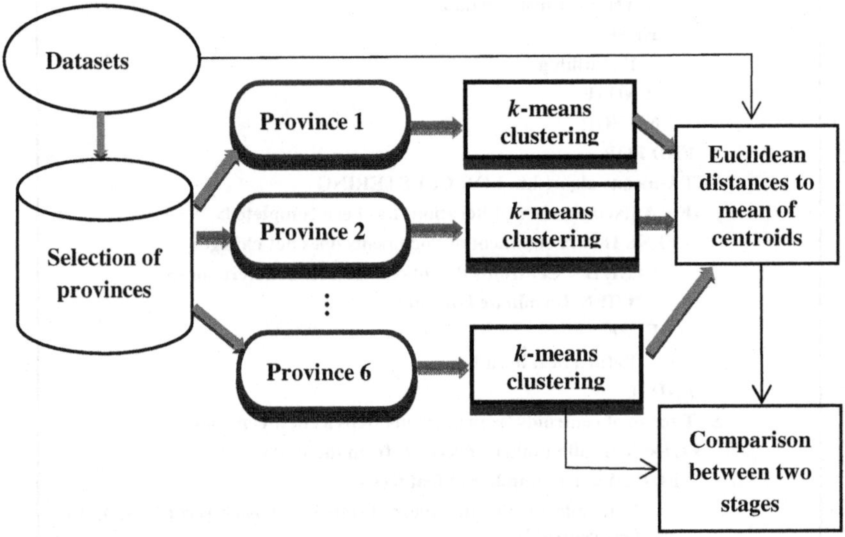

Fig. 3. Conceptual framework of the proposed model

3 Results

The final data set includes 78977 records, which belong to 25 provinces. Five provinces choose among them that have the most requests, including Tehran, Khorasan, Isfahan, Alborz and Fars. The number of requests in each of agencies, respectively is: 13908, 9444, 5642, 4058, and 3720. The *k*-means algorithm is used to cluster datasets in each of the agencies and the results are stored to compare with the result of the next stage. In the second stage margin the data from all of the chosen provinces, then calculate the mean of centroids obtained from the first stage. The clustering method using Euclidean square distances is repeated. The distances between the data and new centroids are calculated and the closest distances consider allocating the data to each cluster. The information obtained by each stage is compared to find useful results. In Tehran 8.15% of data is inserted in cluster 1, 1.21% in cluster 2, 71.86% in cluster 3,

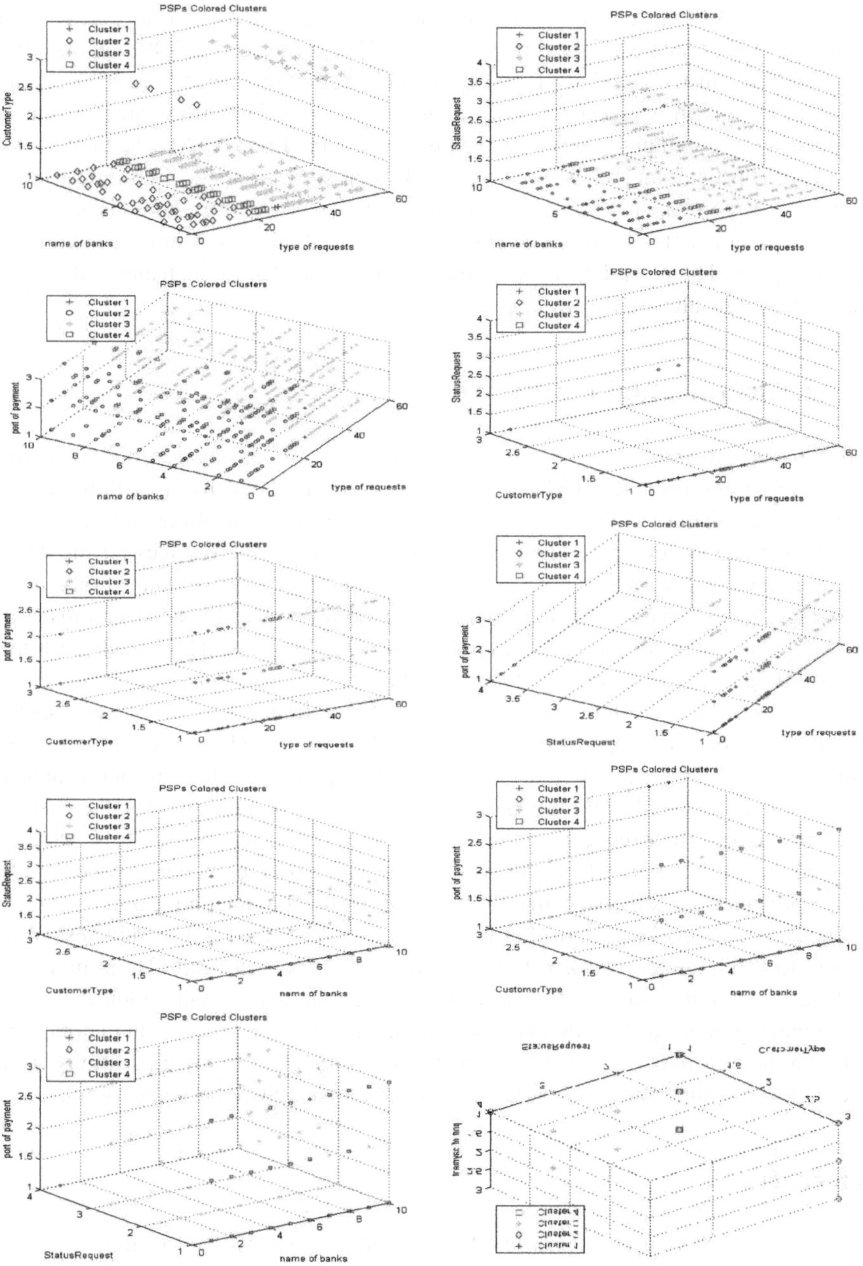

Fig. 4. Clustering Results

and 18.86% in cluster 4. In Khorasan 53.79% of data is inserted in cluster 1, 16.52% in cluster 2, 4.45% in cluster 3, and 25.22% in cluster4. In Isfahan 13.59% is inserted in cluster 1, 5.14% in cluster 2, 51.74% in cluster 3, and 29.51% in cluster 4. In Alborz 16.31% of data is inserted in cluster 1, 6.31% in cluster 2, 33.49% in cluster 3, and 43.87% in cluster 4. In Fars 1.21% of data is inserted in cluster 1, 72.65% in cluster 2, 21.78% in cluster 3, and 4.35% in cluster 4. The information gained from the second stage is stored and 21.01% of data is inserted in cluster 1, 20.27% in cluster 2, 40.75% in cluster 3, and 17.91% in cluster 4, then separate them according to their provinces, in comparison with previous stage the below information is obtained. 84.36% of data inserted in the same cluster and the most similarity among them is the type of request. The most dissimilarity among them is related to the name of the bank. The clustering results are shown in Fig. 4.

4 Conclusion and Future Work

In this paper, we introduce the two-stage framework of information obtained from Payment Systems Provider (PSPs) companies, using the *k*-means algorithm and Euclidean square distances. These companies' new duties are controlling and monitoring the payment due to new Iran central banks' rule. Supplying the services for the whole of the country is one of the most challenging topics for these organizations. After clustering dataset's base in their provinces into some groups by *k*-means clustering algorithm, the mean of centroids is applied as new centroids. Then, each customer requests assigned to a cluster, which has the closest Euclidean square distances. Finally the results of the two algorithms are compared. Dissimilarities between two stages, show the differences in type of requests, ports of payment and customer type in various provinces to the mean of centroids. The result can explain that in each province, according to the strengths and weaknesses of banks in providing services, the customers change and fit their requests to the easiest and the most reliable services, which is provided. As an example, if in a province bank "A" provides better service in telephone ports, their customers change their way of payment to telephone ports.

The *k*-means clustering algorithm has two main problems, (1) The slowness of algorithm to converge and work with a large number of datasets and (2) too much dependence to the initializations, so it causes to get caught in a local minimum.

The initial points lead to different local optimum, so it is the best reason to notice and start with the best initial points. However, for determining such initial partitions, using meta-heuristic optimization such as Genetic Algorithm and SOM could be useful.

References

1. Kohonen, T.: Self-organizing maps, vol. 30. Science & Business Media (2001)
2. Bezdec, J., et al.: Detection and characterization of cluster substructure. SIAM J. Appl. Math. 40, 339–372 (1981)
3. Kaufman, L.R., Rousseeuw, P.J.: Finding groups in data: An introduction to cluster analysis. John Wiley & Sons Inc., Hoboken (1990)

4. Linde, Y., Buzo, A., Gray, R.M.: An algorithm for vector quantizer design. IEEE Transactions on Communications 28(1), 84–95 (1980)
5. Abdou, H., Pointon, J., El-Masry, A.: Neural nets versus conventional techniques in credit scoring in Egyptian banking. Expert Systems with Applications 35(3), 1275–1292 (2008)
6. Chen, C.-M., et al.: Breast Lesions on Sonograms: Computer-aided Diagnosis with Nearly Setting-Independent Features and Artificial Neural Networks 1. Radiology 226(2), 504–514 (2003)
7. Desai, V.S., Crook, J.N., Overstreet, G.A.: A comparison of neural networks and linear scoring models in the credit union environment. European Journal of Operational Research 95(1), 24–37 (1996)
8. Lee, T.-S., Chen, I.-F.: A two-stage hybrid credit scoring model using artificial neural networks and multivariate adaptive regression splines. Expert Systems with Applications 28(4), 743–752 (2005)
9. Lee, T., Jeng, D.S.: Application of artificial neural networks in tide-forecasting. Ocean Engineering 29(9), 1003–1022 (2002)
10. Malhotra, R., Malhotra, D.: Evaluating consumer loans using neural networks. Omega 31(2), 83–96 (2003)
11. Šušteršič, M., Mramor, D., Zupan, J.: Consumer credit scoring models with limited data. Expert Systems with Applications 36(3), 4736–4744 (2009)
12. Tsai, C.-F., Lu, Y.-H.: Customer churn prediction by hybrid neural networks. Expert Systems with Applications 36(10), 12547–12553 (2009)
13. MacQueen, J.: Some methods for classification and analysis of multivariate observations. In: Proceedings of the Fifth Berkeley Symposium on Mathematical Statistics and Probability, Oakland, CA, USA (1967)
14. Barbakh, W.A., Wu, Y., Fyfe, C.: Review of clustering algorithms. In: Barbakh, W.A., Wu, Y., Fyfe, C. (eds.) Non-Standard Parameter Adaptation for Exploratory Data Analysis. SCI, vol. 249, pp. 7–28. Springer, Heidelberg (2009)
15. Han, J., Kamber, M., Pei, J.: Data mining, southeast asia edition: Concepts and techniques. Morgan Kaufmann (2006)
16. Jain, A.K.: Data clustering: 50 years beyond K-means. Pattern Recognition Letters 31(8), 651–666 (2010)
17. Jain, N.C., Indrayan, A., Goel, L.R.: Monte Carlo comparison of six hierarchical clustering methods on random data. Pattern Recognition 19(1), 95–99 (1986)
18. Jain, A.K., Dubes, R.C.: Algorithms for clustering data, vol. 6. Prentice-Hall, Englewood Cliffs (1988)
19. Frigui, H., Krishnapuram, R.: A robust competitive clustering algorithm with applications in computer vision. IEEE Transactions on Pattern Analysis and Machine Intelligence 21(5), 450–465 (1999)
20. Hoover, A., et al.: An experimental comparison of range image segmentation algorithms. IEEE Transactions on Pattern Analysis and Machine Intelligence 18(7), 673–689 (1996)
21. Shi, J., Malik, J.: Normalized cuts and image segmentation. IEEE Transactions on Pattern Analysis and Machine Intelligence 22(8), 888–905 (2000)
22. Anderberg, M.R.: Cluster analysis for applications, DTIC Document (1973)
23. Rasmussen, E.M.: Clustering Algorithms. In: Information Retrieval: Data Structures & Algorithms, pp. 419–442 (1992)
24. Salton, G., Buckley, C.: Global text matching for information retrieval. Science 253(5023), 1012–1015 (1991)
25. Bhatia, S.K., Deogun, J.S.: Conceptual clustering in information retrieval. IEEE Transactions on Systems, Man, and Cybernetics, Part B: Cybernetics 28(3), 427–436 (1998)

26. Iwayama, M., Tokunaga, T.: Cluster-based text categorization: a comparison of category search strategies. In: Proceedings of the 18th Annual International ACM SIGIR Conference on Research and Development in Information Retrieval. ACM Press (1995)
27. Arabie, P., Wind, Y.: Marketing and social networks, vol. 171, p. 254. SAGE FOCUS EDITIONS (1994)
28. Oehler, K.L., Gray, R.M.: Combining image compression and classification using vector quantization. IEEE Transactions on Pattern Analysis and Machine Intelligence 17(5), 461–473 (1995)
29. Ester, M., et al.: A density-based algorithm for discovering clusters in large spatial databases with noise. In: Proceedings of the 2nd International Conference on Knowledge Discovery and Data Mining (1996)
30. Hofmann, T.: Probabilistic latent semantic analysis. In: Proceedings of the Fifteenth Conference on Uncertainty in Artificial Intelligence. Morgan Kaufmann Publishers Inc. (1999)
31. Blei, D.M., Ng, A.Y., Jordan, M.I.: Latent dirichlet allocation. The Journal of Machine Learning Research 3, 993–1022 (2003)
32. Agrawal, R., et al.: Automatic subspace clustering of high dimensional data for data mining applications, vol. 27. ACM (1998)
33. Belkin, M., Niyogi, P.: Laplacian eigenmaps for dimensionality reduction and data representation. Neural Computation 15(6), 1373–1396 (2003)
34. Rekik, A., et al.: A k-means clustering algorithm initialization for unsupervised statistical satellite image segmentation. In: 2006 1st IEEE International Conference on E-Learning in Industrial Electronics, pp. 11–16 (2006)
35. Zhou, X., Shen, Q., Wang, J.: K K-means clustering algorithm based on particle swarm in image classification. Journal of Chinese Computer Systems 29(2), 333–336 (2008)
36. Steinhaus, H.: Sur la division des corp materiels en parties. Bull. Acad. Polon. Sci. 1, 801–804 (1956)
37. Lloyd, S.: Least squares quantization in PCM. IEEE Transactions on Information Theory 28(2), 129–137 (1982)
38. Shindler, M., Wong, A., Meyerson, A.W.: Fast and accurate k-means for large datasets. In: Advances in Neural Information Processing Systems 24, NIPS 2011 (2011)
39. Vidya, V.: A Hs-Hybrid Genetic Improved Fuzzy Weighted Association Rule Mining Using Enhanced Hits Algorithm. Journal of Agricultural & Biological Science 9(6) (2014)
40. Poikolainen, I., Neri, F., Caraffini, F.: Cluster-Based Population Initialization for differential evolution frameworks. Information Sciences 297, 216–235 (2015)
41. Elbasiony, R.M., et al.: A hybrid network intrusion detection framework based on random forests and weighted k-means. Ain Shams Engineering Journal 4(4), 753–762 (2013)
42. Mor, M., Gupta, P., Sharma, P.: A Genetic Algorithm Approach for Clustering. International Journal of Engineering & Computer Science 3(6) (2014)
43. Yaghini, M., Soltanian, R., Noori, J.: Paper: A Hybrid Clustering Method Using Genetic Algorithm With New Variation Operators. International Journal of Industrial Engineering & Production Management 21(2) (2010)
44. Durand, D.: Risk elements in consumer instalment financing. NBER Books (1941)
45. Altman, E.I.: Financial ratios, discriminant analysis and the prediction of corporate bankruptcy. The Journal of Finance 23(4), 589–609 (1968)
46. Meyer, P.A., Pifer, H.W.: Prediction of bank failures. The Journal of Finance 25(4), 853–868 (1970)

47. Sinkey, J.F.: A multivariate statistical analysis of the characteristics of problem banks. The Journal of Finance 30(1), 21–36 (1975)
48. West, R.C.: A factor-analytic approach to bank condition. Journal of Banking & Finance 9(2), 253–266 (1985)
49. Niyagas, W., Srivihok, A., Kitisin, S.: ECTI Transaction on Computer and Information Technology 2(1) (2006)
50. Kumar, M.V., Chaitanya, M.V., Madhavan, M.: Segmenting the Banking Market Strategy by Clustering. International Journal of Computer Applications 45 (2012)
51. Zarandi, S., et al.: Ranking banks using K-Means and Grey relational method. Management Science Letters 4(10), 2319–2324 (2014)

An Artificial Intelligence Approach to Nutritional Meal Planning for Cancer Patients

Richard Fox and Yuliya Bui

Department of Computer Science
Northern Kentucky University
Highland Heights, KY USA
foxr@nku.edu, buijulia@gmail.com

Abstract. It is well-known that cancer patients undergoing chemotherapy suffer from nausea at a time when it is important for them to receive proper nutrition to fight not only the cancer but the impact that chemotherapy has on their body. Yet due to the debilitating nature of both the cancer and the chemotherapy treatment, such patients often have little energy to spend on food selection and preparation. In this paper, an artificial intelligence approach is taken to generate meal plans for users who may or may not be suffering from nausea on any particular day. Specifically, a variety of preferences are applied to evaluate meal components (e.g., dinner entrée, dinner side dish, lunch entrée) in an attempt to generate a balanced, healthy and palatable meal plan for the user.

Keywords: Artificial Intelligence, Planning, Nutrition, Chemotherapy.

1 Introduction

Chemotherapy treatment is one of the most common cancer treatments. The medication stops or slows cancer cells from spreading, destroying existing cancer cells [1]. While chemotherapy may be the best way to combat cancer, it has potentially devastating side effects that can debilitate the patient. Often, patients become overly fatigued, have reactions of bleeding, swelling, skin and nerve ending changes, as well as hair loss and memory changes. At a time when the patient needs energy to fight off the cancer, the patient's body is being subjected to a chemical treatment that leaves the body weakened. Proper nutrition should be employed to help the body adjust [2].

Unfortunately, maintaining adequate nutrition during chemotherapy is challenging because of these and other side effects. Aside from nausea (which can lead to vomiting, diarrhea and/or constipation), the patient could also suffer from changes in their throat and mouth making the consumption of food harder as well as a change in taste leading to a lack of desire to eat. Changes in urination and the development of anemia may further cause patients to alter their eating habits [3].

A nutritional tool that can factor in such symptoms while also weighing the benefits of maintaining nutrition could be extremely useful. Building upon previous artificial intelligence (AI) research in planning, this paper describes the MARY system.

R. Silhavy et al. (eds.), *Artificial Intelligence Perspectives and Applications*,
Advances in Intelligent Systems and Computing 347, DOI: 10.1007/978-3-319-18476-0_22

MARY, `Meal Arranging Rule sYstem`[1], uses general nutritional information and user-specified preference information to generate a daily meal plan. MARY takes into account a number of factors:

- Nutritional information (desired caloric intake, limitations on sodium, fat)
- User preferences of food items and food combinations
- Recency of previously selected foods
- Foods that are easier to prepare and consume when a patient is suffering from nausea and other chemotherapy symptoms

This paper is organized as follows. Section 2 provides background of planning approaches and of previous meal planning systems. Section 3 describes the MARY system. Section 4 provides example runs of MARY. Finally, section 5 consists of conclusions and future work.

2 Background

Planning and design are two facets of the same problem. Design deals with the selection and spatial configuration of physical components and planning requires selection and temporal sequencing of actions [4]. Planning and design utilize constraints to limit the amount of possible sequences to consider. Preference knowledge can be applied to improve performance and/or select superior designs/plans.

AI research has spent a considerable amount of time investigating the planning/design problem. Research is generally classified into three types of problems. In novel planning, there is little to no prior knowledge available to solve the given problem other than applying analogous knowledge. Reactive planning is often used in real-time situations where a robot or monitoring system must modify the current plan based on changes in the environment. Routine design/planning is a type of problem in which the domain and task at hand are well-understood such that there is expert knowledge readily available. Meal planning would fall into this category [4,5].

Solutions for a routine form of planning can vary from simple operator selection based on goal/sub-goal decomposition, as found in early AI systems like STRIPS, NOAH, and ABSTRIPS, or via constraint propagation [6], routine design [5] and case-based reasoning [7]. Routine design based on plan decomposition uses an explicitly encoded hierarchy of component parts/plan steps in the domain. Each component/plan step contains one or more plans. These plans have concrete actions that when applied will generate a design of a component or generate the sequence of steps needed to accomplish the plan step. Each one of these plans has knowledge available to determine the likelihood of that plan's success. Failure handling and redesign knowledge can be applied to fix a plan which either violates constrains or fails to meet the overall goals.

---

[1] This system is named in honor of Mary Cupito, a colleague who passed away recently while battling cancer. It was her idea and inspiration that led to this research.

In case-based reasoning, a library of previous solutions, cases, is available. The reasoner, when given a new problem, examines the library for the most closely matching previous case. Upon retrieving this case, the reasoner must compare the differences between the current case and the retrieved case. To resolve any differences, a series of transformation rules are applied. Once all differences are accounted for, the new solution is attempted. If it works, the new solution is stored as another case and if not, either the retrieved case is further manipulated (repaired) or another case is retrieved. With case-based reasoning, the library grows over time as more and more solutions are obtained. Case-based reasoning is a common solution to routine planning/design problems. In fact, the earliest use of case-based reasoning was with the recipe generating system CHEF [8]. CHEF started with a collection of recipes and given new user specifications, used a previous recipe to generate a new one, adding to the library as successful new recipes were generated.

Several recent systems have tackled nutritious meal planning using other AI approaches. DIligenS applies fuzzy constraint satisfaction to reason over nutritional meal choices [9]. WitF (What's in the Fridge?) helps users decide about meal planning by combining case-based reasoning like CHEF with the WordNet [10] ontology to reason over food components. Other approaches to recipe recommendations use similarity measures based on ingredients [11] and user ratings [12]. Yet another approach is to use genetic algorithms to search for the closest matching meal to given dietary needs using rule-based assessment for a fitness function [13].

A variation of routine design through plan-step decomposition is to use plan step generation, plan step assessment, and plan step assembly. Plan steps are organized in a search space and generated by searching the space for those steps deemed suitable to the given problem. Plan step assessment is used with the generator to supply some initial indication of the plan step's utility. Plan step assembly selects one plan step to fulfill a given need and then reapplies plan step assessment so that the impact of selecting that plan step can be propagated across the other available plan steps. Meal Planner [4], uses this algorithm where individual plan steps are meal components (e.g., ham sandwich, Caesar salad, cheeseburger), each of which can be used to fulfill one or more meal components (e.g., dinner entrée, lunch entrée, dinner side dish).

Meal Planner evaluates all of the possible meal components against the user's daily nutritional goals and taste preferences. Next, a single meal component is selected to fulfill the most significant meal portion (e.g., dinner entrée). With the item selected, the impact of selecting this item is propagated by rescoring all remaining items. In Meal Planner, rescoring is based on eliminating items that could appear in two meal portions (for instance, a Caesar Salad could be in both a lunch and dinner, if selected for one, it is removed from the other) and on its nutritional impact. An item high in sodium may cause other items with high sodium to be re-evaluated as less desirable. This pattern of evaluation, selection and assembly repeats until all meal portions are filled or the daily meal has run out of calories available. Figure 1 illustrates the routine planning approach of plan-step generation, assessment and assembly with the more specific tasks as applied in Meal Planner provided in smaller font.

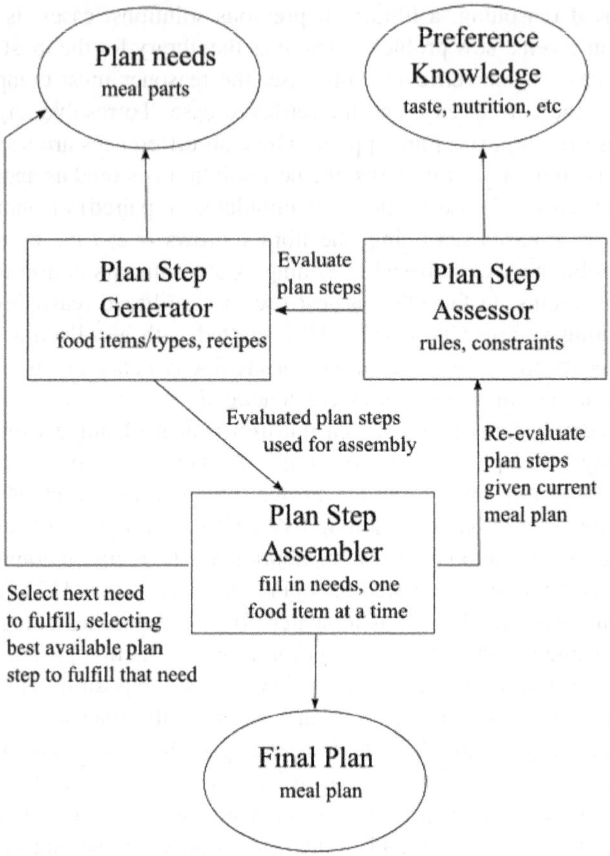

Fig. 1. Routine Design for Meal Planning

The research reported here describes MARY, a system based on Meal Planner. MARY goes beyond Meal Planner and the other systems cited above by factoring in the impacts of chemotherapy side effects on the planning process by taking into account the impact of nausea, ease of preparation and consumption on meal planning.

3 MARY

The Meal Planner system provides a set of meals to fulfill the user's needs for one or more days at a time. The system takes into account nutritional information to generate the meal components but nutrition is primarily limited to a calorie count. Additionally, in Meal Planner it is assumed that the user *wants* to eat. For a cancer patient undergoing chemotherapy, the desire to eat food is often eliminated because of nausea, sores in the mouth and throat, and changes to the taste buds. Yet, the patient must eat to keep up one's strength. The MARY system takes a broader view of meal planning by factoring in the notion that the user may not actually want to eat but must eat something. An overview of MARY's architecture is shown in figure 2.

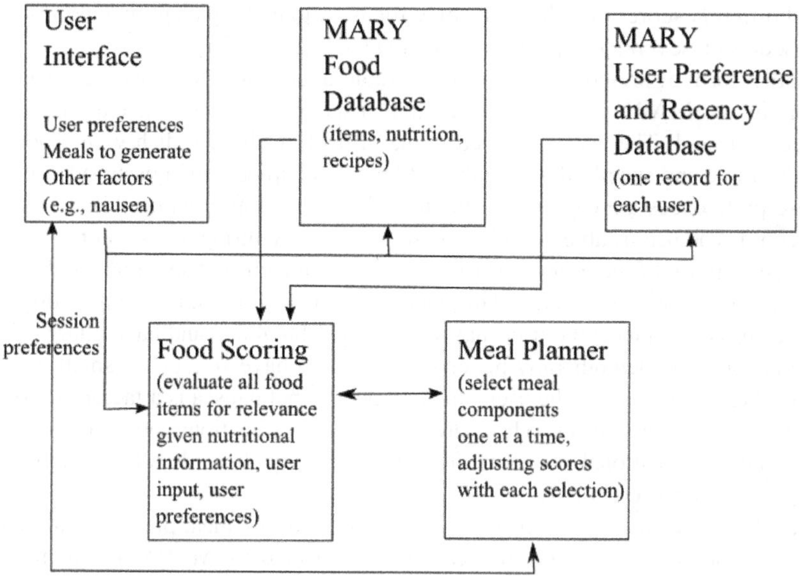

Fig. 2. MARY's Architecture

The main components of MARY are a user interface (discussed in section 3.1), two databases (discussed in section 3.2) and two AI modules (discussed in section 3.3). The user interface is a simple web-based portal to obtain user information including general preferences and meal-specific factors. The two databases are a food database containing a permanent collection of food items and recipes, used to form the basis of a generated meal, and a user database containing user preference information and recency information. The artificial intelligence components are a food scoring component and the planner to select, one at a time, meal components.

3.1 The User Interface

MARY's user interface serves three purposes. A new user must present information about his or her preferences and nutritional goals. The user fills out a series of on-line forms providing name, caloric intake goal, and scoring metrics. Metrics are used to adjust the food scoring algorithm. Metrics are the significance of nutrition (calorie counting, sodium, fat), to what extent taste and user preferences should be taken into account, and whether recently eaten food items should be discarded.

After specifying these initial values, the user next is able to tailor food preferences. All food items and recipes receive an initial score of 5. The user can adjust any individual score to indicate a greater like (up to 10) or dislike (down to 1) of that food item/recipe. The user can also specify a preference for food groups. Food groups include vegetable, fruit, beef, chicken, pasta, spicy food and tomato-based. As an example, a user may indicate that he or she likes spicy foods by increasing the default score for the spicy food group from 5 to 8. This saves the user from having to modify several different scores of individual spicy foods. A diabetic patient may need to lower the scores on several groups like pasta, bread, cereal and fruit.

Another form of preference is that of desirable/undesirable food pairings. This type of score allows MARY to promote or demote a second food item once an initial item has been selected. The user may specify that a beef or chicken dish with a potato is often desirable and specify that pizza should not be coupled with a pasta side dish. All of these preferences are captured in the user database, stored within this user's database record. The user is able to update these preferences during any session.

The second use of the user interface is to specify the current session's goals. These goals can simply be to generate a daily meal plan. However, the user can modify what is expected. For instance, the user may wish to skip breakfast and have a larger lunch or alternatively, a chemotherapy patient may wish to have five or six small meals in place of three large meals. The user can also specify that nausea is a factor to take into account for this day's meals. There are three nausea selections: none, nauseas, and nauseas ignore nutritional concerns. This last category discards all nutritional concerns when scoring food items.

The third use of the user interface is to interact with the meal plan as meal components are generated. As each new meal item is selected by MARY, the item is displayed in a drop-down box and daily nutritional values are updated and displayed. The box will display the top 10 rated food items for the given meal component (e.g., dinner entrée). The user has the ability to change the selected item to one of the other items listed in the drop-down box. As the user makes an adjustment, all other food items are rescored and new nutritional information is presented.

3.2 MARY's Databases

MARY utilizes two databases, a collection of food items/recipes and user information. The food item/recipe database contains one entry per food item/recipe. Recipes are not stored as actual recipes (step-by-step preparation and cooking instructions) but instead a list of ingredients.

Each food item/recipe is stored with nutritional information based on one serving: calories, amount of sodium, fat, fiber, protein, cholesterol and carbohydrates. The serving size is used to compute the item's impact on the daily nutritional count. For instance, an item whose serving size is 300 calories would account for 12% of a person's diet (assuming 2500 calories per day). If this item has considerably more than 12% of a person's daily sodium allowance (2400 milligrams), the item's score might be lowered if nutrition is a concern of the user. Food items/recipes also have scores of how easy they are to eat/prepare when nauseas. The ease of eating and preparing items were taken from various publications regarding chemotherapy side effects.

The food/recipe database also specifies categories of foods so that food items/recipes can be considered in groups. Food groups include vegetable, fruit, beef, chicken, pasta, spicy food and tomato-based, among others.

The second database consists of user records. This database contains one entry per user and stores all of that user's preferences. There are three forms of preferences. First is the set of scoring metrics, used by the scoring algorithm to weigh the importance of nutrition, the impact of nausea, the need for high caloric but easy-to-consume food, and recency. The metrics are filled in during the user's initial session with MARY but can be altered during any session if the user desires. For instance, on a bad day, the impact of nausea may be far more significant than a desire to limit recently eaten items.

The second set of information stored in the user database is recency information. Each time a meal plan is accepted, the items in that plan are time stamped and placed into the database. The score of any recently selected item will be decreased if the user has specified a preference against recently eaten items. The user can flush the recency entries, and for any session the user can modify the recency metric if the user decides that recency should or should not be taken into account.

The last set of information stored in the user database is the set of user specific food preferences, both for individual items and groups of items. Individual food/recipe or food group preferences allow the user to adjust the default score of 5 to another value. Once adjusted, the new score remains in the database unless updated. The user also specifies food combinations that the user would either desire or dislike. For each food item and/or food group, any other food items or groups can be listed as "compatible" or "incompatible".

3.3 MARY's Rules for Food Evaluation and Selection

Food items/recipes are generated with an initial score as recorded in the user database (the user-specified preference, or 5 for the default). Nutritional data for every food item/recipe is normalized based on the item's serving size versus the desired daily caloric intake. Once normalized, the initial preference score for each item is adjusted upward or downward based on fat, sodium, calories, and fiber (cholesterol, carbohydrates and protein are not currently factored into scoring). The amount of adjustment is predicated on the user-specified nutritional metric. That is, if the user values nutrition more, the adjustment is greater and if the user does not care about nutrition, the adjustment has less (or no) impact.

Scores are further adjusted if the user has indicated nausea. In such cases, scores for food items/recipes specified as easy to eat and prepare when nauseas will be increased while the scores of food items/recipes indicated as challenging to eat or prepare will be decreased. Recency is then taken into account. If the item is found on the recently eaten list, then the item's score is further decreased if recency is a concern of the user. Each of these adjustments is weighed based on the user's initial metrics for their importance combined with whether the user has indicated that he or she is nauseas for the day.

With food items/recipes now scored, the planner selects the most important meal component for the day (e.g., dinner entrée). The top choice is selected and the top 10 rated items are provided in a drop-down box. The user can override the item selected by MARY with any other item in the box.

Once an item has been selected for the first meal component, the scorer updates all remaining food item/recipe scores using two additional factors. First, given the user's preferences of compatible and incompatible foods, some food scores will be raised and others lowered. Additionally, nutritional scores are modified based on the amount of calories, sodium and fat existing in the selected item. These nutritional adjustments ensure that the user is held within his or her stated nutritional goals (e.g., 2500 calories and 2400 milligrams of sodium). The nutritional score adjustment is predicated on the user's desire for a nutritious meal. This can be overridden with a low metric for nutrition or by selecting the nauseas ignore nutritional concern option. The scoring algorithm is illustrated in figure 3.

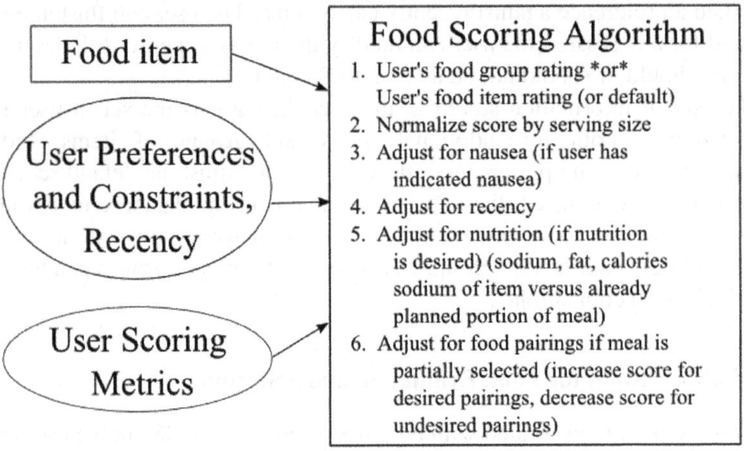

Fig. 3. MARY's Scoring Algorithm

With scores modified, MARY resumes with the planner, selecting the second most important meal component. Again, once an item is selected, scores are adjusted. The process repeats until either a full day's meals have been generated or the meal components selected have used up all of the daily calories, fat and/or sodium. If the planner is unable to complete the meals for the day, MARY lets the user know. When done, the user interface displays the selected meal and its nutritional content. If the user approves of the meal, the individual food items/recipes are noted in the user database record to indicate that these items have been consumed most recently.

4 Example Run Traces

To demonstrate the utility of MARY, this section offers a couple of examples. Assume that the user is a cancer patient who has good days and bad days with respect to chemotherapy side effects. As described in section 3.1, MARY has selections for normal meal selection, nauseas and nauseas with no consideration for nutrition. One might assume that on good days the user will operate MARY in normal mode. On a bad day, nausea mode will be selected. In this mode, MARY takes into account that the meal components should be limited to those that can be easily prepared and eaten. The user, for instance, might have specified a preference in such cases for a few food items like gelatin, broth, pudding, and herbal tea [2]. Any food items deemed too challenging to eat or prepare are downgraded. On the worst days, a chemotherapy patient would prefer not to eat at all. In such a case, MARY throws out any notion of obtaining nutritious meals and selects foods that the user has selected as preferably even when nauseas. In such a case, scores do not reflect nutrition at all. What follows are two examples, one in normal mode and one with nausea selected. The final mode, nauseas with no consideration for nutrition is omitted because this selection generates predictable results in that MARY generates meals consisting only of those items with high preference by the user with respect to a desire to eat when nauseas.

For these examples, assume the user has specified a strong desire to omit recently eaten items with a secondary desire of maintaining nutrition. Further, the user has specified a preference for fruit and items in the chicken group and a slight preference for spicy food while having a preference against oranges specifically and against items in the fish group. Other items are left at the default values. Food combination choices are also left at the default values.

For the first run, the user has selected no nausea. Dinner is selected as an entrée of stuffed peppers, a cherry tomato and cucumber salad, Tuscan vegetable soup and strawberries. For lunch, the first item would be stuffed peppers again but because of recency, this item is ignored and the top item selected is (meat-based) lasagna with a side order of avocado salad with apples and a fruit smoothie as a beverage. Finally, for breakfast, MARY generates a banana, but the user selects the second item oatmeal with a banana on the side. The top-rated beverage is cranberry juice. Orange juice was listed as a choice but the score was greatly downgraded because of the user's preference against oranges.

In the second example, the user selected nausea but is still interested in nutrition but without recently eaten items. MARY generates the following meals. For dinner, the top choice is grilled chicken breast and a baked potato. These are deemed acceptable in spite of nausea because of the user's preference for chicken when nauseas along with a desire for a pairing of chicken and baked potato. A second side of soup broth with chicken would be ranked highly but is demoted because it contains chicken, which now appears in the recency list. Instead, vegetable soup is generated as the top choice. Plain yogurt is selected for a dessert and water for a beverage. Lunch consists of a baked potato and an apple with a beverage of ginger ale. Notice the repetition of the baked potato in spite of recency. Here, MARY has little choice because most lunch items were rejected due to nausea, so recency wins out over nausea in this case. Breakfast consists of fruit cocktail and cottage cheese along with water.

5 Conclusions

Nutritional meal planning is an important part of every person's day. With a cancer patient undergoing chemotherapy, there are many added challenges including the impact of the chemotherapy medication on the patient and the need for the patient to maintain his or her strength. The MARY system combines a user interface, two database and two artificial intelligence components to generate a user's daily meal plan.

The selection of meal components combines nutrition, user preferences for individual and groups of food, recency, and the impact of chemotherapy side effects.

Work continues on MARY. MARY's limited number of food items is being expanded to include a greater variety. Currently, the system is tuned to a single user. The next step is to include multiple users in the user database. The scoring mechanism continues to be tweaked as experiments are performed. Additional nutritional elements will be included such as cholesterol and carbohydrates. Currently, recency is treated as a Boolean value, either an item has been eaten recently or it has not. Another change is to provide a range for recency such as one day, two days, within the past week. Finally, MARY is intended to be a public tool made available on the web. It is hoped that patient feedback will help improve other aspects of the system.

References

1. Warburg, O.: On the origin of cancer cells. Science 123, 309–314 (1956)
2. Rivadeneira, D.E., Evoy, D., Fahey, T.J., Lieberman, M.D., Daly, J.M.: Nutritional support of the cancer patient. CA: A Cancer Journal for Clinicians 48, 69–80 (1998)
3. Greene, D., Nail, L.M., Fieler, V.K., Dudgeon, D., Jones, L.S.: A comparison of patient-reported side effects among three chemotherapy regimens for breast cancer. Cancer Practice, Vol 2, 57–62 (1994)
4. Cox, M., Fox, R.: Nutritional Meal Planning Using a Generic Task Routine Decision Making Algorithm. In: The Proceedings of the 2000 Association of Management/Int'l. Assoc. of Management International Conference, vol. 18(3), pp. 107–112 (2000)
5. Brown, D.C., Chandrasekaran, B.: Expert systems for a class of mechanical design activity. In: Knowledge Engineering in Computer-aided Design, pp. 259–282 (1985)
6. Hendler, J., Tate, A., Drummond, M.: AI planning: Systems and techniques. AI Magazine 11(2), 61 (1990)
7. Riesbeck, C.K., Schank, R.C.: Inside case-based reasoning. Psychology Press (2013)
8. Hammond, K.J.: CHEF: A Model of Case-based Planning. In: Proceedings of the Fifth National Conference on Artificial Intelligence, pp. 267–271 (1986)
9. Tom, M.: Computational intelligence using Fuzzy Multicriteria Decision Making for DIligenS: Dietary Intelligence System. In: Fuzzy Systems (FUZZ-IEEE), pp. 1–7 (2012)
10. Zhang, Q., Hu, R., MacNamee, B., Delany, S.J.: Back to the future: Knowledge light case base cookery. In: ECCBR 2008, The 9th European Conference on Case-Based Reasoning, Workshop Proceedings, pp. 239–248 (2008)
11. van Pinxteren, Y., Geleijnse, G., Kamsteeg, P.: Deriving a recipe similarity measure for recommending healthful meals. In: Proceedings of the 16th Int'l. Conference on Intelligent User Interfaces, pp. 105–114 (2011)
12. Freyne, J., Berkovsky, S.: Intelligent food planning: personalized recipe recommendation. In: Proceedings of the 15th Int'l. Conference on Intelligent User Interfaces. ACM (2010)
13. Gaál, B., Vassányi, I., Kozmann, G.: Application of Artificial Intelligence for Weekly Dietary Menu Planning. In: Vaidya, S., Jain, L.C., Yoshida, H. (eds.) Advanced Computational Intelligence Paradigms in Healthcare. SCI, vol. 65, pp. 27–48. Springer, Heidelberg (2007)

E2TS: Energy Efficient Time Synchronization Technique in Large Scale Wireless Sensor Network

K. Nagarathna[1] and J.D. Mallapur[2]

[1] Visvesvaraya Technological University,
Belagavi, Karnataka 590018, Republic of India
nagarathnavtu@gmail.com
[2] Dept. of Electronics & Communication Engg. BEC
Basaveshwar Engineering College, Bagalkot, Republic of India

Abstract. Time Synchronization is one of the critical successful factor for performing optimal data dissemination process in wireless sensor network. The literature archives shows that there are two types of research attempts e.g. i) first types are very standard and considered as benchmarks, and ii) second types of research work are enhancement work of prior types. However, the levels of enhancement have not witnessed any robust benchmarks in time synchronization principle most recently. The proposed investigation have identified that there are few studies that have evaluated the impact of time synchronization algorithms on energy efficiency. Hence, we present a novel algorithm E^2TS-Energy Efficient Time Synchronization that lay emphasis on modelling a control message for performing time synchronization using node-to-node interaction by minimizing the transmission latency and tuning up the difference between the local clocks. With low storage complexity, E^2TS also ensures that energy required for performing time synchronization is very minimal. Designed in Matlab, the outcome of E^2TS is compared with available benchmarks to find more energy efficient with considerably less error occurrence with increasing number of hops. Hence, the E^2TS hold suitable to perform time synchronization for large wireless sensor network.

Keywords: Clock Drift, Time Synchronization, Clock Synchronization, Wireless Sensor Network.

1 Introduction

With the increasing demands of monitoring the data and events, sensors are increasingly deployed in the area of wireless communication system. In an area, where it is infeasible for the human to collect the data over a larger period of time in unfriendly environmental situation, wireless sensor networks are adopted. It comprises of various miniature size of electronic device called as sensor node that can perceive the physical and environmental data most effectively [1]. Some of the applications of wireless sensor network in this regards are like habitat monitoring, environmental monitoring, industrial appliances monitoring. Although some applications in wireless sensor network like rainfall detection, precipitation detection etc doesn't call for much accuracy,

R. Silhavy et al. (eds.), *Artificial Intelligence Perspectives and Applications*,
Advances in Intelligent Systems and Computing 347, DOI: 10.1007/978-3-319-18476-0_23

but certain other applications like industrial appliances and healthcare applications calls for highest degree of accurate and precise information. Such precision in information is often affected if the sensor network works on large size and amount of data collected are huge. One of the way to find such precision is the uniqueness and freshness of the data being received by the aggregator node. Very often such captured data and time stamped by each nodes at the time of forwarding it to the other nodes (aggregator node or sink). The problems occur when irrelevant time stamping is done owing to faulty hardware clocks in the sensor nodes. It is also quite possible that hardware clocks differ in oscillation frequency from one to another node and in such situation, the defective time stamping is done. The adverse effect of this is overhead and data redundancy leading to significant degradation in performance of data aggregation in wireless sensor network. Hence, the method by which the drift in local clocks is rectified (or minimized) is called as time (or clock) synchronization [2]. An effective time-synchronization leads to process only the unique data and thereby highest packet delivery ratio, throughput, and lowered transmission delay. However, the area of time synchronization is shrouded by another prominent factor called as energy issues, which means energy required to performed time synchronization. Although various studies in past have focused on energy issues in wireless sensor network but very few studies have witnessed it for the reason of time synchronization.

Therefore, the proposed study introduces a technique by which the system can jointly address the time synchronization and energy efficiency. The primary objective of this paper is i) to discuss modelling of logical clock component, ii) to introduce the formulation of control message for time synchronization, iii) to introduce an energy efficient error minimization technique for large scale wireless sensor network. Finally, the outcomes are benchmarked with available standards of time synchronization. Section 2 discusses about the existing research work. Section 3 discusses about the proposed system and illustrates it is based on its architecture. Section 4 discusses about the research methodology followed by Simulation & Results discussion on Section 5. Finally Section 6 makes some concluding remarks.

2 Related Work

Shahzad et al. [3] have presented a framework that can reduce errors in time synchronization while ensuring cost effectiveness too. The authors have used the framework for interchanging the conventional Reference Broadcast Synchronization and Time Sync Protocol for Sensor Network using threshold-based approach. The outcomes were evaluated using throughput and energy mainly. The protocol gave better switching facilities but bears the similar flaws of errors in conventional time synchronization techniques. Bheemidi and Sridhar [4] have applied a technique that continuously screens the clock drift and automatically rectify the errors in local clock difference. Albu et al. [5] have performed a simulation study to investigate time synchronization technique on IEEE 1588 standard. Ikram et al. [6] have presented a proposal for minimizing errors in the radio-clocks using real-time sensors. Tang et al. [7] have presented a protocol to accomplish higher energy efficiency on MICAz nodes. The protocol is designed based on duty cycle of MAC on multiple channels. Sanchez et al.

[8] have presented a technique of time synchronization on body area network. The study performs prediction of skew factor in clock as well as cross-validates the errors in predictions. Shanti and Sahoo [9] have proposed a time slot assignment technique using spanning tree. The authors have performed evaluation of clock offset and performed simulation study using flooding technique. Chauhan and Awasthi [10] have proposed a clock synchronization protocol where the technique is used for rectifying the skew factor in the clock. Rao et al. [11] have also investigated on the energy efficiency factor while working on the time-synchronization. The authors have performed the simulation study using where cumulative energy is recorded. Simon [12] has investigated ring topology and the issues associated with the time synchronization problems. The authors have experimented on real-time sensor nodes where the outcomes have been evaluated considering the synchronization errors. Lenzen et al. [13] have proposed a scheme for scattering the information pertaining to local clocks to neighboring nodes using time-stamping on the MAC layer. Another unique study have been proposed by Li et al. [14] who have introduced a time synchronization protocol that self-adjust itself based on diffusion approach. Vonlopvisut et al. [15] have performed amendment to the conventional time synchronization technique and tested in for underwater mobile network. Zhao et al. [16] have introduced a time synchronization technique while investigating the election mechanism of cluster head in wireless sensor network. Most recently, Leva et al. [17] have proposed a scheme to compensate the drift and skew factor to enhance the time synchronization technique.

3 Proposed System

The prime aim of the proposed study is to formulate a condition for optimal decision based routing by adopting the principle of large scale and multi-hop wireless sensor network. Our prior work [18] has already reviewed various studies in time synchronization and has discussed about the research challenges. This paper has presented a novel technique called as Energy-Efficient Time Synchronization (E^2TS) that is focused majorly on large scale networks. The proposed system posses following contribution:

- To design a simulation test-bed for large scale wireless sensor network considering member node, aggregator node, and base station and empirically model a logical clock with all possible parameters of local clocks.
- To model a decision-based routing by designing a control message with less storage complexity. Such modelling also permits node-to-node compensation of clock drift for large network. The control messages are used reference node to receive, transmit, and update the time factor to balance the difference between clock times between the two sensors.
- To apply a standard radio-energy model [19] for evaluating the energy consumption in the nodes owing to time synchronization. The system also considers time stamping of message on MAC layer for both source and destination sensor nodes.

The proposed method accomplishes minimization of errors in time synchronization by interchanging the control message among the source and destination nodes via intermediate nodes (multihop). The design principles of the message does the following job e.g. i) it will request for finding reference node and index it, ii) it will request for performing time-synchronization process, and iii) it will send response for the successful receiving of time synchronization information from other nodes in the network.

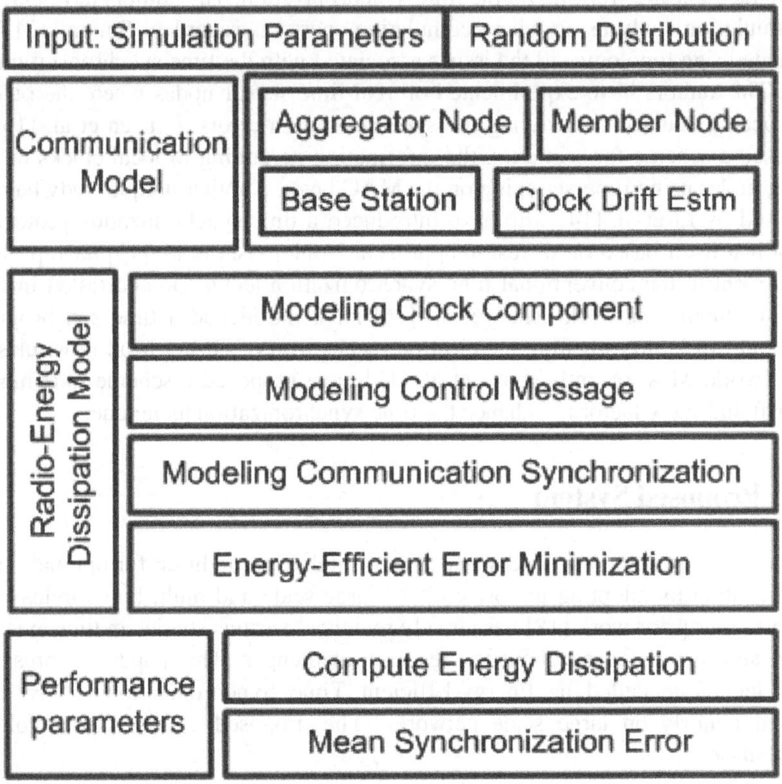

Fig. 1. Schema of E^2TS Design

4 Research Methodology

The design principle of the proposed model targets to accomplish the highest precision for time synchronization in large scale wireless sensor network with an assurity of energy efficiency for any cluster-based topologies. The essential components of the design principles includes following components:

1. **Modelling Clock Component**: All the sensor mote is considered to possess a local clocks with frequency ω_t (t), principle time t, principle frequency ω_o, clock frequency offset as $\Delta\omega$, clock frequency drift as σ_ω, and arbitrary clock error process as err_ω (t).

Considering that local clocks don't posses any extrinsic hardware issues, the cumulative oscillator frequency can be now mathematically represented as

$$\omega_t(t) = \omega_o + \Delta\omega + \sigma_\omega + err_\omega(t) \quad (1)$$

Considering at an ith event, the test-observation time for the hardware clock O_i (t) can be now represented as,

$$O_i(t) = O_i(0) + \frac{1}{\omega_o}\int_0^t \omega_i(t).dt \quad (2)$$

$$O_i(t) = O_i(0) + \frac{1}{\omega_o}\int_0^t (\omega_o + \Delta\omega + \sigma_\omega + err_\omega(t)).dt \quad (3)$$

Eq.(2) represents the oscillator frequency with respect to preliminary reference time setting t value as 0, whereas the second component of eq.(2) shows the consecutive time screened. Substitution of ω_t (t) from eq. (1) in eq. (2) finally results in eq. (3). For easiness in computation, the proposed design considers zero value for error in clock process and drift in clock frequency. Hence, eq. (3) now becomes

$$O_i(t) = O_i(0) + (1 + (\frac{\Delta\omega}{\omega_o})).t \quad (4)$$

Eq. (4) can be further normalized as,

$$O_i(t) = C_i^o + C_i^s.t \quad (5)$$

In the above eq. (4), the first component is clock offset and second component is clock skew. The system considers *r* as optimal rate of clock drift under the condition (1-r) $\leq C^s_i \leq$ (1+r).

2. **Modelling Control Message**: The design principle of the proposed study also considers modelling a particular control message. The control message is formulated by involving parameters e.g. message ID, source node ID, intermediate hop ID, destination node ID, time factors of local clocks (time for transmitting, receiving, time instant for transmitting the reply message and reference time). The time factors of the proposed system is designed based on the time-stamping mechanism followed in MAC layer in both source and destination node. The method also introduces a novelty by incorporating a technique to explore and balance the transmission delay that is frequent in large scale wireless sensor network. This mechanism will also assist in optimally reducing the errors in time synchronization for the proposed method.
3. **Modelling Communication Synchronization**: The proposed system considers multi-hop wireless sensor network. The reference node receives the request message from the source node via intermediate nodes. If the intermediate node is not found to be synchronized, it forwards the request message to its neighbor nodes in a periodic manner. The process continuous till reference node doesn't receive the request message. The request message also contains the time-stamp information.

The process also maintains lesser amount of time complexity as very few amount of request pertaining to time synchronization spreads in the network. The prime reason behind this principle is that in a dense and larger wireless sensor network, the nodes are in multi-hop mode of communication for which reason the synchronization happens quite faster among the proximity (or neighboring) nodes. The system also ensures lower storage complexity as sensors maintains memory for only ID of reference node and client ID, with timestamps.

4. **Energy Efficient Error Minimization Technique**: Every node checks the new synchronization message in the request message. Based on the receiving of the request message, the reference node as well as aggregator node performs transmitting the reply message on that direction. Transmission delay is evaluated after the reference node forwards the reply. Clock drift, skew, and offset are estimated in this process and is subjected to condition for minimizing errors. If the updated value of the skew is found to be either lower or greater than unit value, the source node will ask for timeslot information in their request message to be forwarded to reference node. In case of difference of global time with local time, a sensor evaluates its local time and checks its equivalence with the global time. If the difference is found to be significantly higher, the node will begin forwarding the request for time synchronization or else it is not required thereby saving computational resources and maintaining ongoing data aggregation process. Such minimization of error drastically controls the energy dissipation which the node requires to do in conventional system. One of the interesting way, the energy efficiency is ensured was by selection of reference node, which should have highest residual energy to perform time synchronization in large scale wireless sensor network. Once the reference node is about to depletes off its threshold-point energy (T_h > 0Joules), the selection of the new reference node is carried out. Once the new reference node is selected, the old reference node (that still has certain amount of residual energy) is treated as common node that can assist in further time synchronization process. Hence, network lifetime is increased. In order to further minimize the load of processing time synchronization, the broadcasting frequency is highly minimized to save unwanted forwarding of request for time synchronization. Once the node is time synchronized, it is capable enough to reply the request for other nodes.

5 Simulation and Result Discussion

The proposed study is designed in Matlab considering 25-1000 nodes considering 1000 rounds of iterations. The nodes are distributed in random fashion in simulation area of 1500 x 1200 m^2 with 0.5 Joules of initialized energy and considered to have transmission delay of 5-150 milliseconds. The design principles of the wireless environment were done based on MAC layer of 802.15 standards.

The outcome of the proposed study is evaluated using synchronization errors in milliseconds. For better benchmarking process, the proposed system have been compared with the standard framework of time synchronization proposed by Elson et al. [20], Ganeriwal [21], and Maroti [22]. From the section 2, it was seen that there are abundant research work on time synchronization algorithms being proposed, but very few are found to be benchmarked out. Our findings from Google scholar citation shows that all the existing studies have very irregular and very low evidence of comparative analysis and moreover, we select Elson et al. [20], Ganeriwal [21], and Maroti [22] as it was found to be technically adopted by various researchers in last decade making the three studies are most standard research milestone till date. Table 1 shows the evidence behind this rationale.

Table 1. Extent of citation for the considered benchmarks

Available Benchmarks	Cited by
Elson et al. [20]	2396
Ganeriwal [21]	1874
Maroti [22]	1758

Figure 2 exhibits the mean error of time synchronization for proposed system as well as available benchmarks of time synchronization. It can be seen that proposed system has exponentially lower mean errors of synchronization as compared to the work of Elson [20](Error=30.216ms) and Ganeriwal [21](Error=21.621ms). However, proposed system also yields improved outcome as compared to Maroti [22] (Error=0.596ms).

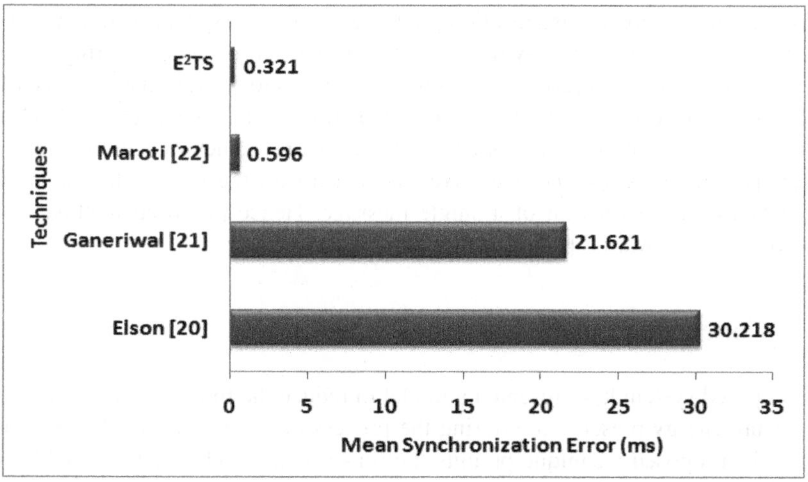

Fig. 2. Performance Comparative Analysis of Mean Synchronization Error

It was also necessary to investigate the behaviour of the proposed system with increasing number of hops. Figure 3 shows the simulation study that was carried out to analyze the effect of increasing number of connection (or hops) to the energy factor. The outcome shows that Elson [20] and Ganeriwal [21] model have pretty high error in time synchronization as compared to Maroti [22]. However, our proposed system has even outperformed Maroti [22] model in observation carried out where increasing number of simulation iterations were studied closely.

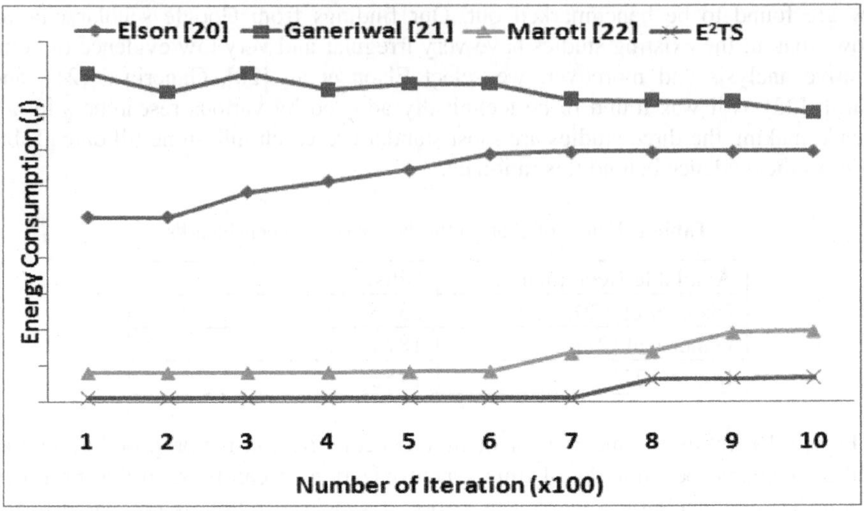

Fig. 3. Performance Analysis of Energy Dissipation with Iteration

Basically, the reasons of energy consumptions in the wireless sensor network are many, but the proposed system only emphasizes the time synchronization as one of the prominent reason of energy dissipation from the sensor nodes. During the analysis, it is found that Elson [20] model uses all the nodes to forward the message of time synchronization leading to higher energy depletion. Whereas Ganeriwal [21] model uses only two set of control message to forward time synchronization information. Maroti [22] model uses one exclusive message to do the same job, and proposed model takes only a fraction of a single message. Hence proposed method is better revised version of Maroti [22] model. E^2TS

6 Conclusion

The proposed system has presented a model to reduce the time synchronization errors and ensure energy preservation during the process in large scale wireless sensor network. The proposed technique permits the sensor node to choose the suitable condition to transmit control messages for requesting time synchronization. The design of such minimizing the drifts is based on standard deviation as well as skew factor from local to global clock. The outcome of the proposed system is compared with the most standard models of time synchronization till date to find proposed system has better controls over error minimization and its design principle optimally controls the amount of message to be transmitted for performing time synchronization to conserve enough amount of energy in the reference node.

References

[1] Fraden, J.: Handbook of Modern Sensors: Physics, Designs, and Applications. Technology & Engineering, p. 678. Springer Science & Business Media (2010)

[2] Hu, F., Cao, X.: Wireless Sensor Networks: Principles and Practice. Technology & Engineering, p. 531. CRC Press (2010)

[3] Shahzad, K., Ali, A., Gohar, N.D.: ETSP: An Energy-efficient Time Synchronization Protocol for Wireless Sensor Networks. In: IEEE-22nd International Conference on Advanced Information Networking and Applications – Workshops (2008)

[4] Bheemidi, D.R., Sridhar, N.: A Wrapper-Based Approach to Sustained Time Synchronization in Wireless Sensor Networks. In: Proceedings of 17th International Conference on Computer Communications and Networks, ICCCN 2008 (2008)

[5] Albu, R., Labit, Y., Gayraud, T., Berthou, P.: An Energy-efficient Clock Synchronization Protocol for Wireless Sensor Networks. In: Wireless Days (WD), pp. 1–5 (2010)

[6] Ikram, W., Stoianov, I., Thornhill, N.F.: Towards a Radio-Controlled Time Synchronized Wireless Sensor Network: A Work in-Progress Paper. IEEE EFTA Conference, Bilbao, Spain (2010)

[7] Tang, L., Sun, Y., Gurewitz, O., Johnson, D.B.: EM-MAC: A Dynamic Multichannel Energy-Efficient MAC Protocol for Wireless Sensor Networks. In: International Symposium on Mobile Ad Hoc Networking and Computing, p. 23 (2011)

[8] Sanchez, D.S., Alonso, L., Angelidis, P., Verikoukis, C.: Secure Precise Clock Synchronization for Interconnected Body Area Networks. EURASIP Journal on Wireless Communications and Networking. Article ID 797931, 14 (2011)

[9] Shanti, C., Sahoo, A.: TREEFP: A TDMA-based Reliable and Energy Efficient Flooding Protocol for WSNs. In: IEEE International Symposium on World of Wireless, Mobile and Multimedia Networks, pp. 1–7 (2011)

[10] Chauhan, S., Awasthi, L.K.: Adaptive Time Synchronization for Homogeneous WSNs. Intech Open-International Journal of Radio Frequency Identification and Wireless Sensor Networks (2011)

[11] Rao, A.S., Gubbi, J., Ngo, T.: Energy Efficient Time Synchronization in WSN for Critical Infrastructure Monitoring, pp. 314–323. Springer (2011)

[12] Simon, G.: Efficient time-synchronization in ring-topology wireless sensor networks. In: IEEE International Instrumentation and Measurement Technology Conference (I2MTC), pp. 958–962 (2012)

[13] Lenzen, C., Sommer, P., Wattenhofer, R.: PulseSync: An Efficient and Scalable Clock Synchronization Protocol. Institute of Electrical and Electronics Engineers (2014)

[14] Li, M., Zheng, G., Li, J.: Clock Self-Synchronization Protocol based on Distributed Diffusion for Wireless Sensor Networks. International Journal of Future Generation Communication and Networking 7(5), 11–22 (2014)

[15] Vonlopvisut, P., Annur, R., Wuttisittikulkij, L., Chridchoo, N.: Enhancement of MU-Sync: A Time Synchronization Protocol for Underwater Mobile Networks. In: The 29th International Technical Conference on Circuit/Systems Computers and Communications (2014)

[16] Zhao, B., Li, D., Yang, D., Zhao, J.: A Low-power Cluster-based Time Synchronization Algorithm for Wireless Sensor Networks. Journal of Computational Information Systems 10(14), 6323–6330 (2014)

[17] Leva, A., Terraneo, F.: High-precision synchronisation in Wireless Sensor Networks with no tuning in the field. In: The International Federation of Automatic Control, Cape Town, South Africa (2014)

[18] Nagarathna, K., Mallapur, J.D.: Article: An Investigational Analysis of Different Approaches and Techniques for Time Synchronization in Wireless Sensor Network. International Journal of Computer Applications 103(5), 18–28 (2014)
[19] Emary, I.M.M.E., Ramakrishna, S.: Wireless Sensor Networks: From Theory to Applications. Computers, p. 799. CRC Press (2013)
[20] Elson, J., Girod, L., Estrin, D.: Fine-grained network time synchronization using reference broadcasts. ACM SIGOPS Oper. Syst. Rev. 36(SI), 147–163 (2002)
[21] Ganeriwal, S., Kumar, R., Srivastava, M.B.: Timing-sync protocol for sensor networks. In: Proc. 1st Int. Conf. SenSys, pp. 138–149 (2003)
[22] Maróti, M., Kusy, B., Simon, G., Lédeczi, Á.: The flooding time synchronization protocol. In: Proc. 2nd Int. Conf. SenSys, pp. 39–49 (2004)

A Brief Survey on Event Prediction Methods in Time Series

Soheila Mehrmolaei[1] and Mohammad Reza Keyvanpourr[2]

[1] Science and Research Branch, Islamic Azad University, Qazvin, Iran
s.mehrmolaei@qiau.ac.ir
[2] Alzahra University, Tehran, Iran
keyvanpour@alzahra.ac.ir

Abstract. Time series mining is a new area of research in temporal data bases. Hitherto various methods have been presented for time series mining which the most of an existing works in different applied areas have been focused on event prediction. Event prediction is one of the main goals of time series mining which can play an effective role for appropriate decision making in different applied areas. Due to the variety and plenty of event prediction methods in time series and lack of a proper context for their systematic introduction, in this paper, a classification is proposed for event prediction methods in time series. Also, event prediction methods in time series are evaluated based on the proposed classification by some proposed measures. Using the proposed classification can be beneficial in selecting the appropriate method and can play an effective role in the analysis of event prediction methods in different application domains.

Keywords: Time series, Time series mining, Event prediction, Methods.

1 Introduction

Mining large temporal data sets is an active area of research. Time series is a kind of temporal data and time series mining is the process of mining large data sets with time dimension for the purpose of knowledge identification and discovery [1]. It has also provided a research opportunity for detecting and predicting event-based knowledge [2]. Events can be defined as real world occurrences that unfold over space and time [3]. Event E_t is an existence which occurs with effective and significant changes in time series values at time t. In a time series, the definition of an event depends on the objective of application. For example, rain, storms, profit, and loss of stock are the events in the real world. The purpose of time series mining is classified into four groups: prediction, description, explanation and control [4]. Event prediction which is one of the main tasks in time series mining is an important research topic. Time series mining for event prediction is used in different application areas of medicine [5], seismology [6, 7], astronomy [8], finance [9, 10], crime and criminal investigations [11], weather [12, 13], etc. Thus, event prediction has always an important role in

R. Silhavy et al. (eds.), *Artificial Intelligence Perspectives and Applications*,
Advances in Intelligent Systems and Computing 347, DOI: 10.1007/978-3-319-18476-0_24

people's life and helps them make apt decisions in different of application areas. Beside, in the real world, time series behavior is always nonlinear. Thus event prediction in time series is problem. In recent years, various time series mining methods for event prediction have been proposed in different of application areas. Study of related researches is descriptive variety and multiplicity of event prediction methods in time series. Time series mining for event prediction is always required to employ an efficient method in this field. Hence, it is necessary to propose a general classification of the existing methods for event prediction in time series. This study attempted to represent a general collection and evaluation for these methods. The rest of the paper is organized as follows: Section 2 defines event prediction in time series. In Section 3, event prediction methods in time series are classified based on the input type of behavior. Evaluation of event prediction methods in time series according to this classification is presented in Section 4. Section 5 presents the conclusion.

2 Event Prediction in Time Series

Recently, the increasing use of temporal data, in particular time series data, has initiated various research and development attempts in the field of data mining. Time series is an important class of temporal data objects, which can be easily obtained from scientific and financial applications, daily temperature, weekly sales totals, and prices of mutual funds and stocks. A time series is a collection of observations which is made chronologically. The nature of time series data is as follows: large in data size, high dimensionality and necessity of continuous updating [14]. Mining is the final goal for discovering hidden information or knowledge from data. Time series mining is knowledge extraction from the data with time dimension. Knowledge discovery in data set is rapidly in progress and is essential for practical, social and saving fields. Knowledge discovery is the process of identification of reputing, novel, understandable and inherently useful patterns in data [15]. Extracted knowledge can be event. Event is an existence which occurs with important and effective changes in time series values at point *t*. An earthquake is defined as an event in the seismic time series in real world [7]. Events can be instantaneous or have duration. As mentioned earlier, one of the main goals of time series mining is prediction, which is achieved by combining some techniques such as machine learning, statistics, database and etc. Event prediction is a form of data analysis that can be used to extract models and predict future data trends. Event prediction problems are very similar to time series prediction problems [16]. In the last decade, different methods of learning have been used for Predicting of meaningful events in time series, the most common of which are decision trees, artificial neural networks, and support vector machines briefly introduced below. Decision tree induction is free from parametric assumptions and generates a reasonable tree by progressively selecting attributes to branch the tree. Decision tree techniques have been already shown to be interpretable, efficient, problem independent and able to treat large scale applications [9]. However, they are also recognized as highly unstable classifiers in terms of minor perturbations in training data. Fuzzy logic provides advantages in handling these variances owing to the elasticity of fuzzy sets formalism. In Fig. 1, fuzzy decision tree is shown for price prediction in stock market. It is obvious that the three linguistic can be described as Low, med and high. A decision tree is a flow chart like structure, in which each node represents a test on an attribute, each branch represents an outcome of the test and leaf nodes show a classification of an instance.

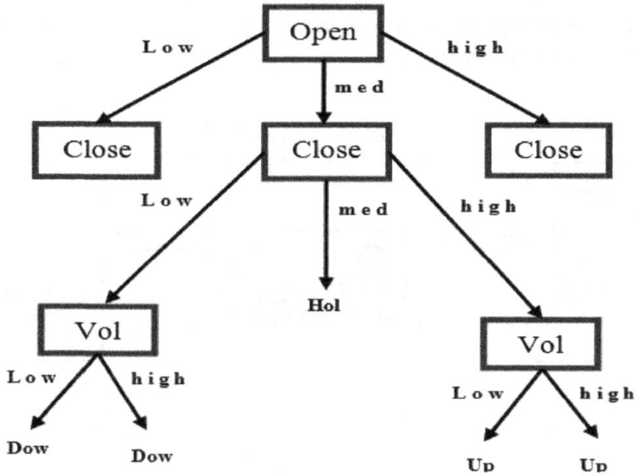

Fig. 1. Fuzzy decision tree for price prediction in stock market

Artificial neural networks (ANNs) are flexible computing frameworks for modeling a broad range of nonlinear problems. One significant advantage of ANN models is their universal approximate nature that can approximate a large class of functions with a high degree of accuracy. ANNs are a multilayer structure widely applied in forecasting. They usually consist of an input layer, an output layer and one or more intervening layers also referred to as hidden layers. The hidden layers can capture the nonlinear relationship between variables. Support vector machines (SVMs) are promising methods for the prediction of financial time series. SVMs use a linear model to implement nonlinear class boundaries through some nonlinear mapping of the input vectors x into the high dimensional feature space. A linear model constructed in the new space can represent a nonlinear decision boundary in the original space. In the new space, an optimal separating hyper-plane is constructed. Thus, SVMs are known as the algorithm that finds a special kind of linear model, the maximum margin hyper plane.

2.1 General Architecture for Event Prediction in Time Series

In general, event prediction in time series involves four main tasks: pre-processing, model derivation, time series prediction and event detection. They form the general architecture layers for event prediction systems in time series. The nature of time series data includes large data size, high dimensionality and strong correlation among features. Pre-processing layer is used to get high quality time series data or select a sub set of features effective in the prediction accuracy. The two major components of the pre-processing layer are feature extraction and class labeling. In model derivation layer, the final model based on type of input behavior and application area is extracted for event prediction in time series. Past values of time series is used for the model derivation. In the layer of time series prediction, future values are forecasted based on the derived model. Finally, in the last layer, events are detected based on the predicted time series and available patterns. The general architecture for event prediction process in time series is shown in Fig. 2.

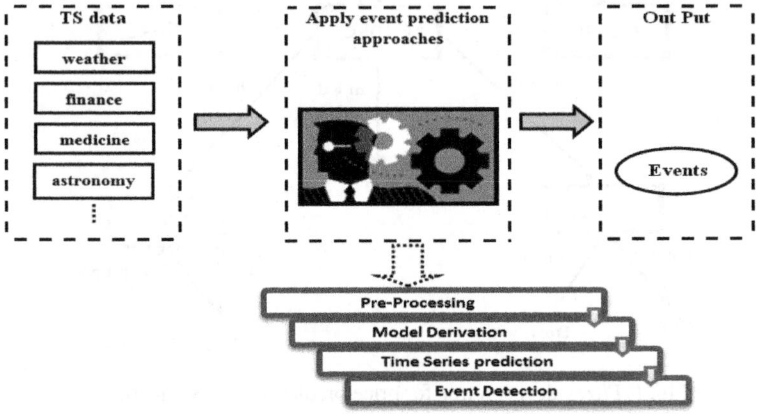

Fig. 2. General architecture for event prediction process in time series

3 Proposed Classification for Event Prediction in Time Series

There are various time series mining methods for event prediction in different applied domains. Review of the related literature is demonstrates the plurality and variety of time series mining methods for event prediction. Consequently, their classification and evaluation are difficult. In this section of the paper, a classification is proposed for event prediction methods in time series. In our point of view, proposed classification basis of the methods in time series is the categorization of type input behavior, which can be linear or nonlinear. Type input behavior can be determined by statistical analysis and using auto correlation function. Hence, the proposed classification categorized the methods into statistical learning methods and machine learning methods. Details will be discussed in this context.

3.1 Statistical Learning Methods

Statistical learning methods are widely used for event prediction in time series in the conditions where input has a linear behavior. In this sub-section, methods are divided into subgroups: seasonal and non-seasonal. Each sub-group is briefly described below.

Seasonal: Many business and economic time series exhibit seasonal and trend variations. Seasonality is a periodic and recurrent pattern caused by factors such as weather, holidays, repeating promotions, and the behavior of economic agents. Statistical learning methods with seasonal components such as seasonal autoregressive integrated moving average are one of the most popular methods which have been widely used over the past three decades and are called SARIMA. SARIMA methods assume that future values of a time series have a linear relationship with the current and past values as well as white noise; therefore, approximations by SARIMA methods may not be adequate for complex nonlinear problems. Forecasting airport passenger traffic has been done by SARIMA methods [17]. The popularity of SARIMA is related to its statistical properties as well as the well-known Box Jenkins methodology in the model building process. Many studies have examined the performance of SARIMA method for predicting air traffic flows [18].

Non Seasonal: Autoregressive is the most common method of linear behavior modeling of time series in statistical learning methods without seasonal component for prediction. In this method, output variable linearly depends on the previous values [19]. Autoregressive integrated moving average method is commonly the method in statistical learning group for the analysis of time series and is called ARIMA. Basically, an ARIMA consists of three parts: an autoregressive (AR) process, a moving average (MA) process and integrated (I) process. In time series forecasting, the method of autoregressive integrated moving average, sometimes called Box Jenkins model, most widely used method for predicting future value. Future value of a variable is calculated by a linear combination of past values and errors by Equation 1:

$$y_t = \theta_0 + \varphi_1 y_{t-1} + \dots + \varphi_p y_{t-p} + \varepsilon_t - \theta_1 \varepsilon_{t-1} - \dots - \theta_q \varepsilon_{t-q} \tag{1}$$

Where y_t is real value at time of t, θ and φ are model coefficients, P and q are integer numbers called order of the autoregressive and moving average, respectively, and ε_t is error. Actually, an ARIMA is the extension of ARMA time series by differencing them with the order of d to ensure the stationary of the time series. Use of integrated process can help remove the Non-stationary section of the time series data and be usually combined with other process. MA is a linear regression of the current value of the series against the current and previous random shocks. ARIMA and SARIMA methods are performed for water level forecasting [13]. Also, many studies have examined the performance of ARIMA method for predicting air traffic flows [18].

3.2 Machine Learning Methods

In the real world, time series behavior is always nonlinear; thus, using statistical learning methods does not have acceptable accuracy for time series analysis. Due to the linearity, statistical learning methods cannot capture any nonlinearity. Linear methods are not always suitable for complex real world problems. Therefore, the new methods of time series mining such as machine learning methods are used to model the behavior of nonlinear data for prediction of significant events. In this study, machine learning methods are investigated from two different aspects. The first aspect is type input data set and the second aspect is type applied approach for prediction.

Categorization Based on Type Input Data Set: According to a first aspect, the related researches are divided into two sub-groups: univariate data set and multivariate data set. Each sub-group is briefly described below. Target classification according to first aspect is shown in Fig. 3.

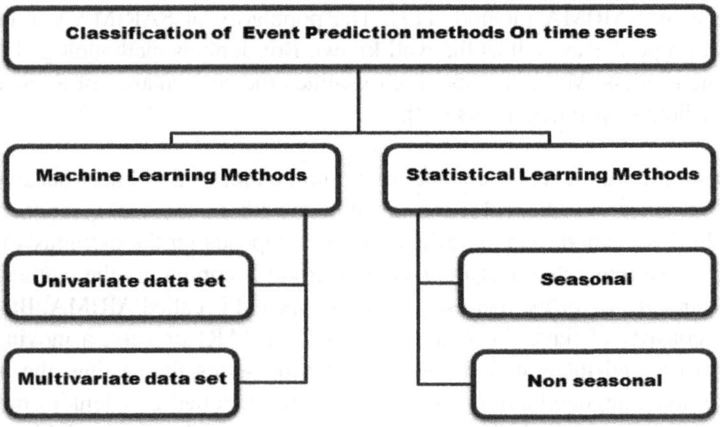

Fig. 3. Classification of event predition methods in time series based on first aspect

Univariate Data Set: Univariate time series refers to a time-series that consists of single observations recorded sequentially through time, e.g. the monthly unemployment rate. There are many researches on univariate time series data for event prediction. SVM is used to predict future direction of stock price index on univariate time series data [20]. Change directions are forecasted for financial univariate time series data using HMM model [21]. Semi-supervised learning with data calibration has been done for long-term forecasting [22]. Forecasting of sales has been done through univariate time series data clustering [23]. Exchange rate has been forecasted by using support vector machines [24]. Electric load forecasting has been done using support vector machines optimized by genetic algorithm [25]. Also, load forecasting has been done by using fixed-size least squares support vector machines on univariate time series data [26].

Multivariate Data Set: Multivariate time series data are widely available in different fields including medicine, finance, bioinformatics, science and engineering. Modeling multivariate time series data accurately is important for many decision making activities. Multivariate models have more parameters than univariate ones. It is difficult to generalize nonlinear procedures to the multivariate case. Commonly, time series has the characteristic of nonlinearity, and consists of multiple variables. However, most of the published papers are concerned with univariate time series other than the multivariate time series. However, multivariate time series often contains more dynamic information of the underlying system than univariate time series. As a result, the research on multivariate time series prediction has drawn an increasing focus. Multivariate time series prediction has been done based on multiple kernel extreme machine learning [27]. Multivariate short-term traffic flow forecasting has been done using time series analysis [28]. Combination of decision tree and clustering has been done for event prediction on multivariate time series data [29].

Categorization Based on Type Applied Approach on Data set: According to a second aspect, related works are divided into four sub-groups: neural network based methods, evolutionary computation based methods, feature based methods and fuzzy logic based methods. Each sub-group is briefly described below.

Neural Network Based Methods: Neural network attempt to solve problems by imitating the human brain. A neural network is a graph-like structure that contains an input layer, zero or more hidden layers, and an output layer. Each layer contains several "neurons" which have weighted connections to neurons of the following layer. A neuron from the input layer holds an input variable. For forecasting models, this input is a previous time series observation or an explanatory variable. There are many researches in different application domains for event prediction by neural networks. Water level forecasting of Dungun River has been done by neural networks [13]. Time series forecasting has been done using a hybrid ARIMA and neural network model [30]. Event Prediction has been done in time series data [16].

Evolutionary Computation Based Methods: For methods based on evolutionary computation, the process of biological evolution is mimicked in order to solve a problem. After an initial population of potential solutions is created, solutions are ranked based on their "fitness." New populations are produced by selecting higher ranking solutions and performing genetic operations of "mating" to produce offspring solutions. This process is repeated over many generations until some termination condition is reached. A novel financial time series forecasting model has been done by clustering, evolving fuzzy decision tree and evolutionary computation algorithm for stocks in TSEC [9]. A time series based temperature model using integrated back propagation with genetic algorithm technique is proposed for prediction [31].

Feature Based Methods: Feature based time series prediction methods work on transforming the sequential data into feature set before handing it to the prediction algorithms. Feature selection and extraction are very common strategies in several research and technical areas. Selecting the most relevant variables is a critical step toward obtaining better performing and more efficient methods. Classical prediction algorithms, such as support vector machine and decision trees, do their prediction based on feature set. Exchange rate has been forecasted by using support vector machines [24]. Electric load forecasting has been done using support vector machines optimized by genetic algorithm [25]. Feature selection has been proposed for time series forecasting [32]. Decision tree classification and forecasting has been done for pricing time series data [33].

Fuzzy logic Based Methods: Due in many of application areas, fuzzy logic combines with statistics based, machine learning based and hybrid methods that accuracy is improved for prediction of behavior indefinite data. However, the environment is uncertain and changes rapidly, we usually must forecast future situations using little data in a short span of time, and it is hard to verify that the data is a normal distribution. Fuzzy logic is handled behavior indefinite data. Evolving and clustering fuzzy decision tree have been proposed for financial time series data forecasting [9]. Fuzzy knowledge discovery has been done from time series data for events prediction [34]. Fuzzy modeling and neural networks is coupled for river flood prediction [35].

4 Evaluation of Event Prediction Methods in Time Series

So far, different methods have been presented for event prediction in time series in various applied domains. An important problem in the assessment and development of methods, are the identification of suitable evaluation criteria and development of the related benchmarks. Generally, there are different criteria that could be considered to describe and evaluate time series mining methods for event prediction. In this section, proposed measures are designated on the base of two factors. Factors are literature review, works of carried out by researchers and theoretical reviews of authors. Hence, performed comparisons on the basis of the mentioned factors can be considered introduction to empirical comparison in upcoming works and the first step in this way.

4.1 Proposed Measures

To assess the aforementioned methods, the following criteria are taken into account, and ranking is done in five different tiers – low, medium, high, simplex, and complex. Each measure is briefly described below.

Accuracy: Accuracy has an important role in the performance evaluation of methods. Higher accuracy of methods leads to reduced error and consequently increased efficiency.

Scalability: Refers to the ability of efficiently to constructing the predictor given large amounts or high dimension of data. If the method has acceptable performance with the increasing dimensions of data, it is scalable.

Flexibility: The existence of different types of data decreases the efficiency of methods. Some methods are not usable on different data types. For example, statistical learning methods are usable on numerical data. Therefore, flexibility is low in statistical learning methods.

Model Selection: The number of potential candidates for multivariate models exceeds its univariate counterpart. Model selection is therefore more complex and lengthier and more susceptible to errors, which then affect prediction.

Computation Complexity (Cc): It is difficult to generalize nonlinear procedures to the multivariate case. Generally, multivariate methods must have a simpler structure than univariate ones, to overcome the additional complexity that is imposed by being multivariate.

4.2 Evaluation of Methods According to Proposed Measures

In this section, efficiency of event prediction methods in time series is evaluated based on the proposed classification. Event prediction methods in time series are evaluated according to the proposed classification based on type input data set by some proposed measures and the results of this evaluation are presented in Table 1. Accuracy is high in seasonal methods than others. Model selection is simplex on seasonal and Non- seasonal methods, since data set is univariate. Flexibility is medium on seasonal methods, since it is applied in seasonal data. Flexibility is low on methods which have been applied in univariate data set, since time series data set is univariate and numerical. It has not been done on different data types. Model selection is difficult on methods which have been applied in multivariate data set, since number of potential candidates for multivariate methods exceeds its univariate counterpart. Thus, it is more susceptible to errors, which then affect prediction. Computation complexity is high on methods which have been applied in multivariate data set, since it is difficult to generalize nonlinear procedures to the multivariate case. Model selection is complex on machine learning methods, because it is applied in univariate and multivariate datasets. Computation complexity is high, since it is applied in multivariate dataset.

Table 1. Evaluation of event prediction methods based on type input data set

Methods	Flexibility	Model Selection	Cc
Statistical learning methods	Low	Simplex	Low
Seasonal	Medium	Simplex	Low
Non seasonal	Low	Simplex	Low
Machine learning methods	High	Complex	High
Univariate data set	Low	Simplex	High
Multivariate data set	High	Complex	High

In this part, event prediction methods in time series are evaluated according to the proposed classification based on type applied approach on data set by some proposed measures and the results of this evaluation are presented in Table 2. Statistical learning methods have higher accuracy for event prediction in time series data with a linear behavior. Also, flexibility is low on them, since they are used for modeling linear behaviors. Flexibility is medium in seasonal sub-group, since it can be applied on numerical data with seasonal component, but it is low in Non seasonal methods, since it can be applied on numerical data without seasonal component. Statistics learning methods are applied to one variable time series. Thus, scalability is low on them. Behavior of most of the time series data is complex and nonlinear in the real world.

Therefore, accuracy is higher in machine learning methods than others, because, it models the nonlinear data behavior. Scalability is high, since data dimensions can be high and increase. Flexibility is higher than others statistical learning methods, since machine learning methods can applied on different data types. Accuracy is high in evolutionary computation than others, because evolutionary computation process is repeated over many generations until some termination condition is reached for problem solving. Fuzzy logic based methods have been predicted behavior indefinite time series data and forecasting is uncertain. Thus, accuracy is medium on fuzzy logic methods.

Table 2. Evaluation of event prediction methods based on type applied approach on data set

Methods	Accuracy	Scalability	Flexibility
Statistical learning methods	Medium	Low	Low
Seasonal	High	Low	Medium
Non seasonal	Medium	Low	Low
Machine learning methods	High	High	High
Neural network based	Medium	High	High
Evolutionary computation based	High	High	High
Feature based	Medium	High	High
Fuzzy logic based	Medium	High	High

In general, according to obtained results of evaluation of methods and due to the nonlinear behavior of most of the time series data in the real world, combination of methods can be a more suitable choice for event prediction. A combination of methods together improves the accuracy of prediction, since it is used advantages of methods together.

5 Conclusion

In this paper, time series mining was presented as a new research for event prediction in the area of temporal data bases. Due to the existence of the multiplicity and variety of methods for event prediction in time series, it is difficult to classify and evaluate them. Thus, it is necessary to classify and evaluate of event prediction methods in time series. Hence, first, in this paper, a classification was proposed for event prediction methods based on input type of behavior in data. Classification was presented based on different two aspects. First aspect was type input data set. Second aspect was type applied approach on data set. Then, the methods were evaluated according to some of the suggested measures. One of the most important problems in time series mining, which is an important and active research field and requires more investigation, is elimination of challenges and improvement of efficiency of algorithms. Results of this research asserted that the proposed classification can help select the suitable prediction method in different areas of application.

References

1. Morchen, F.: Time Series Knowledge Mining. MS. Thesis, Marburg (2006)
2. Rude, A.: Event Discovery and Classification in Space-Time Series. MS.Thesis. National Institute of Technology, The University of Maine (2011)
3. Koohzadi, M., Keyvanpour, M.R.: An analytical framework for event mining in video data. Artif. Intelli. Rev. 41, 401–413 (2012)
4. Shasha, D., Zhu, Y.: High Performance Discovery in Time Series. Techniques and Case Studies, pp. 1–190. Springer, New York (2004) ISBN:0-387-00857-8
5. Soni, J., Ansari, U., Sharma, D.: Predictive Data Mining for Medical Diagnosis. International J. Comp. Appli. 17, 808–816 (2011)
6. Gabarda, S., Cristobal, G.: Detection of events in seismic time series by time – frequency methods. In: Proceeding of 8th Signal Processing, IET, vol. 4, pp. 413–420 (2009)
7. Preethi, G., Santhi, B.: Study on Techniques of Earthquake Prediction. International J. Comp. Appli. 29, 55–58 (2011)
8. Preston, D., Brodleyz, C., Protopapas, P.: Event Discovery in Time Series. In: International Conference on Data Mining, vol. 3, pp. 34–38 (2000)
9. Robert, K.L., Chin-Yuan, F., Wei-Hsiu, H., Pei-Chan, C.: Evolving and clustering fuzzy decision tree for financial time series data forecasting. Expert Systems with Appli. 4, 3761–3773 (2009)
10. Lin, Y., Yang, Y.: Stock markets forecasting based on fuzzy time series model. In: Proceedings of IEEE International Conference on Intelligent Computing and Intelligent Systems, Shanghai, vol. 1, pp. 782–786 (2009)
11. Reuse, H., Joshi, M.J., Rascal, R.: Importance of Data Mining Time Series Technique in Crime and Criminal Investigation: A Case Study of Pune Rural Police Stations. International J. Comp. Applic. 30, 38–42 (2011)
12. Damle, C.: Flood forecasting using time series data mining. MS.Thesis, College of Engineering. University of South Florida (2005)
13. Arbian, S., Wibowo, A.: Time Series Methods for Water Level Forecasting of Dungun River In Terengganu Malayzia. International J. Engin. Science Technology 4, 1803–1811 (2012)
14. Tak-chung, F.: A review on time series data mining. Engin. Applic. Artifi. Intell. 24, 164–181 (2011)
15. Keyvanpour, M.R., Etaati, A.: Analytical Classification and Evaluation of Various Approaches in Temporal Data Mining. In: Thaung, K.S. (ed.) Advanced Information Technology in Education. AISC, vol. 126, pp. 303–311. Springer, Heidelberg (2012)
16. Yan, X.B., Lu, T., Li, Y.J., Cui, G.B.: Research on Event Prediction In Time-Series Data. In: Proceedings of IEEE International Conference on Machine Learning and Cybernetics, Shanghai, pp. 2874–2878 (2004)
17. Lajevardi, S.B., Minaei-Bidgoli, B.: Forecasting Airport Passenger Traffic: The Case of Hong Kong International Airport. In: Proceeding of Aviation Education and Resaerch, pp. 54–62 (2011)
18. Coshall, J.: Time series analyses of UK outbound travel by air. Travel. Research, 335–347 (2006)
19. Anderson, O.D.: The Box-Jenkins Approach To Time seies Analysis. R. A. I. R. O Research Operationelle/Operations Research 11, 3–29 (1997)
20. Kyoung-jae, K.: Financial time series forecasting using support vector machines. Neuro Computing 3, 307–319 (2003)

21. Park, S.-H., Lee, J.-H., Song, J.-W., Park, T.-S.: Forecasting Change Directions for Financial Time Series Using Hidden Markov Model. In: Wen, P., Li, Y., Polkowski, L., Yao, Y., Tsumoto, S., Wang, G. (eds.) RSKT 2009. LNCS, vol. 5589, pp. 184–191. Springer, Heidelberg (2009), J. Pattern. Recog,
22. Haibin, C., Pang-Ning, T.: Semi-supervised Learning with Data Calibration for Long-Term Time Series Forecasting, pp. 1–9. ACM (2008)
23. Sanwlani, M., Vijayalakshmi, M.: Forecasting Sales Through Time Series Clustering. International J. Data Mining Knowledge Manage. Process 3, 39–56 (2013)
24. Kao, D.Z., Pang, S., Bai, Y.H.: Forecasting Exchange Rate Using Support Vector Machines. In: Proceedings of 4th International Conference on Machine Learning and Cybernetics, Guangzhou, pp. 3448–3452 (2005)
25. Hong, W.C.: Electric load forecasting by support vector model. Applied Mathematical Modelling 33 32, 2444–2454 (2009)
26. Espinoza, M., Suykens, J.A.K., De Moor, B.: Load Forecasting Using Fixed-Size Least Squares Support Vector Machines. In: Cabestany, J., Prieto, A.G., Sandoval, F. (eds.) IWANN 2005. LNCS, vol. 3512, pp. 1018–1026. Springer, Heidelberg (2005)
27. Wang, X., Han, M.: Multivariate Time Series Prediction based on Multiple Kernel Extreme Learning Machine. In: Proceeding of International Joint Conference on Neural Networks (IJCNN), Beijing, China, pp. 198–201 (2014)
28. Ghosh, B., Basu, B., Mhony, M.: Multivariate Short-Term Traffic Flow Forecasting Using Time Series Analysis. IEEE Trans. Intelligent Transportation Systems 10, 246–254 (2009)
29. Lajevardi, S.B., Minaei-Bidgoli, B.: Combination of Time Series, Decision Tree and Clustering: A Case Study in Aerolology Event Prediction. In: Proceeding of IEEE International Conference on Computer and Electrical Engineering, pp. 111–115 (2008)
30. Zhang, G.P.: Time series forecasting using a hybrid ARIMA and neural network model. Neurocomputing, 159–175 (2003)
31. Shaminder, S., Pankaj, B., Jasmeen, G.: Time Series based Temperature Prediction using Back Propagation with Genetic Algorithm Technique. International J. Comp. Science 8, 28–32 (2011)
32. Crone, S.F., Kourentzes, N.: Feature selection for time series prediction – A combined filter and wrapper approach for neural networks. Neurocomputing, 1923–1936 (2010)
33. Lundkisi, E.: Decision Tree Classification and Forecasting of Pricing Time Series Data. MS. Thesis, Stockholm, Sweden (2014)
34. Gholami, E., Borujerdi, M.M.: Fuzzy Knowledge Discovery from Time Series Data for Events Prediction. In: Ho, T.-B., Zhou, Z.-H. (eds.) PRICAI 2008. LNCS (LNAI), vol. 5351, pp. 646–657. Springer, Heidelberg (2008)
35. Corani, G., Guariso, G.: Coupling Fuzzy Modeling and Neural Networks for River Flood Prediction. IEEE Trans. Systems, Man, Cybernetics—Part C: Application and Reviews 35, 382–390 (2005)

Integrating Grid Template Patterns and Multiple Committees of Neural Networks in Forex Market

Nikitas Goumatianos[1,2], Ioannis Christou[1], and Peter Lindgren[3]

[1] Athens Information Technology,
19km Markopoulou Ave. P.O. Box 68, Paiania 19002, Greece
{nigo,ichr}@ait.edu.gr
[2] Aalborg University, Aalborg, Denmark, DK-9920
in_ng@es.aau.dk
[3] Aarhus University, Birk Centerpark 15, 7400 Herning, Denmark
peterli@hih.au.dk

Abstract. We present a hybrid framework for trading in Forex spot market by integrating two different technologies: price patterns based on an array of grid template methods and multiple committees of neural networks. This integration is applied in four currency pairs (EUR/JPY, EUR/USD, GBP/JPY and GBP/USD) using data of a 20 min timeframe. In this research we examine two different fusion approaches for Forex trading: the first one is based on price pattern discovery methods and committees of neural networks as independent entities and works by assigning one entity as basic signal provider and the other as filtering system. The second approach is to take price pattern properties (e.g. forecasting power values) together with technical indicators to feed and train the neural networks. Results show that in both approaches, integration of these independent technologies can improve the trading performance by bringing higher net profits and less risk.

Keywords: Forex Market, Committee of Neural Networks, Template Grid Method, Price Pattern Discovery, Trading Strategies.

1 Introduction

Generally, price patterns can be seen in the form of technical analysis as predefined patterns (candlesticks and chart formations) or more extensively in other generic forms not described in technical analysis.

Candlesticks is one form of price patterns which may be used to produce profits at various markets. A lot of researchers have examined the profitability of the candlesticks such as Caginalp & Laurent [1] who found that candlestick patterns of daily prices of S&P 500 stocks between 1992 and 1996, provided strong evidence and high degree of certainty in predicting future prices. In another study, Lee and Jo [2] developed a chart analysis expert system for predicting stock prices. On the other hand, in a paper by Marshall et al [3], the authors found candlestick technical analysis had no value on U.S. Dow Jones Industrial Average stocks during the period from 1992–2002.

R. Silhavy et al. (eds.), *Artificial Intelligence Perspectives and Applications*,
Advances in Intelligent Systems and Computing 347, DOI: 10.1007/978-3-319-18476-0_25

Regarding price patterns in a view of chart formations, numerous methods have been proposed. Zhang et al. [4] worked on pattern matching based on Spearman's rank correlation and sliding window, which is more effective, sensitive and constrainable compared to other pattern matching approaches such as Euclidean distance-based, or the slope-based method. Another study [5] investigated the performance of 12 chart patterns in the EURO/Dollar (5-min mid-quotes) Forex (foreign exchange) market, involving Monte Carlo simulation, using identification methods for detecting local optima. The authors found that some of the chart patterns (more than one half) have predictive power, but unfortunately, when used for creating & applying trading rules they seemed unprofitable. In [6], four different pattern recognition methods designed for trading systems in Forex market were compared. An interesting study of evaluating pre-specified chart formations is the Template Grid method (TG), which involves the rank method for expressing the similarity between two patterns. Wang and Chan [7], used this method for pattern recognition to predict stock prices by detecting so-called "bull flag" formations. A 10x10 Grid was used, with corresponding weights stored in the cells and 20-day fitting data (each of two successive days corresponding to one cell). In a series of papers, Leigh et al. [8], [9], [10], Bo et al. [11], extended the Template Grid method (TG). The common characteristic of these studies is that the algorithm is only capable of evaluating pre-specified chart formations which are already well-known and popular in the technical analysis world.

Besides price pattern methods, another popular approach for predicting financial markets is using neural networks. Generally, in most cases, neural network prediction models demonstrate better performance than other approaches. Yao and Tan [12] developed a neural network model using six simple indicators to predict the exchange rate of six different currencies against the US dollar. The forecasting results were very promising for most currencies except the Yen and were better in comparison than ARIMA-based models. Dunis et al [13] examined the use of Neural Network Regression (NNR) models in forecasting and trading the US Dollar against Euro exchange rate. The authors demonstrated that NNR model gives better forecasting and trading results than ARMA models. Another study on using neural networks to perform technical forecasting of the Forex was done by Yao and Tan [14]. They fed the neural network with time series data and technical indicators such as moving average to capture the underlying rules of movement in currency exchange rates. They report empirical evidence that a neural network model is applicable to the prediction of the Forex market. However, the experiments showed it is not easy to make profits using technical indicators as input to the neural networks. Giles, Lawrence, and Tsoi [15] found significant predictability in daily foreign exchange rates using their neural network models.

Neural Networks have been used in multiple classifier systems, also known as Committee of Neural Networks (CNNs), to decide and vote on a given example, hoping that errors would be canceled out as there are several experts [16]. These structures can be static or dynamic. In static structures, the combiner does not make use of the input. This category includes methods such as ensemble averaging where the outputs of different predictors are linearly combined to produce the output, bagging [17] which is a short of "bootstrap aggregation" and members of committee are trained on different samples (data splits) while the prediction is by averaging all results, boosting using weak and strong learners to achieve higher accuracy [18]. In dynamic structures the combiner makes direct use of the input. Examples of this category are the Mixture of Experts (there is nonlinear combination of expert outputs by means of a single gating network), Hierarchical Mixture of Experts (there are hierarchically arranged gating networks). Research in this area has been initiated by Jacob et al. [19]. Later this approach was extended and improved by Avnimelech [20], Srivastava [21], Lima [22], and others. The biggest disadvantage of the method is its tendency for overfitting due to the complexity of the model.

2 Methods and Application

The main purpose of our research is to combine and integrate two radically different approaches to forex trading so as to maximize profits and reduce risk of the trading strategies. In the first approach, for the price patterns methodology, we developed a novel pattern recognition algorithm using an extended version of the Template Grid method (TG) for pattern matching, to construct more than sixteen thousand new intraday price patterns. Of those patterns, after processing and analysis, we extracted 3,518 chart formations that are capable of predicting the short-term direction of prices. In parallel, we created two groups containing four committees of neural networks to make predictions for the short-term direction of prices, similarly to price patterns. The first group of committees contains neural networks that are fed only by technical indicator values. The second group contains a mix of members (neural networks) of the first group and new members which are fed by the prediction output of the price patterns. Each committee is assigned the task of learning to predict a specific output (for a total of 8 outputs).

Table 1. Predictive Variables for Price Patterns and Neural Networks

Predictive Variables (Outputs) – Changes in pips	Price Patterns	Neural Networks
(Variable 1) Next period lowest price less than	-5	N/A
(Variable 2) Next period highest price greater than	+5	N/A
(Variable 3) Within next 5 periods lowest price less than	-12	-12
(Variable 4) Within next 5 periods highest price greater than	+12	+12
(Variable 5) Within next 10 periods lowest price less than	-25	-25
(Variable 6) Within next 10 periods highest price greater than	+25	+25
(Variable 7) Within next 20 periods lowest price less than	-35	N/A
(Variable 8) Within next 20 periods highest price greater than	+35	N/A

2.1 Discovery of Template Grid Patterns

A Template Grid is two-dimensional table used for capturing chart formations in which each column corresponds to a specific time t (the latest column is the most recent time t_1, the previous columns being $t_1 - tf, t_1 - 2tf, \dots$ where $tf =$ timeframe length). The vertical column is used to present the position of the point of chart formation for a specific time. The lowest point (in a column) represents the lowest price value of the specific timeframe while the highest point represents the highest price value for the same timeframe. Our price pattern system does not use the well-known chart formations described in technical analysis. In initialization mode, it creates thousands of various shapes corresponding to 1 of 4 different in size grid dimensions (10x10, 15x10, 20x15 & 25x15) which have been taken from historical data. As an example, in Fig.1, on the left side, there is a template grid 10x10 which captures a chart formation from historical data to be used as prototype pattern.

To compute the weights in each cell of the grid, we execute the following steps. First, we find the cells that will be assigned weight value 1, which are the cells (in vertical) where the given price value is found together with their vertical neighboring two cells (up and down). Them, we calculate the cell length in vertical position which is $(H - L)/10$, where H, L are the Highest and Lowest close values over the last 10 (=horizontal grid dimension) periods. So, a price is in the first cell of the column (coded as 0) if it is between L and $L + (H - L)/10$, is in the second vertical cell (coded as 1) if price is between $L + \frac{H-L}{10}$ *and* $2L + \frac{H-L}{10}$, *etc.* In general, the position of a cell that will be assigned the value 1 at any given time, given a price value p is computed as $pos = [G_H(p - L)/(H - L)]$ where L, H represent the lowest and highest (close) prices during the selected period and G_H represents the vertical dimension of the grid (here is 10).

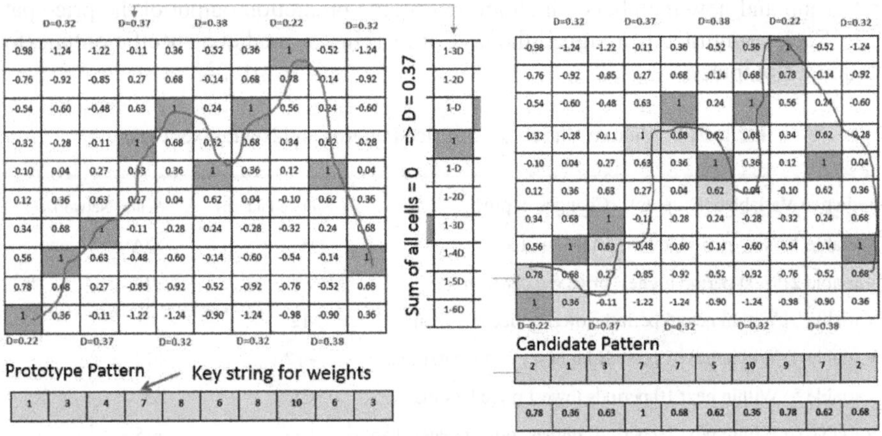

Fig. 1. Prototype Pattern & Candidate Pattern

The dynamic method of calculating weights is rather similar to the method of Wang and Chan [7] with the difference that each grid column has only one cell value equal to "1". The equation of the fourth column is: *1+2(1-D)+2(1-2D) +2(1-3D)+(1-4D) +(1-5D)+(1-6D)= 0* based on the following formula:

$$\text{Sum of 4}^{\text{th}}\text{ column} = 1 + \sum_{j=1}^{G_H - p}(1 - jD) + \sum_{j=1}^{p-1}(1 - jD) = 0$$

where p denotes the cell-position in the column that gets the value "1" (in the example p=7). Solving the above equation, gives D = 0.37. Therefore, the immediate two neighbor cells of the cell with value 1 are assigned the value $1-D=1-0.37=0.63$, the next two have values $1-2D=0.27$, etc. This template grid which contains all weights of the prototype pattern is called template ranker. We define the similarity (rank) of the candidate chart formation (price pattern) with respect to a prototype pattern, as the ratio of the sum of the weights which correspond to each position found on prototype pattern's ranker divided by the length of the pattern, expressed as a percentage. Fig. 1, right side grid, illustrates the above described process. By definition then, $Rank = 100\sum_{i=1}^{G_W} w_i / G_W$ where G_W is the horizontal dimension of the grid. The rank measure attempts to capture the similarity of a candidate chart formation (pattern) to that of the prototype's chart formation. In our research we consider that a price pattern (P_m) belongs to a prototype pattern ($Prot_i$) if rank($Prot_i, P_m$) ≥ 60.

For a set of historical time series data, the "prediction accuracy" of predictive variable i of a given prototype pattern A is:

$$\textit{Prediction Accuracy}(\textit{Prototype A}, \quad \textit{Predictive Variable i}) = 100 \cdot \frac{\text{number of patterns (having rank} \geq 60\text{) future price change is above or below } \textit{the threshold value}}{\text{total number of patterns (having rank} \geq 60\text{)}}$$

The strategy for exploiting the pattern prediction is simple: if the current formed template pattern matches ($Rank \geq 60$) with a prototype template grid whose prediction accuracy is greater than 60% it produces signal either long (if price is above threshold) or short (price below threshold). In case 2 or more in different dimensions prototype patterns produces different signal (at least one short and one long), the systems cancels to open position.

2.2 Multiple Committees of Neural Network

The first group of committees, called Multiple Committees of Neural Networks (MC-NNs) is fed by technical indicator values. Each member (NN) may have different number of inputs. The selected inputs are determined by the help of a genetic algorithm. The pool of the available inputs contains many popular indicators (totally 28 inputs including previous values) such as Acceleration/Decelerator Oscillator, Arron, Commodity Channel Index, Chande Momentum Indicator, Directional Movement Index (+DI) & (-DI), Simple Moving Average (SMA) of 5, 20, 50 and 100 Periods, Price Oscillator of 12/26 periods, Price Oscillator of 20/50 periods, Relative Vigor Index, Relative Strength Index, Vertical Horizontal Filter. The members of the second group of committees, called Multiple Committees of Pattern based Neural Networks (MC-PNNs) are fed mainly by the prediction accuracy (probabilities) of all predictive variables of prototype matching patterns and four inputs from technical indicators (Chande Momentum and Moving Average of 5, 20 and 50 periods). It should be noted that the members of MC-PNNs are fed only when the current template grid matches with prototype pattern which have prediction accuracy greater than 55%. Example inputs 1 &2:

a. (Template Grid 10x10) Prediction Accuracy of predicate variable 1 (price for next period will be less than the threshold): 0.67 (=67%).
b. (Template Grid 20x15) Prediction Accuracy of predicate variable 6 (price within next 10 periods will be greater than the threshold.): 0.72 (=72%).

In the above example, the total inputs are 8 (see table 1, predictive variables are 8) for each template grid and 4 technical indicator inputs.

Regarding the neural training, we selected three forms of propagation training for our experiments: Resilient Propagation Training (RPROP), Levenberg Marquardt (LMA) and Scaled Conjugate Gradient. For the accuracy of Neural Network prediction we used the Mean Square Error (RMSE). The strategy for exploiting the outputs of MC-NNs or MC-PNNs is similar to pattern strategy.

2.3 Integration Model – System Architecture

The first system is evolved by a genetic algorithm (regarding the input selection). The second system uses prototype patterns with the support of four technical indicators as inputs for the neural networks. Both systems are designed based on the knowledge trading framework which involves four different outputs instead of six that the previous system had. Fig. 2 shows in high level the structure of the whole system consisting of six different trading systems (cases A to F), as follows:

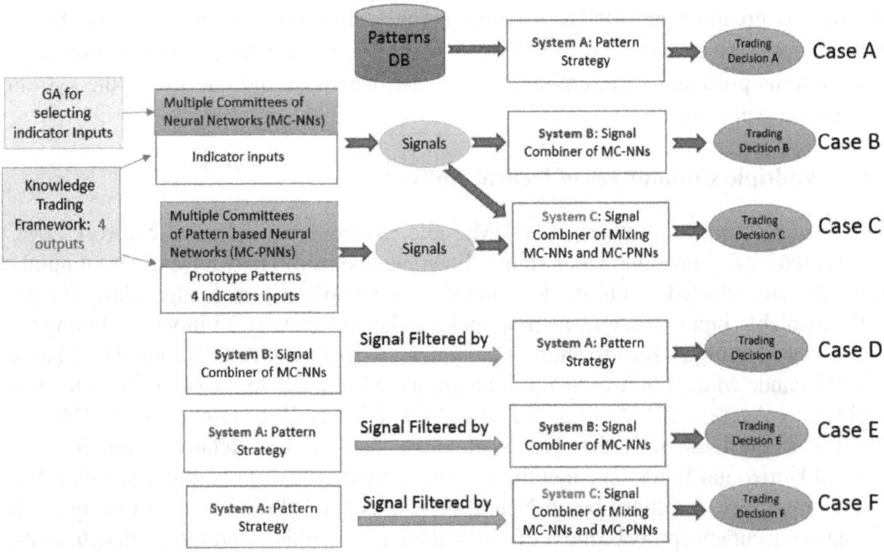

Fig. 2. The Architecture of Trading Systems

Case A: It is the trading pattern strategy (system A) based on array of template grid method.

Case B: Based on the signals produced by the Multiple Committees of Neural networks (MC-NNs), we constructed a signal combiner strategy (system B) for trading decisions (long/short).

Case C: A signal combiner gets both the signals produced by Multiple Committees of Neural networks (MC-NNs) and Multiple Committees of Pattern based Neural networks (MC-PNNs) and produces a trading decision (system C).

Case D: The produced trading decision of the signal combiner strategy of MC-NNs (System B) is filtered by the pattern strategy (System A). If conflict exists, the result is no-trade.

Case E: The produced trading decision of the patterns strategy (System A) is filtered by the signal combiner strategy of MC-NNs (System B). If conflict exists, the result is no-trade.

Case F: The produced trading decision of the patterns strategy (System A) is filtered by the signal combiner strategy of MC-PNNs (System C). If conflict exists, the result is no-trade.

3 Empirical Results

The data setup contains intraday data of 20 minutes time frame for 4 pair-currencies (EUR/JPY, EUR/USD, GBP/JPY, GBP/USD); this dataset starts from Jan 2010 until end of Dec 2013 and contains approximately 220,800 entries. The strategy testing takes place in two different semesters (dataset 1 is the 1st semester of 2013 and dataset 2 is the 2nd semester). Strategy datasets use out-of-sample data.

The trading signals for each case were executed by market orders at the current close prices, while the money management was the 'Fixed Dollar Amount' with initial capital of $20,000 (using 1 lot size contract: USD $100,000), leverage 1:10 (notice that most banks and intermediaries to the Forex market currently offer leverage of at least 1:100) and exposure to margin account 50%. All calculations are based on no profit re-investment, and for net profits the cost of spread commissions was taken into account. No slippage or latency was involved in the calculations. For all strategies we used trailing stop/loss between 35 and 40 pips value. The trading strategies run for all cases are presented in Fig 2. In Tables, 2 to 5, we present the trading results for the 20 min timeframe. The first column data of each table, is the "Pattern Strategy (Case A)". The second column is either "Filter Ptns by MC-NNs (Case E)" or " Filter Ptns by MC-PNNs (Case F)" (depicting the best performing), refers to the strategy where the MC-(P)NNs filters the signals of the pattern strategy. The next two columns show the performance of the MC-NNs (case B) and MC-NNs with MC-PNNs (case C). The last column depicts the strategy where MC-NNs signals are filtered by the Pattern Strategy signals.

Table 2. Average Performance Comparisons: EURO vs. Japanese Yen (EUR/JPY)

Performance Indicators Average of Datasets 1 & 2	Pattern Strategy (Case A)	Filter Ptns by MC-NNs (Case E)	MC-NNs (Case B)	MC-PNNs (Case C)	Filter MC-NNs by Ptn Strategy (Case D)
Net Profit Amount($)	20,165.22	21,435.75	20,998.76	22,917.53	19,672.94
Net Profit %	100.82%	107.18%	104.99%	114.58%	98.36%
Number of Total Trades	394	488	210	231	203
% of Winning Trades	46.85%	48.48%	40.30%	40.93%	41.73%
Sharpe Ratio	2.7	2.7	2.43	2.53	2.44

Table 3. Average Performance Comparisons: EURO vs. US Dollar (EUR/USD)

Performance Indicators Average of Datasets 1 & 2	Pattern Strategy (Case A)	Filter Ptns by MC-NNs (Case E)	MC-NNs (Case B)	MC-PNNs (Case C)	Filter MC-NNs by Ptn Strategy (Case D)
Net Profit Amount($)	12,726.50	16,764.75	7,623.5	8,569.75	8,218,00
Net Profit %	63.63%	83.82%	38.12%	42.84%	41.09%
Number of Total Trades	432	417	95	162	93
% of Winning Trades	59.27%	58.20%	45.84%	46.56%	47.08%
Sharpe Ratio	2.44	3.42	3.98	1.75	1.69

Table 4. Average Performance Comparisons: Pound Sterling vs. Japanese Yen (GBP/JPY)

Performance Indicators Average of Datasets 1 & 2	Pattern Strategy (Case A)	Filter Ptns by MC-NNs (Case E)	MC-NNs (Case B)	MC-PNNs (Case C)	Filter MC-NNs by Ptn Strategy (Case D)
Net Profit Amount($)	22,963.13	23,998.26	13,065.68	9,883.92	11,690.39
Net Profit %	114.81%	119.99%	65.33%	49.42%	58.45%
Number of Total Trades	1,012	1370	141	61	90
% of Winning Trades	50.60%	51.75%	43.71%	40.49%	42.95%
Sharpe Ratio	2.33	3.41	1.91	1.38	1.73

Table 5. Average Performance Comparisons: Pound Sterling vs. US Dollar (GBP/USD)

Performance Indicators Average of Datasets 1 & 2	Pattern Strategy (Case A)	Filter Ptns by MC-NNs (Case E)	MC-NNs (Case B)	MC-PNNs (Case C)	Filter MC-NNs by Ptn Strategy (Case D)
Net Profit Amount($)	18,859.00	19,943.00	14,290.00	14.932.00	14,297.5
Net Profit %	94.29%	99.72%	71.45%	74.66%	71.48%
Number of Total Trades	442	416	239	263	222
% of Winning Trades	53.37%	52.40%	46.34%	46.82%	46.76%
Sharpe Ratio	3.26	3.11	2.54	2.65	2.56

The main points of analyzing trading results from Tables 2 to 5 (20 min timeframe) are the following:

a. Filtering the pattern strategy (case A) by the MC-(P)NNs (case E or F) brings a little higher net profits and decreases slightly the trading risk.
b. The concept of filtering the MC-NNs by the pattern strategy, generally, has not improved the trading performance (only 3 out of 10 run tests brought a little higher net profits).
c. By comparing the trading performance between pattern strategy (case A) and MC-NNs / MC-PNNs (cases B & C), we can conclude that pattern strategy has showed better performance than those of MC-NNs / MC-PNNs. Actually, the net profit was higher in most run tests and the risk was reduced in most run tests.
d. We cannot conclude in a statistically significant sense if the MC-NNs (case B) or the combination of both MC-NNs with MC-PNNs (case C) produces better results.

4 Conclusion and Future Improvements

Generally, results showed that integration of these independent technologies can improve the trading performance by bringing higher net profits and less risk. More specifically, filtering patterns strategy by MC-NNs, case E, brought better results in all cases (higher net profit and 3 of 4 currencies less risk). The hybrid committees of Neural networks (MC-PNNs), case B, brought higher profits in 3 of 4 currency pairs, while the risk did not improved over the MC-NNs. Finally, filtering MC-NNs by pattern strategy, did not improve the performance of MC-NNs. We found that a single method/strategy cannot be expected to bring good results for all kinds of instruments. Our integration approach could solve such issues and is worth further examination.

As for embedding prototype pattern attributes as inputs to the neural networks, it showed that only one template grid could enhance the results in most cases while involving three template grids did not bring any good results. This opens a new room for further investigation and improvements such as (a) involve again three grids at lower timeframes such as 1 min, because the number of training samples is significant larger in such timeframes, (b) consider the price patterns attributes as individual separate inputs and with the help of a genetic algorithm select those attributes that enhance the neural network performance, or (c) instead of involving different neural networks (NNs) for each different in dimension template grid, use only one NN by taking the average of corresponding prototype pattern attributes.

References

1. Caginalp, G., Laurent, H.: The predictive power of price patterns. Applied Mathematical Finance 5, 181–205 (1998)
2. Lee, K.H., Jo, G.S.: Expert system for predicting stock market timing using a candlestick chart. Exp. Sys. Appl. 16, 257–364 (1999)

3. Marshall, B.R., Young, M.R., Rose, L.C.: Candlestick technical trading strategies: can they create value for investors? J. Bank. Fin. 30, 2303–2323 (2006)
4. Zhang, Z., Jiang, J., Liu, X., Lau, R., Wang, H., Zhang, R.: A real-time hybrid pattern matching scheme for stock time series. In: Proc. 21st Australasian Conf. on Database Technologies, vol. 104, pp. 161–170 (2010)
5. Walid, B., Van Oppens, H.: The performance analysis of chart patterns: Monte-Carlo simulation and evidence from the euro/dollar foreign exchange market. Empirical Economics 30, 947–971 (2006)
6. Goumatianos, N., Christou, I., Lindgren, P.: Intraday Business Model Strategies on Forex Markets: Comparing the performance of Price Pattern Recognition Methods. Journal of Multi Business Model Innovation and Technology 2(1) (2013) ISSN:2245-8832
7. Wang, J.-L., Chan, S.-H.: Stock market trading rule discovery using pattern recognition and technical analysis. Expert Systems with Applications 33, 304–315 (2007)
8. Leigh, W., Modani, N., Purvis, R., Roberts, T.: Stock market trading rule discovery using technical charting heuristics. Expert Systems with Applications 23(2), 155–159 (2002)
9. Leigh, W., Purvis, R., Ragusa, J.M.: Forecasting the NYSE composite index with technical analysis, pattern recognizer, neural network, and genetic algorithm: a case study in romantic decision support. Decision Support Systems 32, 161–174 (2002)
10. Leigh, W., Modani, N., Hightower, R.: A computational implementation of stock charting: Abrupt volume increase as signal for movement in New York stock exchange composite index. Decision Support Systems 37, 515–530 (2004)
11. Bo, L., Linyan, S., Mweene, R.: Empirical study of trading rule discovery in China stock market. Expert Systems with Applications 28, 531–535 (2005)
12. Yao, J., Tan, L.: A case study on neural networks to perform technical forecasting of forex. Neurocomputing 34, 79–98 (2000)
13. Dunis, C.L., Williams, M.: Modelling and trading the EUR/USD exchange rate: Do neural network models perform better? Derivatives Use, Trading & Regulation 8, 211–239 (2002)
14. Yao, J., Tan, C.L.: A case study on using neural networks to perform technical forecasting of forex. Neurocomputing 34, 79–98 (2000)
15. Girosi, F., Jones, M., Poggio, T.: Regularization theory and neural networks architectures. Neural Computation 7, 219–269 (1995)
16. Bettebghor, D., et al.: Surrogate Modeling Approximation using a Mixture of Experts based on EM joint Estimation. Structural and Multidisciplinary Optimization 43(2) (2011)
17. Breiman, L.: Bagging predictors. Journal Machine Learning 24, 123–140 (1996)
18. Schapire, R.E.: The Boosting Approach to Machine Learning: An Overview. In: MSRI Workshop (2002)
19. Jacobs, R.A., Jordan, M.I., Nowlan, S.J., Hinton, G.E.: Adaptive mixtures of local experts. Neural Computation 3, 79–87 (1991)
20. Avnimelech, R., Intrator, N.: Boosted mixture of experts: An ensemble learning scheme. Neural Computation 11(2), 483–497 (1999)
21. Srivastava, A.N., Su, R., Weigend, A.S.: Data mining for features using scale-sensitive gated experts. IEEE Transactions on Pattern Analysis and Machine Intelligence 21, 1268–1279 (1999)
22. Lima, C.A.M., Coelho, A.L.V., Von Zuben, F.J.: Hybridizing mixtures of experts with support vector machines: Investigation into nonlinear dynamic systems identification. Information Sciences 177(10), 2049–2074 (2007)

A Comparison of Handwriting Grip Kinetics Associated with Authentic and Well-Practiced Bogus Signatures

Bassma Ghali[1], Khondaker A. Mamun[1,2], and Tom Chau[1]

[1] Bloorview Research Institute, Holland Bloorview Kids Rehabilitation Hospital, 150 Kilgour Road, Toronto, Ontario, Canada and
Institute of Biomaterials & Biomedical Engineering, University of Toronto, 164 College Street, Toronto, Ontario, Canada
[2] Department of Computer Science and Engineering,
Ahsanullah University of Science and Technology, Dhaka, Bangladesh
{bassma.ghali,k.mamun,tom.chau}@utoronto.ca

Abstract. Handwriting biomechanics may bear biometric value. Kinematic and kinetic handwriting characteristics of authentic and forged handwriting samples have been contrasted in previous research. However, past research has only considered pen-on-paper forces while grip kinetics, i.e., the forces applied by writers' fingers on the pen barrel, have not been examined in this context. This study compares multiple grip kinetic features between repeated samples of authentic signatures and skilled forgeries in a sample of 20 functional adult writers. Grip kinetic features differed between authentic and well-practiced bogus signatures in less than half of the participants. In instances where forces differed between authentic and bogus, there was no clear trend in the difference. Forgeries are not necessarily associated with different or more variable grip kinetics. As long as the written text is well-practiced and written naturally, the handwriting kinetics tend to be similar to those of authentic signature writing.

Keywords: Handwriting biomechanics, Grip kinetics, Bogus signatures, Authentic signatures, Variation between signatures.

1 Introduction

Handwriting grip kinetics are the forces applied by the hand on the pen barrel while writing. Using instrumented pens, many studies have considered grip forces for clinical and rehabilitation applications [1,2,6,9,12]. In [9] and [17], the authors employed a pen that measured the six force and torque components (radial, tangential, and normal) on four contact points on the pen using four sensors and moment arms. Other studies [1, 2, 6, 12] used instrumented pens covered with an array of force sensors that measured the normal force applied on the pen barrel while writing in which case there is no restriction in fingers placement.

Grip kinetics are also being investigated in the fields of biometrics and forensics. For example, grip force patterns have demonstrated biometric value in gun

R. Silhavy et al. (eds.), *Artificial Intelligence Perspectives and Applications*,
Advances in Intelligent Systems and Computing 347, DOI: 10.1007/978-3-319-18476-0_26

control applications [15, 16]. Similarly, grip force patterns may also possess discriminatory features for writer identification and forgery detection [5, 11].

Some studies have compared handwriting characteristics between authentic and forged signatures. When kinematic features were compared between authentic samples of a model writer and simulations of 10 other participants, it was found that forgeries were associated with longer reaction time, slower movement velocities, more dysfluencies (reversals of velocity) and higher limb stiffness [19]. A more recent study investigated the kinematic and kinetic differences between authentic signatures and forgeries [3], finding that some forgers exhibited slower movements, multiple pen stops and higher axial forces, while other forgers simulated their normal writing velocity with no hesitations and comparable or even lower pen tip forces. The difference between writing a true versus a deceptive message has also been examined biomechanically, with the latter being associated with a significantly higher mean axial pressure, stroke length and height [13]. These differences were attributed to the higher cognitive load required to forge another person's handwriting or to write a deceptive message.

None of the above studies have considered the grip kinetics of authentic and forged handwriting. Therefore, in this study, we compared the magnitude and dispersion of grip kinetics associated with repeated writing of the participant's authentic signature and a well-practiced bogus signature. An instrumented pen that measured the normal grip forces applied on the pen barrel while writing was utilized for this purpose. The findings of this study can help to inform the use of grip kinetics for writer discrimination and signature verification.

2 Methods

2.1 Participants

Twenty adult participants (8 males; 17 right handed; 27 $\pm$ 6 years of age) were recruited from students and staff of an academic health sciences center. Individuals with known history of musculoskeletal injuries or neurological impairments that can affect handwriting function were excluded from the study. The research ethics board of the health sciences center approved the study and each participant provided informed, written consent.

2.2 Instrumentation

An instrumented writing utensil was used to obtain the grip force signals (the forces applied on the pen barrel by the fingers) through an array of Tekscan 9811 force sensors that covered the pen barrel. The original sensor array consists of 96 (16 $\times$ 6) sensors; however, only the section of the array covering the pen barrel which includes 32 (8 $\times$ 4) sensors was considered. The dimension of each individual sensor was 6.3 $\times$ 7.9 millimeters. The force sensors were systematically calibrated every two or three days during the data collection to derive calibration curves. The array was replaced regularly due to wear and tear. The grip force signals were acquired by computer at 250 Hz via a custom-made interface box. A Wacom Cintiq 12WX digitizing LCD display served as the writing surface and captured the axial force, pen tip position, pen tilt and rotation angles at a frequency of 105 Hz. The various signals were synchronized by data acquisition software. A grounding strap was worn by the participant to reduce noise in the grip force signals. The instrumentation setup is illustrated in Figure 1. For further details regarding utensil construction and the calibration procedure, the reader is referred to [1] and [5].

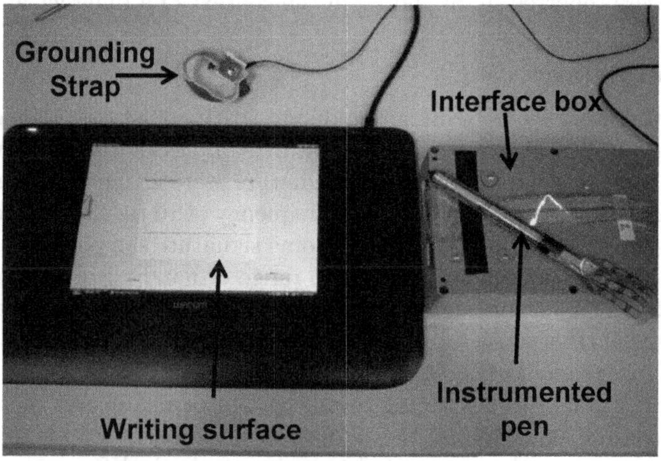

Fig. 1. Data collection instrumentation set-up

2.3 Experimental Protocol

Before the actual data collection sessions, each participant practiced a bogus signature on paper, 25 times each day, for two weeks, in order to become familiar with the signature and to develop idiosyncratic signing patterns. All participants practiced the same bogus signature shown in Figure 2.

Fig. 2. A sample of the bogus signature

For each participant, data collection included 30 sessions spread over 10 days. Three sessions were performed each day (morning, afternoon, and evening). In each session, the participant wrote with the instrumented pen, on the digital writing surface, 20 samples of the well-practiced bogus signature followed by 20 samples of his or her own authentic signature. In each session, the following steps also occurred: customized software was launched to guide the participant through data collection; the force sensors were checked; and pictures of the grip were taken at the midpoint and conclusion of each session. In addition, a 10-second baseline was collected at the beginning of each session to determine the pre-grip value of each sensor, which was used in subsequent off-line data preprocessing.

All data collection took place in a laboratory within Holland Bloorview Kids Rehabilitation Hospital and was completed in 20.4 ± 3.6 days on average depending on participant availability. In total, 600 well-practiced bogus signatures and 600 authentic signatures were obtained from each participant over a total of 30 sessions. All signature samples were considered in this study. The authentic signatures were reported in [5] and [4] in the context of a different analysis.

2.4 Data Pre-processing

Some signature samples were discarded due to writing mistakes, a long pause or a sensor malfunction during writing. Signature samples that were contaminated with extra writing before or after the signature were trimmed accordingly. A Butterworth low-pass filter with a cut off frequency of 10 Hz suppressed high frequency noise from the grip force signals. Some signature samples were discarded because of low frequency noise within the range of handwriting frequencies. In total, 1,522 (12.7%) bogus signatures and 960 (8%) authentic signatures were excluded from further analysis.

The grip force signals of the remaining signature samples were pared down to a zero pre-grip value by subtracting from each sensor reading its corresponding 10-second baseline average from the associated session. The grip force recording from each sensor was then converted to units of physical force via the corresponding calibration curve. These preprocessing steps are further detailed in [5].

2.5 Feature Extraction

Features were extracted from the topographical and functional representations of the grip force signals to facilitate comparison between authentic and bogus signatures. The topographical representation of the kth signature consisted of a grip force matrix, GF_k, which is an 8 by 4 matrix of non-negative real numbers, where each element denotes the force on a sensor, averaged over the duration of the signature. We will use $\widehat{GF_k}$ to denote the grip force matrix whose values have been normalized to [0,1] (e.g., Figure 3B). The mean grip force matrix, GF^p, of participant p is the element-by-element average of all normalized grip force matrices for that participant, i.e., $GF^p = \frac{1}{N_p}\sum_{k=1}^{N_p}\widehat{GF_k}$, where N_p is the number of signatures by participant p. Topographical features are detailed below.

- The 2-dimensional normalized correlation coefficient (NCC_2) between the normalized grip force matrix ($\widehat{GF_k}$) of the kth signature and the mean grip force matrix of the corresponding participant (GF^p) is calculated as follows:

$$NCC_2\left(\widehat{GF_k}, GF^p\right) = \frac{1}{32}\frac{\sum_{i=1}^{8}\sum_{j=1}^{4}(\widehat{GF_k}(i,j) - \overline{\widehat{GF_k}})(GF^p(i,j) - \overline{GF^p})}{\sigma_{\widehat{GF_k}}\sigma_{GF^p}} \qquad (1)$$

where the spatial mean and standard deviation of the grip force matrix are respectively, $\overline{GF_k} = \frac{1}{32}\sum_i \sum_j \widehat{GF_k(i,j)}$ and $\sigma_{\widehat{GF_k}} = \sqrt{\frac{1}{32}\sum_i \sum_j (\widehat{GF_k(i,j)} - \overline{GF_k})^2}$. Likewise, the spatial mean, $\overline{GF^p}$ and standard deviation σ_{GF^p} of the mean grip force matrix for participant p are defined similarly. The NCC_2 feature determined the consistency of the grip shape within each participant.

- The total unnormalized force (TUF) of the kth signature is the sum over all sensors of the unnormalized grip force readings

$$TUF = \sum_{i=1}^{8} \sum_{j=1}^{4} GF_k(i,j) \tag{2}$$

- Grip height of the kth signature is the force-weighted average position of fingers to barrel contact [1], namely,

$$\text{Grip height} = \frac{\sum_{i=1}^{8}(i \sum_{j=1}^{4} \widehat{GF_k}(i,j))}{\sum_{i=1}^{8}\sum_{j=1}^{4} \widehat{GF_s}(i,j)} \tag{3}$$

  The units of grip height are the number of sensors above the proximal edge of the barrel (the end closest to the pen tip) and ranged from 1 to 8.

The functional representation of a signature consisted of the total grip force profile over time, $TGF(t)$, namely,

$$TGF(t) = \sum_{i=1}^{8} \sum_{j=1}^{4} F_{ij}(t) \tag{4}$$

where $F_{ij}(t)$ is the force reading at time t from the sensor on the ith row of sensors above the pen apex and j^{th} sensor strip running longitudinally down the pen barrel (See for example Figure 3A). The total grip force profile of a signature sample is shown in Figure 3C which is the sum of the 32 individual sensors' signals shown in Figure 3A. It can be seen in Figures 3A and 3B that the grip forces are mainly applied on four sensors which are the contact points with the hand. The mean total grip force profile, $TGF^p(t)$, for participant p is the curve obtained by averaging at each sampling instance, all individual signature grip force profiles for that participant, i.e.,

$$TGF^p(t) = \frac{1}{N_p} \sum_{k=1}^{N_p} TGF_k(t) \tag{5}$$

where N_p is the total number of signatures available for the pth participant. Separate mean total grip force profiles were estimated for authentic and bogus signatures. As above, we will denote the amplitude normalized versions of the force profiles as $\widehat{TGF(t)}$ and $\widehat{TGF^p(t)}$. Note that in the functional representation, normalization refers to standardization to 0 mean and unit variance. The total force signals were also time-normalized in a participant-specific manner by re-sampling the signatures to a common length, taken as the average length of the signatures of that participant. Note that bogus and authentic signatures were time-normalized separately.

The functional features used in this study are introduced below.

- The maximum total force of the kth signature is the maximum over time of the total grip force of that signature,

$$TGF_{max} = \max_t TGF(t) \tag{6}$$

- The total force interquartile range (TGF_{IQR}) is the interquartile range of the total grip force profile of each signature.

$$TGF_{IQR} = \gamma_{0.75} - \gamma_{0.25} \tag{7}$$

  where γ_q is the qth quartile of the amplitude distribution of force values in a given force profile $TGF(t)$. In other words, if $D(x)$ is the amplitude distribution of $TGF(t)$, then γ_q is defined implicitly as $q = \int_{-\infty}^{\gamma_q} D(x)dx$.
- The 1-dimensional NCC between the total force profile of the kth signature, $TGF_k(t)$ and the mean total force profile for participant p, $TGF^p(t)$, is given by

$$NCC_1\ (TGF_k, TGF^p) = \frac{1}{T}\frac{\sum_{t=1}^{T}(TGF_k(t) - \overline{TGF_k})(TGF^p(t) - \overline{TGF^p})}{\sigma_{TGF_k}\sigma_{TGF^p}} \tag{8}$$

  where T is the normalized signature duration, $\overline{TGF_k} = \frac{1}{T}\sum_t TGF_k(t)$ is the mean total grip force over the duration of the kth signature and $\sigma_{TGF_k} = \sqrt{\frac{1}{T}\sum_t(TGF_k(t) - \overline{TGF_k})^2}$ is the corresponding standard deviation. The quantities $\overline{TGF^p}$ and σ_{TGF_p} for the pth participant's average temporal force profiles are obtained in like fashion. This feature determined the consistency of the force profile across signatures (either bogus or authentic) of each participant.
- The root mean square error (RMSE) between the normalized total force profile of the kth signature ($\widehat{TGF_k(t)}$) and the normalized mean total force profile of the corresponding participant ($\widehat{TGF^p(t)}$) is defined as

$$RMSE\ (\widehat{TGF_k}, \widehat{TGF^p}) = \sqrt{\frac{\sum_{t=1}^{T}(\widehat{TGF_k(t)} - \widehat{TGF_p(t)})^2}{T}} \tag{9}$$

2.6 Data Analysis

Bogus and authentic signatures were compared on the basis of the aforementioned topographical and functional features. From previous study, it was observed that collection of multiple signatures in each session can introduce/

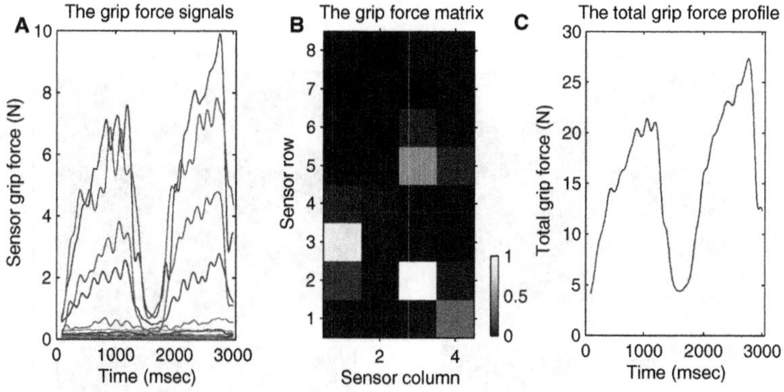

Fig. 3. Different representations of the grip force signals associated with a bogus signature shown in Figure 2. (*'A'*): a functional representation of the grip force signals on each sensor; (*'B'*): a topographical representation of the grip force signals, each square represents a sensor and its value is the normalized average force applied on that sensor over the signature duration; (*'C'*): the functional representation of the total grip force profile over the course of a signature, which is the sum of the signals shown in (*'A'*)

produce within session variability [5], therefore, sessional means and spread of extracted features were compared given their greater stability over corresponding estimates derived from individual signatures.

Given the non-normality of the features based on the Lilliefors test for normality, the non-parametric two-sample Kolmogorov-Smirnov (KS) test was invoked to test the equality of location and shape of the empirical cumulative distribution functions (CDF) of authentic and bogus signatures' features [8]. The Ansari-Bradley (AB) test was deployed to test the equality of dispersion between median-removed versions of the authentic and bogus signatures' features [14]. For these two tests, a significance level of 0.05 was employed and comparisons were performed feature by feature on a per participant basis. In addition, these tests were used to examine the nature of the inequality between the two types of signatures, such that whether the bogus signatures have higher or lower feature values/dispersion compared to the authentic signatures.

3 Results and Discussion

Figure 4 graphically represents the percentage of participants who exhibited statistically significant differences in grip kinetic features between authentic and well-practiced bogus signatures based on the KS and AB tests. The lighter the shading, the greater the percentage of participants showing differences. The left graph depicts differences in the median while the right graph summarizes differences in the interquartile range. For example, in the top-left hand corner of the median graph, 10% of participants exhibited lower NCC_2 values while writing the bogus signatures, according to the KS test. Given the overall dark shading

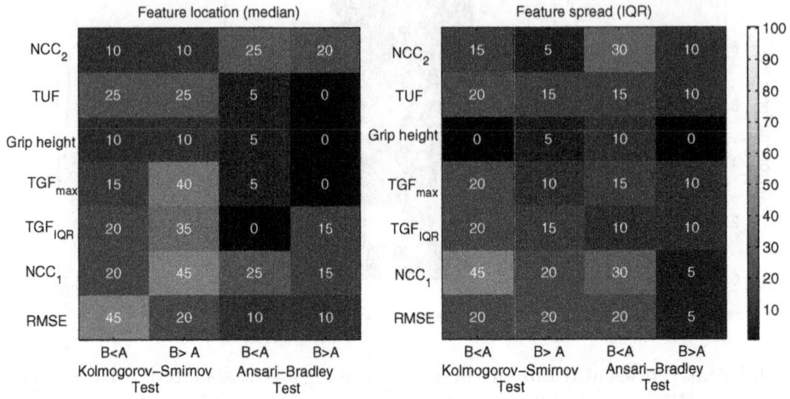

Fig. 4. Heat maps depicting the percentage of participants for whom significant kinetic differences arose between authentic and bogus signatures. Feature medians comparison is shown (*left*) and feature spread comparison is shown on the (*right*). (*Dark coloring*) denotes low percentages. Features are specified on the (*vertical axis*) while the nature of the difference (higher versus lower) between authentic (A) and well-practiced bogus (B) signatures appears on the (*horizontal axis*). (*Each numerical overlay*) corresponds to the percentage of participants showing the specified difference.

of the graphs, we observe that generally, a minority of participants exhibited significant kinetic differences between authentic and bogus signature writing.

In the following, we discuss some specific observations arising.

1. Bogus signatures were not necessarily associated with a different or more variable grip shape when compared to that of authentic signatures. Likewise, total average force and grip height were generally comparable between bogus and authentic signatures.
2. 40% of the participants applied a significantly higher maximum total grip force while writing the bogus signature. This finding is consistent with the findings in [19] and [18], who reported higher pen tip pressure secondary to increased limb stiffness for handwriting that demanded greater cognitive processing such as forging a signature. Nonetheless, in [3], the author contends that higher forces can also be associated with authentic signatures as was the case in 15% of our participants. While these papers only considered the axial pen tip force, we know that axial and grip forces are strongly correlated [1,7].
3. Just over a third of the participants had higher TGF_{IQR} values while writing the bogus rather than authentic signature. Since participants wrote the bogus specimens before the authentic signatures, the recent finding of diminishing grip force variability over a 10 minute writing session might in part explain this observation [12]. Nonetheless, in [19], it was reported that forgeries are associated with less variation in pen tip pressure, which would

appear to be the case in 20% of our participants. Some of the observed difference may have been due in part to the dissimilar demands of authentic and bogus signatures, as the extent of pen force variation has been attributed to the level of task complexity [10].

4. Just under half of the participants had higher NCC_1 values with bogus signature writing. This is corroborated by the lower RMSE values for the same participants for bogus signature writing. Both findings indicate a higher consistency of the force profile while writing the bogus signatures. A similar observation was reported in [19] where variances of spatial and kinematic variables in repeated forgeries were smaller than the variances of these variables for repeated samples of authentic writing.

Overall fewer than 50% of the participants and features had any significant differences in grip kinetic features between the bogus and authentic signature writing. This finding suggests that the two-week practice period was sufficient for participants to become familiar with the bogus signature, to the point that grip kinetics were akin to those of skilled handwriting [13].

4 Conclusions

In this paper, we compared the handwriting grip kinetics associated with repeated samples of a well-practiced bogus signature and authentic signatures for 20 adult participants. The magnitude and the extent of variability of multiple grip kinetic features were compared. In general, only a few participants exhibited any significant difference in grip kinetic features between writing authentic and bogus signatures. These differences were not consistent across participants. The kinetic variability associated with the well-practiced bogus signatures did not exceed that of the authentic signatures. These findings suggest that it is feasible to use bogus signatures to investigate the intra- and inter-subject variability of the handwriting grip kinetic profile and to identify discriminatory features for writer discrimination and signature verification purposes.

Acknowledgements. The authors would like to acknowledge the Canada Research Chairs program, and the Natural Sciences and Engineering Research Council of Canada for supporting this research financially. The authors would also like to acknowledge Ms. Nayanashri Thalanki Anantha, Ms. Jennifer Chan, Ms. Laura Bell, and Ms. Sarah Stoops for their assistance with the data collection for this study.

References

1. Chau, T., Ji, J., Tam, C., Schwellnus, H.: A novel instrument for quantifying grip activity during handwriting. Arch. Phys. Med. Rehabil. 87(11), 1542–1547 (2006)
2. Falk, T., Tam, C., Schwellnus, H., Chau, T.: Grip force variability and its effects on children's handwriting legibility, form, and strokes. Biomech. Eng. 132, 114504 (2010)

3. Franke, K.: Analysis of authentic signatures and forgeries. In: Geradts, Z.J.M.H., Franke, K.Y., Veenman, C.J. (eds.) IWCF 2009. LNCS, vol. 5718, pp. 150–164. Springer, Heidelberg (2009)
4. Ghali, B., Mamun, K., Chau, T.: Long term consistency of handwriting grip kinetics in adults. Biomech. Eng. 136(4), 041005 (2014)
5. Ghali, B., Anantha, N.T., Chan, J., Chau, T.: Variability of grip kinetics during adult signature writing. PLoS One 8(5), e63216 (2013)
6. Hermsdörfer, J., Marquardt, C., Schneider, A., Fürholzer, W., Baur, B.: Significance of finger forces and kinematics during handwriting in writer's cramp. Hum. Mov. Sci. 30(4), 807–817 (2011)
7. Herrick, V., Otto, W.: Pressure on point and barrel of a writing instrument. Exp. Educ. 30(2), 215–230 (1961)
8. Hill, T., Lewicki, P.: Statistics: methods and applications: a comprehensive reference for science, industry, and data mining. StatSoft Inc., Tulsa (2006)
9. Hooke, A., Park, J., Shim, J.: The forces behind the words: development of the Kinetic Pen. Biomech. 41(9), 2060–2064 (2008)
10. Kao, H., Shek, D., Lee, E.: Control modes and task complexity in tracing and handwriting performance. Acta Psychol. 54(1), 69–77 (1983)
11. Koppenhaver, K.: Forensic document examination: Principles and practice. Humana Press Inc., Totowa (2007)
12. Kushki, A., Schwellnus, H., Ilyas, F., Chau, T.: Changes in kinetics and kinematics of handwriting during a prolonged writing task in children with and without dysgraphia. Res. Dev. Disabil. 32(3), 1058–1064 (2011)
13. Luria, G., Rosenblum, S.: Comparing the handwriting behaviours of true and false writing with computerized handwriting measures. Appl. Cogn. Psychol. 24(8), 1115–1128 (2010)
14. Reimann, C., Filzmoser, P., Garrett, R., Dutter, R.: Statistical data analysis explained: applied environmental statistics with R. John Wiley & Sons, Chichester (2008)
15. Shang, X., Veldhuis, R.: Grip-pattern verification for a smart gun. Electron. Imaging 17, 011017 (2008)
16. Shang, X., Veldhuis, R.: Grip-pattern verification for smart gun based on maximum-pairwise comparison and mean-template comparison. In: Proc. 2nd IEEE Int. Conf. Biom.: Theory, Appl. and Syst., Washington, USA, September 29-October 1, pp. 1–5. IEEE Computer Society (2008)
17. Shim, J., Hooke, A., Kim, Y., Park, J., Karol, S., Kim, Y.: Handwriting: Hand-pen contact force synergies in circle drawing tasks. Biomech. 43(12), 2249–2253 (2010)
18. Van Den Heuvel, C., van Galen, G., Teulings, H., van Gemmert, A.: Axial pen force increases with processing demands in handwriting. Acta Psychol. 100(1), 145–159 (1998)
19. Van Galen, G., Van Gemmert, A.: Kinematic and dynamic features of forging another person's handwriting. Forensic Doc. Exam. 9, 1–25 (1996)

A Timetabling Applied Case Solved with Ant Colony Optimization

Broderick Crawford[1,2,3], Ricardo Soto[1,4,5], Franklin Johnson[1,6], and Fernando Paredes[7]

[1] Pontificia Universidad Católica de Valparaíso, Valparaíso, Chile
{broderick.crawford,ricardo.soto}@ucv.cl
[2] Universidad Finis Terrae, Santiago, Chile
[3] Universidad San Sebastián, Santiago, Chile
[4] Universidad Autónoma de Chile, Santiago, Chile
[5] Universidad Central de Chile, Santiago, Chile
[6] Universidad de Playa Ancha, Valparíso, Chile
franklin.johnson@upla.cl
[7] Escuela de Ingeniería Industrial, Universidad Diego Portales, Santiago, Chile
fernando.paredes@udp.cl

Abstract. This research present an applied case of the resolution of a timetabling problem called the University course Timetabling problem (UCTP), the resolution technique used is based in Ant Colony Optimization metaheuristic. Ant Colony Optimization is a Swarm Intelligence technique which inspired from the foraging behavior of real ant colonies. We propose a framework to solve the University course Timetabling problem effectively. We show the problem and the resolution design using this framework. First we tested our proposal with some competition instances, and then compare our results with other techniques. The results show that our proposal is feasible and competitive with other techniques. To evaluate this framework in practice way, we build a real instance using the case of the school of Computer Science Engineering of the Pontifical Catholic University of Valparaíso and the Department of Computer Engineering at Playa Ancha University.

Keywords: Ant Colony Optimization, Swarm Intelligence, University Course Timetabling Problem.

1 Introduction

The timetabling problems are commonly faced by many institutions as schools and universities. The basic problem is defined as a set of events that must be assigned to a set of timeslot of a way that all the students can attend to all of their respective events. With the reservation of which hard constraints necessarily must be satisfied, and soft constraints that deteriorate the quality of the generated timetabling. Of course, the difficulty can vary in any particular case of the UCTP. [10]. The problem difficulty depends on many factors and in addition the assignment of rooms makes the problem more difficult. Many techniques have

R. Silhavy et al. (eds.), *Artificial Intelligence Perspectives and Applications*,
Advances in Intelligent Systems and Computing 347, DOI: 10.1007/978-3-319-18476-0_27

been used in the resolution of this problem. We can find evolutionary algorithms, tabu-search, constraint programming and genetic algorithms [2]. We present the resolution using an Ant Colony Optimization (ACO) algorithm through the implementation of Hypercube framework (HC). ACO is a Swarm Intelligence technique which inspired from the foraging behavior of real ant colonies. The artificial ants seek the solutions according to a constructive procedure as described in [9]. This ACO exploits an optimization mechanism for solving discrete optimization problems in various engineering domain [8]. We establish the representation for the problem to be solved with ACO, generating an appropriate construction graph and the respective pheromone matrix associated. In the following sections we present the UCTP problem, and the ACO Metaheuristic. Later we present the experimental results. Finally the conclusions of the work appear.

2 University Course Timetabling Problem

The UCTP is an adaptation of an original timetabling problem presented initially by Paechter in [13,12]. It consists of a set of events E and must be scheduled in a set of timeslots $T = \{t_1, ..., t_k\}$ ($k = 45$, they correspond to 5 days of 9 hours each), a set of rooms R in which the events will have effect, a set of students S who attend the events, and a set of features F required by the events and satisfied by the rooms. Each student attends a number of events and each room has a maximum capacity.

We present below a mathematical formulation of the problem. The simplest formula for this problem can be described as a problem of binary integer programming numbers in which the variable $X_{ij} = 1$ if course i is assigned to the classroom j is equal to 0 otherwise . The time in which a classroom can take in a day is divided into k periods:

$$\sum_{i=1}^{m}\sum_{j=1}^{n} c_{ij}x_{ij} \tag{1}$$

$$\sum_{j=1}^{n} x_{ij} = 1, \forall i \tag{2}$$

$$\sum_{i \in P_k}^{n} x_{ij} \leq 1, \forall i, \forall k \tag{3}$$

$$x_{ij} \in \{0,1\} \forall i \forall j \tag{4}$$

where P_k is the set of all courses offered in the period k, and C_{ij} is the cost of assigning the course i to the classroom j. The first constraint ensures that each course is assigned to one classroom. The second in a given period, in most courses offered i during the period k is assigned to each classroom. Implicitly each course is assigned to an only classroom.

A feasible timetable is one in which all the events have been assigned to a timeslot and a room so that the following hard constraints **H[1-3]** and soft constraints **S[1-3]** are satisfied.

- **H1:** No student attends more than one event at the same time.
- **H2:** The rooms must be sufficiently great for all students who attend a class and to satisfy all the features required by the event.
- **H3:** Only one event per each room at any timeslot.

In addition, all possible generated timetables are penalized for the number of soft constraint violated. these constrint appear next:

- **S1:** A student has a class in the last slot of the day.
- **S2:** A student has more than two classes in a row.
- **S3:** A student has exactly one class on a day.

Feasible solutions are always considered to be superior to infeasible solutions, independently of the numbers of soft constraint violations (SCV). One feasible solution is better than another, if it minimizes the SCV.

3 Framework for UCTP

According to the constraints presented in the previous section and the characteristics of the problem, we can now consider the option to design an effective scheme for the UCTP. We have to decide how to transform the assignment problem (to assign events to timeslots) into an optimal path problem which the ants can solve [3] and then optimize the problem.

We propose the following: an instance of the problem is received as input, then it assigned events to a timeslot, later a matching algorithm [14,11] is used for makes the assignation from rooms to each one of events associated to timeslot. In this point a solution is complete, but a low quality one. Then a local search algorithm [4] is applied that improves the quality of the solution and gives as final optimized result

3.1 Using Max-Min Ant System

Ant colony optimization is a metaheuristic algorithm based on a graph representation in which a colony of artificial ants cooperate in finding good solutions to discrete optimization problems [9]. The ants travel through the construction graph starting from an initial point and selecting the nodes which travel according to a probability function that is given by the pheromone and heuristic information of the problems.

We choose the Max-Min Ant System (MMAS) algorithm to solve the UCTP. The Max-Min Ant System is one of the best performing ACO algorithms [16]. MMAS can easily be extended by adding local search algorithms.

Contruction Graph: One of the main elements of the ACO metaheuristic is to model the problem on a construction graph [7,5], that way a trajectory through the graph which represents a problem solution. In this formulation of the UCTP is required to assign each one of $|E|$ events to $|T|$ timeslots. Where direct representations of the construction graph is given by $E \times T$; given this graph we can then establish that the ants travel throughout a list of events choosing timeslot for each event. The ants follow one list of events. For each event, the ants decide timeslot t, each event is a single time in a timeslot, thus in each step an ant chooses any possible transition as showed in figure 1.

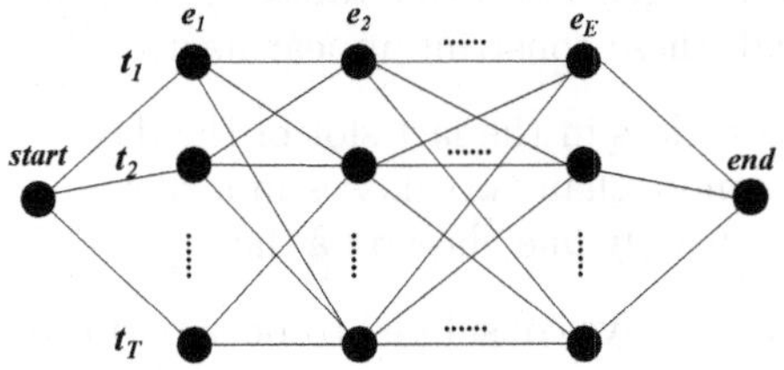

Fig. 1. Construction graph: For each event e, the ant chooses a timeslot t

Now we present the probabilistic function. This function adapting to MMAS according to HC and allows the ants travel through the construction graph selecting a path. We use the probability function, defined in [7]. This function directly depends on the pheromone information τ, and the importance is determinate for the parameter α, and the heuristic information η is determinate for β, for the possible path for a k ant.

$$p_{ij}^{k} = \frac{[\tau_{ij}]^{\alpha} [\eta_{ij}]^{\beta}}{\sum_{l=0}^{T} [\tau_{il}]^{\alpha} [\eta_{il}]^{\beta}}, j \in \{1, ..., T\} \tag{5}$$

The pheromone matrix represents the pheromones in the path where the ants travels, and indicates the absolute position where the events must be placed. With this representation the pheromone matrix the pheromone does not depend on the partial assignments.

Heuristic Information and Pheromone Update: We define as heuristic information a function that calculates a weighted sum of a set of the soft and hard constrains in each assignation. This function has a high computational cost [14]. In the hypercube framework the pheromone trails are forced to stay in the interval $[0, 1]$. We represent computationally the evaporation of pheromone and in addition the amount of pheromone in the best ant path through the graph, once is completed a tour. The pheromone update rule for MMAS to UCTP is as follows:

$$\tau_{ij} = \rho\tau_{ij} + (1 - \rho)\Delta\tau^{kbest} \tag{6}$$

where ρ is a rate of evaporation $\rho \in \,]0, 1]$. And $\Delta\tau$ it is associated with quality of the current solution of ant *kbest*. We can use an updating pheromone strategy considering the quality of timetabling solution:

$$\Delta\tau^{kbest} = \frac{1}{fitnessfunction(kbest)} \tag{7}$$

where *fitnessfunction* determines the quality of the solution of *kbest* ant, according to the SCV.

3.2 UCTP-MMAS Algorithm

The general structure of the algorithm is presented, in which some modifications added to the ones presented in [14,16]. A new assignation values to τ_{max},τ_{min}, a new pheromone update rules.

Algorithm 1. UCTP-MMAS Algorithm

```
Input: Problem Instance I
initialize feromone values τ_max and τ_min
calculate dependencies between events e_i, i ∈ {1, ..., E}
sort Events according dependencies
repeat
  for a = 1 to m do
    construction process of ant a
    for i = 1 to |E| do
      chooser timeslots t_j according to probabilities p_i,j for de event e_i
      storing partial route for k_a ant
    end for
    apply matching algorithm for assign rooms
    select the best solution for iteration
  end for
  applying local serach to best solution according to the fitness
  select the best global solution
  apply pheromone update for k_best ant
until the termination condition is satisfied (iterations or time)
Output: An optimized solution for I
```

Only the solution that causes the fewest number of hard constraint violations is selected for improvement by the Local Search. The pheromone matrix is updated only once by each iteration, and the global best (k_{best}) solution is used for the update. The description is illustrated in algorithm 1. This algorithm use it a matching function to associate events with rooms. To optimize the solution uses 1OPT, 2OPT and 3OPT Local Search algorithm [1].

4 Comparisons

The algorithm was implemented and submitted to a series of tests. The behavior of the proposed framework was observed in the resolution of the UCTP.

Instances of the UCTP are structured using a generator[1]. This generator allows generating classes of small, medium instances which reflect varied timetabling problems. In addition it was used a series of 20 instances created for International Timetabling Competition[2], these instances are made with the same generator. The instances with different parameters are presented in the following table.

Table 1. Parameter for the small and medium instances

Parameter	small	medium
Number of events	100	400
Number of rooms	5	10
Number of features	5	5
Features by room	3	3
Usage percentage	70	80
Number of students	80	200
Events Maximum for students	20	20
Maximum students per event	20	20

We firstly studied the best parameters configuration using the small instances. The UCTP-MMAS was tested without local search, making an evaluations with different ants numbers m and with different evaporations factors ρ, the parameters of $\alpha = 1$, number on attempts = 10 and a maximum time by attempt = 90 seconds for all the tests. The results are in table 2:

Table 2. Evaluation of parameters m and ρ using small1.tim

Parameter	m			ρ		
Values	5	10	20	0.2	0.5	0.8
SCV	17	16	16	15	13	17
Seconds	6.79	7.46	6.06	7.11	8.1	6.79

According to table 2, we observe that the best results are obtained using the parameter $m = 20$, obtaining an evaluation of 16 in 6.06 seconds. And for the case of evaporation factor, the best value is $\rho = 0{,}5$ in 8.1 seconds.

4.1 Comparative Results

Table 3 presents comparative results between the solutions obtained for different instances[3] the UCTP solved with different techniques such as Simulated annealing (SA), Advanced Search (AS) and Simulated Annealing with Local Search (SA-LS). These algorithms are compared according to SCV.

[1] http://www.dcs.napier.ac.uk/~benp
[2] http://www.or.ms.unimelb.edu.au/timetabling
[3] http://www.idsia.ch/Files/ttcomp2002/

Table 3. Number of SCV obtained with International Timetabling Competition instances

Algorithm	com01	com02	com03	com04	com05	com06	com07	com08	com09	com10
SA	45	25	65	115	102	13	44	29	17	61
AS	257	112	266	441	299	209	99	194	175	308
SA-LS	211	128	213	408	312	169	281	214	164	222
UCTP-MMAS	240	133	204	426	406	179	261	204	157	263
Algorithm	com11	com12	com13	com14	com15	com16	com17	com18	com19	com20
SA	44	107	78	52	24	22	86	31	44	7
AS	273	242	364	156	95	171	148	117	414	113
SA-LS	196	282	315	345	185	185	409	153	281	106
UCTP-MMAS	268	212	341	329	172	234	371	124	245	101

For these instances and compared with the other solutions, the UCTP-MMAS presents two characteristics to evaluate. First, it has the capacity to generate feasible solutions for these instances. These instances are difficult because they are from competitions Timetabling. Second, the quality of the generated solutions is very low compared with to Simulated Annealing, which has the best found historical results for these instances, but in comparison with the other instances it does not present great difference. It is not possible to decide if a technique is better than other, since the differences in results can be explained by different external agent.

Table 4 presents the comparison for the small and medium instances for the algorithm for UCTP with HC and local search (UCTP-MMAS) and MMAS pure (MMAS-p).

Table 4. It present the SCV obtained with small and medium instances for UCTP-MMAS and MMAS-p

Algorithm	small1	small2	small3	medium1	medium2
UCTP-MMAS	**0**	**4**	**1**	**138**	**186**
MMAS-p	3	6	3	152	250

We can observe for these instances that the UCTP-MMAS present a superiority in the quality of the generated solutions (smaller SCV). We can say the our proposed improves the quality of the ant algorithm applied. Table 5 presents the comparison with other ACO algorithm such as Ant Colony System algorithm of Krzysztof Socha (ACS) and to algorithm based on Random Restart Local Search (RRLS).

According to the results of UCTP-MMAS performs better than the other algorithms for small and medium instances, improving in all tested instances. only in the medium2 instance was surpassed by ACS.

Table 5. Results obtained with small and medium instances

Algorithm	small1	small2	small3	medium1	medium2
UCTP-MMAS	**0**	**4**	**1**	**138**	186
ACS	1	3	1	195	**184**
RRLS	11	8	11	199	202

4.2 Practical Case

To test this project on a practice way, we implement a the UCTP-MMAS with 2 real cases. We created an instance of the Pontificia Universidad Católica de Valparaíso (PUCV) and specifically for the school of Informatics Engineering. We implemented a tool in C language, to enter the courses, semester, assistants and assistantship, and indicate the times to the week that are dictated and his characteristics. In addition the number of rooms and their characteristics are entered to him. The system generates an instance introducing a factor of correlation between the events, generated an instance with the same format as competition instances. Stored this information, the algorithm is ready to be used. Table 6 presents the characteristics for the PUCV instance.

Table 6. Characteristics of UCV instance

Characteristic	**value**
Rooms and lab	9
Events	194
Total Attending	600
Features	5
maximum events by student	8
maximum students by event	20-45

Before using the instance it was necessary to correct some parameters of the algorithm implemented, since for the instance of PUCV the number of timeslot that they are used are 40 and not 45 like for other problems of the UCTP. In addition we create an adaptation to the soft constraint.

The instance was executed using a number of ants m= 20, evaporation factor ρ= 0.5. Time to local search 100 seconds, total time by reboots = 900 seconds, number of reboots = 10. The best solution was obtained approximately in 600 seconds with an evaluation of SCV = 0, which implies that the algorithm generated a complete timetable feasible and with the best possible quality.

For the quality of the obtained solution, it can be inferred that the generated instance that simulate the hour load of a semester of the school of computer science engineering had a low degree of correlation between courses of different semesters, thus a high performance solution was obtained.

We also implemented a real instance of Universidad de Playa Ancha (UPLA). This instance has 40 timeslots and their characteristics are presented in table 7:

Table 7. UPLA instance characteristic

Characteristic	value
Rooms and lab	22
Events	112
Total Attending	154
Features	5
maximum events by student	8
maximum students by event	40

The instance was executed using a number of ants $m=15$, evaporation factor $\rho=0.01$, total time by reboots = 600 seconds, number of reboots = 2. The best solution was obtained approximately to the 270 seconds with an evaluation of SCV = 0, which implies that the algorithm generated a feasible solution. This occurs because the instance is very simple, given the high number of rooms available, the few events to program and the low number of attendants.

5 Conclusion

In this research we have presented a formal model in order to apply the Hypercube framework to solve the University course timetabling problem (UCTP) making use of Max-Min Ant System, an efficient model was generated to solve instances of this problem creating good construction graph and a good pheromone matrix. We presented the test result made for the UCTP-MMAS. We observed that the UCTP-MMAS presented good results for instances of small and medium. Although the results were of low quality for the instances of the competition, we emphasize the fact that our approach always generates feasible solutions and for instances of normal difficulty have a good evaluation. We applied our algorithm to solve a real instance to the school of Computer Science of the PUCV and UPLA, for which created a feasible solution, this validates the use of a technique useful in real applications. As future work, we hope to improve the proposed algorithm and develop a suitable interface to apply the algorithm to other real instances and integrate the constraints of teachers in future instances. In addition we will try to integrate our algorithm with Autonomous Search [15,6].

Acknowledgments. Franklin Johnson is supported by Postgraduate Grant PUCV 2014, Broderick Crawford is supported by Grant CONICYT/FONDECYT/REGULAR/1140897, Ricardo Soto is supported by Grant CONICYT/FONDECYT/INICIACION/11130459, And Fernando Paredes is supported by Grant CONICYT/ FONDECYT/REGULAR/1130455

References

1. Abuhamdah, A., Ayob, M., Kendall, G., Sabar, N.: Population based local search for university course timetabling problems. Applied Intelligence, 1–10 (2013)
2. Babaei, H., Karimpour, J., Hadidi, A.: A survey of approaches for university course timetabling problem. Computers and Industrial Engineering (2014) (in press)
3. Blum, C., Dorigo, M.: Hc-aco: the hyper-cube framework for ant colony optimization. In: Proc. MIC 2001-Metaheuristics Int. Conf., pp. 399–403 (2001)
4. Blum, C., Dorigo, M.: The hyper-cube framework for ant colony optimization. Trans. Sys. Man Cyber. Part B 34(2), 1161–1172 (2004)
5. Blum, C., Roli, A., Dorigo, M.: Hc-aco: The hyper-cube framework for ant colony optimization. IEEE Transactions on Systems, Man, and Cybernetics–Part B, 399–403 (2001)
6. Crawford, B., Soto, R., Castro, C., Monfroy, E.: Extensible CP-based autonomous search. In: Stephanidis, C. (ed.) Posters, Part I, HCII 2011. CCIS, vol. 173, pp. 561–565. Springer, Heidelberg (2011)
7. Dorigo, M., Di Caro, G.: The ant colony optimization meta-heuristic. In: New Ideas in Optimization, pp. 11–32. McGraw-Hill Ltd., UK (1999)
8. Dorigo, M., Gambardella, L.M.: Ant colony system: A cooperative learning approach to the traveling salesman problem. IEEE Transactions on Evolutionary Computation (1997)
9. Dorigo, M., Stutzle, T.: Ant Colony Optimization. MIT Press, USA (2004)
10. ten Eikelder, H.M.M., Willemen, R.J.: Some complexity aspects of secondary school timetabling problems. In: Burke, E., Erben, W. (eds.) PATAT 2000. LNCS, vol. 2079, pp. 18–27. Springer, Heidelberg (2001)
11. Johnson, F., Crawford, B., Palma, W.: Hypercube framework for ACO applied to timetabling. In: Bramer, M. (ed.) Artificial Intelligence in Theory and Practice. IFIP, vol. 217, pp. 237–246. Springer, Boston (2006)
12. Paechter, B.: Course timetabling evonet summer school (2001), `http://evonet.dcs.napier.ac.uk/summerschool2001/problems.html`
13. Paechter, B., Rankin, R.C., Cumming, A., Fogarty, T.C.: Timetabling the classes of an entire university with an evolutionary algorithm. In: Eiben, A.E., Bäck, T., Schoenauer, M., Schwefel, H.-P. (eds.) PPSN 1998. LNCS, vol. 1498, pp. 865–874. Springer, Heidelberg (1998)
14. Socha, K., Knowles, J., Sampels, M.: A $\mathcal{MAX}-\mathcal{MIN}$ ant system for the university course timetabling problem. In: Dorigo, M., Di Caro, G.A., Sampels, M. (eds.) ANTS 2002. LNCS, vol. 2463, pp. 1–13. Springer, Heidelberg (2002)
15. Soto, R., Crawford, B., Monfroy, E., Bustos, V.: Using autonomous search for generating good enumeration strategy blends in constraint programming. In: Murgante, B., Gervasi, O., Misra, S., Nedjah, N., Rocha, A.M.A.C., Taniar, D., Apduhan, B.O. (eds.) ICCSA 2012, Part III. LNCS, vol. 7335, pp. 607–617. Springer, Heidelberg (2012)
16. Stützle, T., Hoos, H.H.: Max-min ant system. Future Gener. Comput. Syst. 16(9), 889–914 (2000)

Visualization of Semantic Data

Martin Žáček, Rostislav Miarka, and Ondřej Sýkora

Department of Computers and Informatics, University of Ostrava
{martin.zacek,rostislav.miarka}@osu.cz,
ondrej.sykora@outlook.com

Abstract. The main goal of the Semantic Web is to direct the current syntactic web on the path to the Semantic Web. The vision of the Semantic Web is to interpret information on the web to be readable and machine-interpretable. Therefore, the article focuses on the creation of an instrument for visualizing semantic data on the basis of the identified advantages and disadvantages of existing applications for visualization, which are currently available.

Keywords: Semantic web, RDF model, Schema RDF, URI, SPARQL, graph, tool.

1 Introduction

There are several tools for visualization of semantic data; therefore we will focus only on editors of the RDF (Resource Description Framework) model or their applications, which provide only data visualization.

These applications are basically divided into two groups. One group of applications used to control text commands or enable editing of the file only in a text editor (e.g. Morla or BrownSauce) [1] [2] and the second group used a graphical user interface. For a greater user comfort it was decided that further work will consider only applications with a graphical user interface.

2 RDF Model

RDF data can be viewed in three representations: as a graph, triples or XML representation. The graph view is the simplest and most understandable for the user. The trio is best available for application software, which uses the triples as an input to their operations. The XML version is suitable for the transmission of RDF data between computers. In terms of logic, these three forms of view are equivalent. [3] The graph view of RDF data is the simplest form possible of its modeling and is often used for simple and intuitive display of an RDF model. The triples constituting assertion can be understood as part of an oriented graph with labeled nodes where edges represent a named relationship between two sources. The sources are marked in the graph as nodes. An RDF graph consists of a plurality of RDF triples. The orientation of edge of entity is routed from node to the object node. The RDF directly represents only binary

R. Silhavy et al. (eds.), *Artificial Intelligence Perspectives and Applications*,
Advances in Intelligent Systems and Computing 347, DOI: 10.1007/978-3-319-18476-0_28

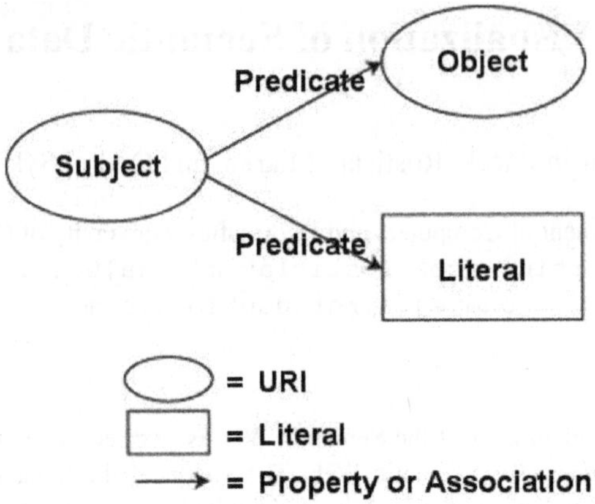

Fig. 1. An RDF triple of RDF model [6]

relationships. [4] Displaying using a directed graph can run assertion consecutively and the subject of one assertion may be the subject of another assertion. [5]

2.1 URI

Data sources in the semantic web applications, or in RSS or FOAF, are represented by identifiers (URI - Uniform Resource Identifier). URI is a text string with a defined structure used to unambiguously identify web sources. Data representation using identifiers is a simple way to refer to data found on the web. Despite frequent usage of identifiers with RDF when representing data source, data representation using an identifier is not compulsory. In fact, identifiers do not have to be present in RDF at all.

http://www.w3.org/1999/02/22-rdf-syntax-ns#type

Fig. 2. An example of URI string

Definition of URI-link
URI-link is a UNICODE string which

- Does not contain specified control characters
- Can contain arbitrary fragment identifier (separated by #)

An advantage of an RDF model is that it also concerns a graph model, which means that any data represented by RDF can be graphically represented with oriented edges. The graph can be also written as a set of triples.

2.2 RDF Triples

RDF provides a universal and flexible way to break down complex knowledge into individual elements. These elements are called RDF triples. The triples are used to graphically represent knowledge using an oriented graph.

Each edge in the graph represents a relation between two objects. For instance, a node "Nicolas Cage" connected by an edge "starred in" with a node "Lord of War" means that actor Nicolas Cage acted in film Lord of War. Each fact represented in the graph is thus defined by the following three parts:

- Subject
- Predicate
- Object

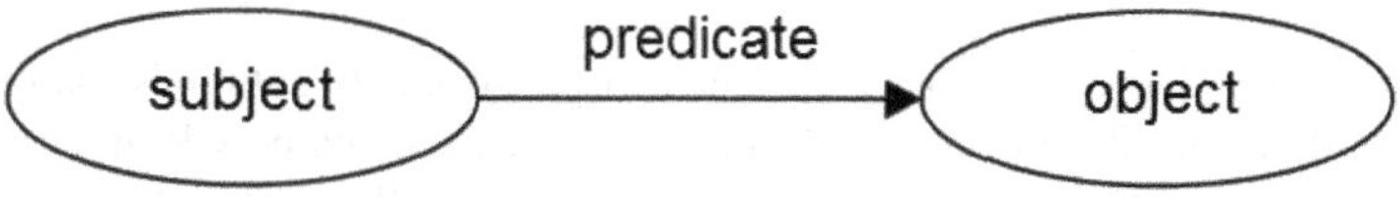

Fig. 3. The RDF triple

A subject in RDF is formed by either URI or an blank node. Sources marked using an blank node are marked as anonymous sources. It concerns sources that are not identifiable by RDF. Predicates in RDF are formed by identifiers of sources representing a relation between the subject and object. Object can be represented by identifiers, blank nodes, or as text strings alias literals.

Literals
A literal in RDF represents a simple text string. It can be used to represent an object in an RDF triple. Unlike names or identifiers, which serve as representatives or real sources, literals are mere text data. They can be used, for example, to name people or as ISBN for books.

Blank Nodes
A blank node in RDF is a node which does not have any defined source identifier or literal. Blank nodes are also marked as anonymous nodes. According to RDF specification, blank nodes can be used as a subject or object in an RDF triple. Blank nodes graph can be found in an RDF only if we want to point out existence of a thing without its accurate identification.

Syntax
As RDF itself is an abstract format providing only the way of notation of resource information, it does not have any defined syntax. In order to give RDF a notation legible by humans, we can use various languages, but the most used is the XML markup language thanks to its versatility and popularity. W3C also defined XML/RDF, which is a particular syntax taking bases from XML. [7]

3 Current Tools for Data Visualization

IsaViz

The IsaViz editor was created as an output of postgraduate work by Emmanuel Pietrgo for the W3C consortium at Massachusetts Institute of Technology (MIT). This modeling tool written in Java offers attractive processing of RDF graph representation without any experimental conception. RDF graph visualization is performed by engine Grphviz (Graphviz is an open-source6 software for representing structured information using a graph).

IsaViz is equipped with a number of functionalities, such as full-text search, easy navigation in the graph (rule, move). Control elements are, however, optimized for mouse control, which makes them too small for touch control.

RDFAuthor

RDFAuthor is graphic editor based on the Java platform. Its author is Damian Seer, who got familiar with RDF in one of his preceding projects, namely RDF text browser called BrownSace. Although this RDF editor was initially started to be developed only for the MacOS X operating system, a bit later was released a version for Windows - thanks to the Java multiplatform language. This made it possible to include this editor in the set of the tested programs.

Individual control elements are in contrast to untraditional colorful processing of RDF graph representation, but individual information in the graph is quite clear. Control elements were big enough for touch control. However, when testing this editor, creating relations between objects was not possible with no apparent reasons.

RDF Gravity

The tool RDF Gravity (RDF Graph Visualization Tool) was included in the test set due to low number of found RDF graphic editors. It does not represent a full-featured RDF editor, but rather a tool for displaying RDF graphs. It is also necessary to point out that the program could not be initially started up due to the following security alert:

„java.lang.SecurityException: com.sun.deploy.net.JARSigningException: Found unsigned entry in resource: http://semweb.salzburgresearch.at/apps/rdf-gravity/jws/colt.jar“

When eliminating this error, the software proved to be quite a sophisticated implementation for displaying information in an RDF graph and might inspire development the output visualization tool.

RDFet

RDFed was developed as a result of a diploma thesis by Helena Šestáková at Masaryk University in Brno. [8] The tool was created at the time when most of the mentioned tools did not exist. The tool can be used to display RDF files. Unfortunately, it does not support graph editing. RDFed is implemented in Java and RDF data processing uses RDF API. [9]

Protégé

Protégé is one of the most sophisticated editors of ontologies and knowledge bases that are currently available. Its development took place under Stanford and Manchester University, but a large number of it parts in the form of plug-in modules was developed by independent developers. Protégé supports creation, editing, and export of ontologies and knowledge bases in all basic formats (RDF, RDFS, OWL, etc.). [10] A highly complex user interface of the application is divided into several tabs representing different views of the displayed data.

WebOnto

WebOnto is an ontology editor available in a form of a java applet. Thus it offers a possibility to browse and edit ontologies on the web. WebOnto was developed as part of projects PatMan, HCREMA and Endrich, and currently it is used in projects PlanetOnto nebo ScholOnto.

4 Evaluation of the Tools

One of the objectives was to create a suitable tool for visualization of semantic data which could be used not only as an educational tool. In order that such a tool could be usable, it should meet certain criteria.

The most important criterion is RDF data editing itself. Although low number of the found tools resulted in including tools that only display data, such tools can contribute the development of the final application, for example in proposing the graphic look of an RDF graph.

The look of the RDF graph representation itself is also very important for us. When analyzing the tools, we came across RDF graph visualization that was very unclear (e.g. RDFAuthor). Such visualization can be very disturbing in education and heaps of colorful objects very badly legible.

The application should be very easy to control, both on a computer and projection screen, touch screen respectively. Thus its control elements should be big enough. Another selected criterion is intuitive control. Although this quality is difficult to evaluate, it is important that the application offered simple and clear control.

Last but not least, the tested tools offered to export the final RDF graph into various file formats. Graph export as an image is a much appreciated part of an application, e.g. if you need to insert the graph into your work in a form of an image. Export into an RDF XML file is handy when you open a file of a different format (OWL) and you need to save it as an RDF.

RDF Gravity had a bit of problems with it start-up caused by a security exception of the Java language. This error was later eliminated by changing one of the application libraries. RDFAuthor, on the other hand, did not allow to add an edge into the RDF graph, which made the editor mostly unusable. Therefore the tow applications were finally marked as non-functional.

Table 1. Results of programs [12]

	IsaViz	RDF Author	RDF Gravity	Protégé	RDFed
RDF editing	Yes	No	No	Yes	No
Clear RDF graph	Yes	Yes	Yes	No	No
Touch control	No	Yes	No	No	No
Intuitive control	3/5	4/5	5/5	2/5	1/5
XML, IMG export	Yes	Yes	Yes	Yes	No
Functional	Yes	No	No	Yes	Yes

5 Implementation of a Tool for Visualization of Semantic Data

5.1 Requirements

As the results of the analysis of current tools for RDF visualization and editing show, no tool meets all criteria that have been set. Thus the primary objective of this paper was to develop a new graphic editor that would meet all the set criteria best. [12]

Partial Objectives

1. RDF graph visualization
   (a) Using standard RDF graph marks
   (b) Support of touch control
2. Export of RDF into other formats
   (a) Export into an image

5.2 Implementation

Before proposing the application itself, it was necessary to define principles to be followed. Therefore all the criteria that have been set to be met were broken down in details.

The main requirement on the developed tool was its support of work with RDF files. As it will concern an RDF editor, visualization will be obviously supplied with editing data of an open RDF graph.

Another criterion focused on certain clarity of the visualized graph. Thus visualization will be displayed in a standard form of ellipsis for graph nodes and arrows to represent graph edges in the direction of the edge orientation.

One of the criteria was the possibility of easy touch control, it is necessary to carefully consider layout and size of the elements. Many tools for work with RDF data are primarily mouse-control oriented, thus they provide only small control elements. New applications with the possibility of touch control should dispose of control elements of a larger size.

Concerning controlling the application, there is one more criterion, namely intuitive control. It is not easy to decide whether control is intuitive or not due to a very subjective judgments. Nevertheless, we can claim that if the control is similar to most of applications that the user has worked with, our application control will be intuitive for him. As the application will be developed for the Windows operating system, it has been set that the application will be controlled using a standard system main menu.

Next fundamental functionality was the criterion of RDF graph data export into selected image formats. The PNG format was selected as the principal export image format.

5.3 Tools

Before starting the development, we had to define which tools to use during its development. As mentioned before, the application will be developed for the Windows operating system. In particular, it will be developed using Windows Forms API included in Microsoft .NET Framework. The development will be performed in the development environment Visual Studio (version 2012). The whole application will be written in a high-level object-oriented language C#.[12]

The following frameworks will also be used:

- Diagram.NET – to simplify work with RDF graph representation.
- dotNETRDF – to simplify work with adding, removing, and editing of RDF graph elements.
- Force-Directed drawing algorithm – for the purposes of calculating the layout of elements in an RDF graph.

5.4 Architecture of the Application

The application (Fig. 4) has been divided into three modules. The main part is for visualization of and manipulation with an RDF graph. This part is located in a separate project and designed as an independent control element. In addition, this part consists of a diagram which is responsible for visual mediating of an RDF graph. The diagram also contains a document that contains the RDF itself. The document includes all methods for with an RDF graph.

The second part is responsible for launching the application and displaying a form and all control elements. The form also includes a control element of a diagram type from the previous part of the application. Control elements from this part of the application control the control element of the diagram. This solution secures reusability of the diagram element in other parts of the application, or in other applications respectively.

The third part of the application is a project which serves as a framework for work with RDF data. The project is used to upload, save and edit RDF files in the document environment in the first part of the application. [12]

5.5 Manipulation with RDF

The principal task was to solve how to manipulate RDF data within the application. This was solved by using the dotNETRDF library. The dotNETRDF library was developed by Rob Vesse, Ron Michael Zettlemoyer, Khalil Ahmed, Graham Moore, and Tomasz Pluskiewicz. The library is available under license creative commons, particularly by its branch CC BY 3.0. The library disposes of an extensive documentation and wide user base. The library also lays basis for other useful tools for work with RDF files.

(https://bitbucket.org/dotnetrdf/dotnetrdf/wiki/UserGuide/Tools). [11]

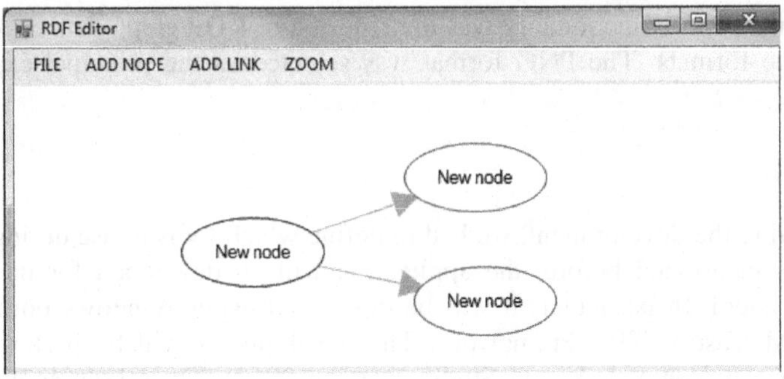

Fig. 4. RDF Editor [12]

6 Export

One of the most important functions of the editor is the possibility to export an open graph into several formats. The main format is the RDF format. However, there is a possibility to export RDF graphs into various image formats, such as PNG, BMP, JPEG or GIF (Fig. 5).

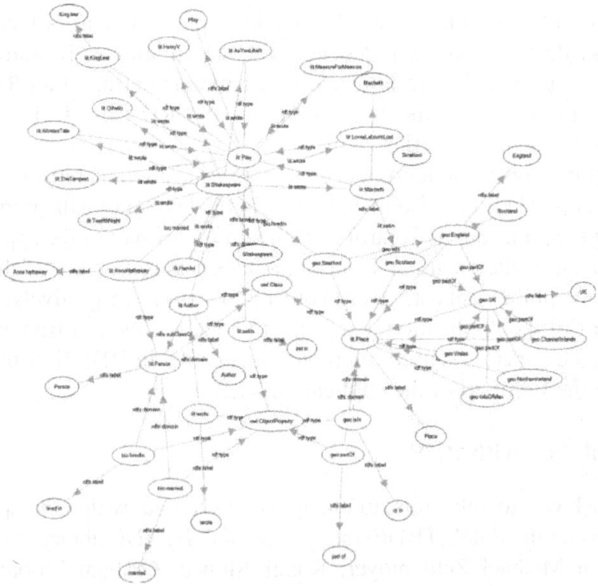

Fig. 5. Export of RDF Editor [12]

7 Conclusion

The aim of this work was to develop an RDF graphical editor which would be usable both for work with RDF graph and as an educational tool for lessons on semantic data. This need led to the development of simple tool, which was designed to meet the set criteria best.

Next step in the development of this editor is querying an RDF graph using the SPARQL language. In the case of further development of the editor, possibilities are open. A good choice to enhance its functionality would be adding functions that would deal with graph editing from the ontology point of view. Another suitable functionality would be implementation of full-text search within the graph structure or a possibility to edit more RDF graphs simultaneously in the environment of one workspace, similarly to that in tool Protégé.

Acknowledgments. The research described here has been financially supported by University of Ostrava grant SGS/PřF/2015. Any opinions, findings and conclusions or recommendations expressed in this material are those of the authors and do not reflect the views of the sponsors.

References

1. Steer, D.: "BrownSauce" (2007), `http://brownsauce.sourceforge.net/`
2. Marchesini, A.: "Morla" (2007), `http://www.morlardf.net/`
3. Hjelm, J.: Creating the semantic Web with RDF: Professional. Wiley, New York (2001)
4. Manola, F., Miller, E.: R. P. W. Recommendation (2004), `http://www.w3.org/TR/2004/REC-rdf-primer-20040210/`
5. Houben, G.-J.: Web Information Systems: Wis Data, Semantic Web, RDF(S) (2008), `http://wwwis.win.tue.nl/~houben/wis/`
6. Štencek, J.: Užití sémantických technologií ve značkovacích jazycích (2009), `http://vse.stencek.com/`
7. W3C, RDF/XML Syntax specification (Revised) (2004), `http://www.w3.org/TR/REC-rdf-syntax/`
8. Šestáková, H.: Editor of RDF metadata (in Czech), thesis, Fakulta informatiky, Masarykova univerzita, Brno (2000)
9. S. University: Sergey Melnik: RDF API Draft (2001), `http://infolab.stanford.edu/~melnik/rdf/api.html`
10. S. University: Protégé (2014), `http://protege.stanford.edu/`
11. Vesse, R.: dotNetRDF (2014), `http://www.dotnetrdf.org/`
12. Sýkora, O.: Visualization of semantic data (in Czech), thesis. Faculty of Science, University of Ostrava, Ostrava (2014)

Lexical Similarity Based Query-Focused Summarization Using Artificial Immune Systems

Sulabh Katiyar and Samir Borgohain

Department of Computer Science & Enginnering,
National Institute Of Technology, Silchar, 788010, India

Abstract. Query Focused Summarization has been explored mostly with statistical or graph based methods which haven't utilised semantic similarity between words. Graph Based methods which use sentence to sentence comparisons do not utilize lexical relations between words fully due to entailing complexity of finding relationship among all words. Lexical Chaining Methods which are used in Generic Text Summarization systems also utilize only a limited set of word types such as nouns. They do not utilize the full potential of Semantic Similarity measures by overlooking sentence to sentence comparisons. We propose a novel method for Query Focused Summarization which makes full use of semantic relationships between sentences arising out of relationships between their constituent words by using Artificial Immune Systems to compare the sentences thereby reducing the complexity. Experiments show the potential of the approach to be used in situations with large input data.

Keywords: Lexical Analysis, Semantic relationships, Artificial Immune Systems, Automatic Text Summarization.

1 Introduction

In Text Summarization [1],[2],[3] a source text is condensed into a more compact version of itself without losing the salient information of the text. The document summaries may be generic or query relevant. The former captures the central ideas of the document and the latter reflects the relevance of document to a user-specified query. The automatic text summarization involves two steps [4]:

- Building an intermediate source representation from the source text.
- Summary generation: Formation of a summary from the source representation.

Text summarization may yield two types of summaries: abstractive and extractive [2], [1]. The former is produced by expressing salient features of the text with a new set of sentences and can serve as its substitute or an analysis. The extractive summaries are produced by selecting important sentences from the text and can be produced easily. Extractive summarization is the focus of the

R. Silhavy et al. (eds.), *Artificial Intelligence Perspectives and Applications*,
Advances in Intelligent Systems and Computing 347, DOI: 10.1007/978-3-319-18476-0_29

paper and the term summarization will imply extractive summarization henceforth. A number of techniques have been used for building intermediate source representation [3]. One group of techniques is statistical approaches where word frequencies, location of text in document and cue words are used to determine the most important concepts within a document [2], [1]. In these approaches [2] text units are treated indifferent to the semantic information they represent. On the other extreme, approaches try to analyze the discourse structure of the text [5]. One of the shortcomings in these approaches is that they suffer from the problem of large computational requirement and usage of domain specific knowledge. In between the two extremes, entity based approaches [2] use inherent semantic relationships among constituent words. Lexical Chaining [6] is one such method where lexical relations between words or word sequences are analyzed. The author [6] envisaged that lexical chains of words can be built with the help of Rogets Thesaurus. The underlying features include utilizing the inherent semantic relationships between the words that are grouped together under different hierarchies and categories. Unfortunately, the idea could not materialize due to lack of availability of a machine readable version of Rogets Thesaurus. Based on the lexical analysis paradigm, Barzilay [7] constructed the lexical Chains using WordNet. Relationships between the words are calculated using distance between the positions of words and the shape of path which connects them in the WordNet hierarchy. Silber [8] improved Barzilay's model by enhancing the computational complexity to O(n) in the number of nouns in the text. Graph based ranking approaches as in [9] and [10] are another type of entity based approaches which represent text units as nodes in a graph and establish relationships between them such as collocation, co-reference, etc.

1.1 Query-Focused Summarization

The goal of Query-focused summarization is to extract a summary from a collection of documents based on a query. It has been addressed mostly with statistical methods. Graph Based Centrality Methods based on Page Rank Algorithm [9] have been developed. Notable are LexRank Algorithm proposed in [10] where cosine similarity between sentences is used to find the ones with highest importance in the text. TextRank Algorithm as proposed in [11] uses content overlap for finding relations between sentences. These methods have been further improved upon in [12] and [13]. Some other recent works include usage of Manifold Ranking [14], probabilistic ranking of sentences using relevance, coverage and novelty as in [15], matrix factorization approaches as in [16]. With supervised learning methods query-specific domain knowledge is utilized to guide the selection of sentences as in [17]. Some attempts have been made to do the job by utilizing discourse analysis methods as in [18] and by utilizing Latent Semantic Analysis methods as in [19].

1.2 Artificial Immune Systems

The primary role of the biological immune system is to eliminate antigens (foreign substances) from human bodies with the help of antibodies. A number of immune-inspired algorithms [20], [21], [22], [23] like negative selection, positive selection, idiotypic network, clonal selection have been developed by the research community which closely approximate the biological immune system. Neil Jerne [24] proposed the idiotypic network theory. According to his theory, antibodies are not isolated but communicate with each other to form a large scale network. Farmer [25] proposed the mathematical formulation of the idiotypic model. Concentration of antibodies changes as

$\frac{dx_i}{dt} = c[(Antibodies\, recognized) \pm (I\, am\, recognized) + (Antigen\, recognised)] - (deathrate)$.

where c is rate constant. First term is stimulation from recognizing other antibodies whereas second term represents positive or negative selection from being recognized by other antibodies. The third term is stimulation from being able to recognize the antigen and last term is the constant death rate to which each antibody is subjected.

2 Proposed Model

2.1 Approach

The lexical chains methods which identify the important concepts focused only on nouns [6]. Although these methods are able to select the most important topics based on the assumption that the most important sentences represent those topics. But, the assumption of selecting important sentences from lexical chains may not always be true. The amount of salient information extracted from a text may vary greatly from chain to chain. In addition, there may be many sentences which contain some information about lot of concepts and may carry more compressed information as well. These sentences will be semantically related to many sentences by virtue of common information content but may not contain a lot of information, in particular, about any one of the concepts. Hence these may not be selected as representatives of any chain. Also a lot, though not all, of the sentences which are related to other sentences are also generally the ones which carry the flow of information through the text via the constructs of elaboration, generalization, negation, critique, co-reference, etc. These are the types of sentences which discourse analysis methods wish to find. In essence finding the sentences which are related to a lot of other sentences takes us away from entity based extraction methods towards discourse analysis methods, which is the long term plan in the field of text summarization [26]. A simple solution might be to calculate similarities between the constituent sentences of the text based on lexical similarities of the words they contain and select those sentences which are 'most popular' in the text. But this approach, though promising conceptually, suffers from a drawback in terms of complexity involved. Comparing all sentences with each other require time of the order $O(n^2)$ in the number of sentences in the text whereas the lexical chain methods, originally proposed in [7], have been improved such that building lexical chains can be done in a time linear in number of nouns in the text [8].

We propose an algorithm which calculates the most important sentences of the text using lexical relations between their constituent words. Instead of comparing all sentences with each others at once, we compare each sentence with a limited number of chosen sentences which are of higher importance and carry more information. These sentences are selected iteratively by utilizing Artificial Immune systems for selection of sentences to be compared in a particular step of the process. Our algorithm uses the Idiotypic nature of Artificial Immune systems and Clonal Selection principle. Complexity of system is linear in number of sentences. Steps for selecting the appropriate metric for extracting suitable sentences and their extraction are subsumed into one common step.

2.2 Architecture

Similarities between words are calculated using WordNet dictionary which is described in [27]. There are six basic WordNet relations, viz. Synonymy (similar senses), Antonymy (opposing senses), Hyponymy (subordination or sub-name, as in 'tree' being a hyponym of 'plant') and its inverse Hypernymy, Meronymy (being a part of something, as in 'ship' being a part of 'fleet') and its inverse Holonymy, Troponymy (being a manner of doing something, as in 'whisper' being a manner of 'speak') and entailment (being usually succeeded by something, as in 'marriage' being entailed by 'divorce'). We used these relations to give scores, normalized between 0 and 1, to each relation between two words. The score is calculated as Wu-Palmer similarity [28] in which the score denotes how similar two word senses are, based on the depth of the two senses in the WordNet taxonomic hierarchy and the depth of their Least Common Subsumer (most specific ancestor node). A score of 1 means concepts are the same. Score of 0 never occurs because Least Common Subsumer never has a depth of zero in Word Net Hierarchy.

Summaries are generated using either Maximal Marginal Relevance Principle [29] or using Global Selection Approaches. In summaries extracted using Maximal Marginal Relevance Principle, diversity of sentences is maximized by a Greedy Approach so that sentences which are similar to the ones already selected are discarded. In summaries where sentences have been extracted using Global Selection Approaches, the most important sentences are selected with indifference towards the amount of redundancy in information they bring into the summary.

Following three steps constitute the architecture:

1. Preprocessing Stage: The whole text is broken down into constituent sentences. Sentences are Tokenized and Brill's style Part-of-Speech Tagger is used to extract Nouns, Adverbs, Adjectives and Verbs for each sentence forming a Word Tuple for each sentence. Tuples for the sentences are called Antibodies. Tuple formed for Query is called Antigen. The collection of Antibodies and Antigens are stored in a list called Text Input Pool.

2. Calculation of Affinities: Antibodies are extracted iteratively from the Text Input Pool and their affinities are computed with the Antigen and the Antibodies present in System. System is the list of Antibodies constituting our Artificial Immune System. These antibodies represent the most important sentences derived so far. Similarity between two antibodies, or between the antigen and an antibody is calculated as the summation of similarities between their constituent words as described below:
   - Words in the two Tuples are added to LIST_1 and LIST_2 respectively.
   - Lexical Similarities are calculated among word pairs from the two lists.
   - The words constituting the word pair with the highest similarity score are deleted from both the lists. The score is added to the affinity measure between the two tuples (two antibodies or the antigen and an antibody). The procedure is repeated until no more word pairs are available.

   The similarity measures calculated are added to the concentration of the antibodies present in System, as mentioned in the Algorithm subsequently.
3. Addition of Antibody to System: Antibodies in the system represent those sentences which have the most information pertinent to the query. So when the antibody is added upon processing in the previous step, it is added as:
   - If the goal is to maximize Diversity then the among the antibodies already present in the system, the one with the lowest concentration is deleted. Then the incoming antibody is added to system. This ensures that antibodies present in system represent most diverse information while maintaining maximum relevance to the query by eliminating less popular information with respect to the query.
   - If the Goal is to maximize popularity then antibody is added before the decision is taken to select the candidate antibody for elimination. This ensures that if the incoming antibody isn't more popular than the least popular antibody already present, it will be deleted since it doesn't bring substantial information.

Algorithm for the Model

```
1.Perform Preprocessing; Get antibodies & antigen in Text Input Pool.
2.Initialize the System:
  2.1.Antigen is present.
  2.2.No antibody is Present.
  2.3.Upper Limit for Antibody Count is set.(say 'x')
  2.4.Each Antibody has same starting concentration.
3.Perform following steps until no antibody remains to be added into the System.
  3.1.Randomly select an antibody to be added to the System.
      Delete it from Text Input Pool.
    3.1.1.Calculate the affinity with antigen and antibodies already present.
    3.1.2.Add to incoming antibody's concentration proportionate to affinity
          with antigen.
    3.1.3.Change Incoming Antibody's concentration further as follows: If,
      3.1.3.1.Goal is to Maximize Diversity:
              Subtract from its concentration, amounts proportionate to its
```

```
            affinity with Each Antibody already present in System.
    3.1.3.2.Goal is to extract most popular sentences:
            Add to its concentration, amounts proportionate to affinity
            with each Antibody already present in System.
  3.1.4.For each Antibody already present in System, Add to its concentration
        amount proportionate to its affinity with incoming antibody.
  3.1.5.Addition of Incoming Antibody:
    3.1.5.1.If the Goal is to Maximize Diversity:
            [a].Delete the antibody with lowest concentration in System.
            [b].Add the Incoming antibody.
    3.1.5.2.If the Goal is to select most popular sentences:
            [a].Add the Incoming antibody.
            [b].Delete the antibody with lowest concentration in System.
```

Following points need to be mentioned about the algorithm:

- Having a limit on the number of antibodies present in the system as mentioned in Step 2.3 one of the two ways to eliminate antibodies with low concentration. The other way is to eject the antibodies whose concentration decreases below a certain level. This is not desirable as the antibodies entering the system with low concentration may not get fair chance of survival.
- Step 3.1.5.1. is a way to ensure only those antibodies which have a lot of exclusive information survive. This is unlike the case in Step 3.1.5.2 where the most popular antibodies are to be retained even if they bring in redundancy.
- Change in concentration of Antibodies already present in system changes at each iteration as:
  $\frac{d(conc(AB_p))}{dt} = conc(AB_p) + \theta \; aff(AB_p, AB_{in})$
  Changes in concentrations of incoming antibodies follow, for Maximizing Diversity & Maximizing Popular Sentences, respectively:
  $\frac{d(con(AB_{in}))}{dt} = con(AB_{in}) + \alpha \; aff(AB_{in}, AN) - \beta \; aff(AB_p, AB_{in})$
  $\frac{d(con(AB_{in}))}{dt} = con(AB_{in}) + \alpha \; aff(AB_{in}, AN) + \gamma \; aff(AB_p, AB_{in})$
  where AB_{in}, AB_P and AN are Incoming Antibody, Antibodies Already Present and Antigen respectively and θ, α, β and γ are proportionality constants whose values are calculated experimentally.
  Since the affinities are symmetric, θ is much greater than β or γ.
- Since every sentence is compared to a maximum 'x' (see Step 2.3 above) number of sentences, the complexity is O(n) in the number of sentences.

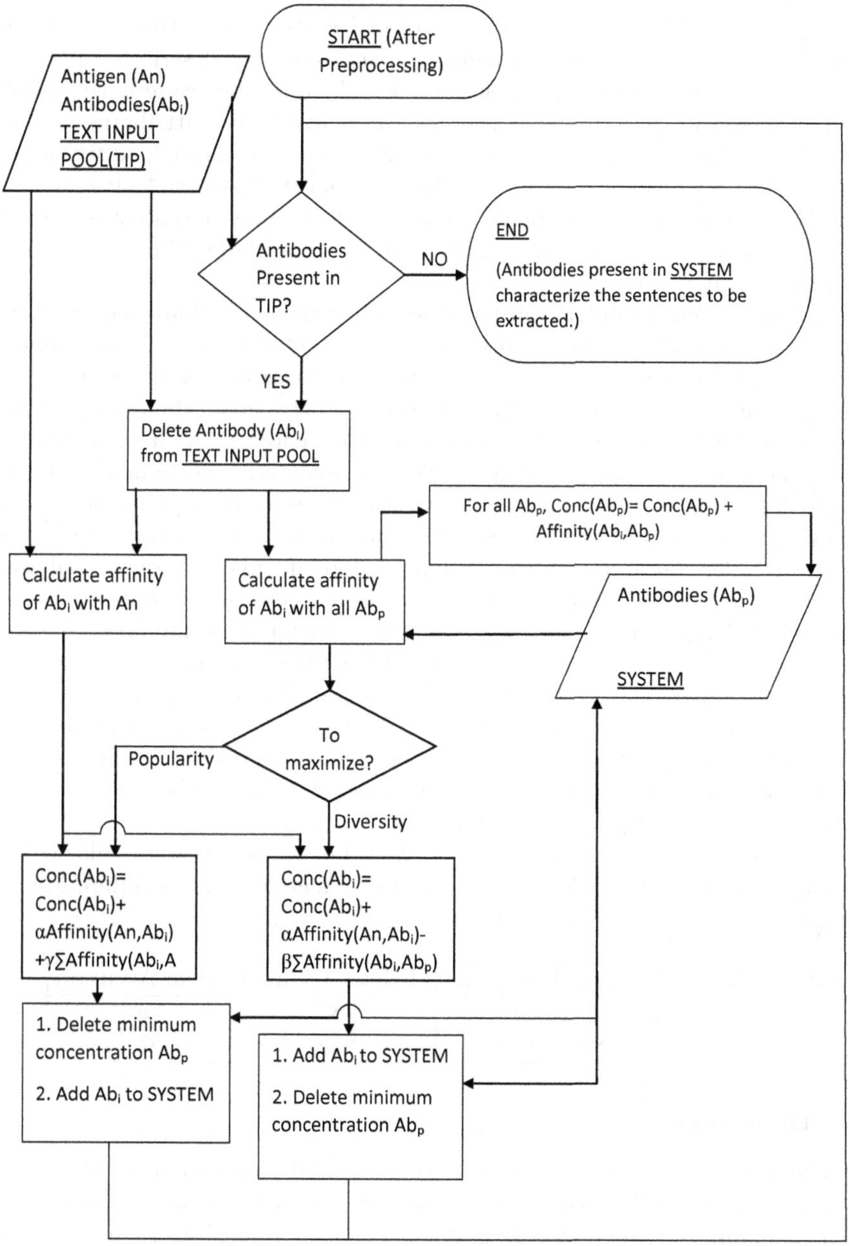

Fig. 1. Flow chart of the Algorithm

3 Experiments and Results

We implemented the system on a data set of Questions and Answers collected from Students Feedback Forms in which students answered descriptive questions about a teachers performance in classrooms. The questions were fairly specific yet descriptive, in the sense that they clearly defined what attributes teachers were to be judged on. The reviews were collected online through survey forms. Two queries were given and data was collected from 80 students. A little over 400 sentences were found for each query response. The number of sentences required in the summary was 40 which was the number of antibodies in the system at any point of time.

For comparison, we took as our baseline system the method involving comparison among all sentences in the text and subsequent selection of most popular sentences. In the baseline system, similarity measures between sentences were calculated in the same way as similarity measure between antibodies in our system by comparing representative tuples for the sentences with each other to obtain the most important sentences. Then a predetermined number (number of sentences required in summary) of sentences are selected from the list of sentences and their importance scores. The summaries were generated with both the greedy selection approach to minimize redundant information by following Maximal Marginal Relevance (MMR) Principle and also by following Global Selection (GS) Approach by selecting the sentences with highest importance scores. Then results on our system were compared with this baseline system so as to find out the changes with respect to baseline system. The percentage of sentences retained by our system as compared with the sentences extracted by the baseline system is the score of the system. It depicts the extent to which our system is able to retain the properties of the baseline system, which is the system ideal in terms of results but impractical in terms of complexity.

The scores obtained are depicted in table below. Scores were collected by running our system multiple times. Hence the 'range of scores' column. Median and Mean score values are also given.

Extraction Type	Range of Accuracy	Median Accuracy	Mean Accuracy
MMR	69% - 88%	78%	79%
GS	83% - 91%	85%	87%

3.1 Discussion

- The data sets contained source text generated as response to the query. Data sets generally used for most applications are those where query pertinent summary is generated from documents containing generic information. Performance is expected to rise for such systems because of availability of sentences containing redundant information which will increase chances of desirable sentences during random selection.
- Performance is expected to rise in bigger data sets (in which size of data sets is much larger than the summary) for the same reason mentioned above.

– In MMR Approach a lot of sentences which appeared in Baseline Approach were not selected because the information they represent was similar to the sentences already present. This has led to the following observations:
  1. Accuracy of GS approach was more as evidenced by median and mean scores. Popular sentences (those selected in baseline approach) which appear later during random sentence selection step in MMR would not be selected because of the presence of common information with the sentences already present.
  2. Variation in range of score values is more in MMR than GS Approach. Different sentences would appear as substitutes of the sentences selected in baseline approach by virtue of commonality in information. Since maximizing diversity is the motive, only one of the many sentences representing similar information would be retained.

4 Conclusion and Future Work

The results seem promising given that sample data set was very small and it is expected that recall rate will improve with the increase in size of sample data as it would lead to better chances for desirable sentences to be introduced into the system during processing. It seems that results of Sentence extraction using Global Selection strategy will show more significant improvement. If the recall rates can be brought up to more than 90% in bigger data sets then the system will have scored significant advantages as the entailing reduction in complexity will make it feasible to use direct sentence comparison methods in query-focused text summarization. Furthermore, query expansion approaches can be used in our system before the sentence selection step in the algorithm to identify concepts in the query which can be used to discard sentences that do not relate to query topics, thereby reducing the number of sentences being compared.

References

1. Lloret, E., Palomar, M.: Text summarisation in progress: a literature review. Artificial Intelligence Review (2012)
2. Mani, Maybury: Advances in automatic text summarization. MIT Press (1999)
3. Nenkova, A., McKeown, K.: A survey of text summarization techniques. Mining Text Data, pp. 43–76. Springer US (2012)
4. Jones, K.S.: What might be in a summary? Information Retrieval, 9–26 (1993)
5. Marcu, D.: Discourse trees are good indicators of importance in text. In: Advances in Automatic Textsummarization, pp. 123–136 (1999)
6. Morris, J., Hirst, G.: Lexical cohesion computed by thesaural relations as an indicator of the structure of text. Computational Linguistics 17(1), 21–43 (1991)
7. Barzilay, R., Elhadad, M.: Using lexical chains for text summarization. In: Advances in Automatic Text Summarization, pp. 111–122 (1999)
8. Silber, H.G., McCoy, K.F.: Efficient text summarization using lexical chains. In: Proceedings of the 5th International Conference on Intelligent User Interfaces. ACM (2000)

9. Page, Brin, Motwani, Winograd: PageRank citation ranking: Bringing order to the web. Technical report, Stanford University (1998)
10. Erkan, G., Radev, D.R.: LexRank: Graph-based lexical centrality as salience in text summarization. Journal of Artificial Intelligence Research 22 (2004)
11. Mihalcea, R., Tarau, P.: TextRank: Bringing order into texts. In: Proceedings of EMNLP 2004, Barcelona, Spain, pp. 404–411 (2004)
12. Zhao, L., Wu, L., Huang, X.: Using query expansion in graph-based approach for query-focused multi-document summarization. Information Processing and Management (45), 35–41 (2009)
13. Ramesh, A., Srinivasa, K.G., Pramod, N.: SentenceRank - A graph based approach to summarize text. In: Applications of Digital Information and Web Technologies (ICADIWT). IEEE (2014)
14. Cai, X., Li, W.: Mutually reinforced manifold-ranking based relevance propagation model for query-focused multi-document summarization. IEEE Transactions on Audio, Speech, and Language Processing 20(5), 1597–1607 (2012)
15. Luo, W., Zhuang, F., He, Q., Shi, Z.: Exploiting relevance, coverage, and novelty for query-focused multi-document summarization. Knowledge-Based Systems 46, 33–42 (2013)
16. Canhasi, E., Kononenko, I.: Weighted archetypal analysis of the multi-element graph for query-focused multi-document summarization. Expert Systems with Applications 41(2), 535–543 (2014)
17. Berger, A., Mittal, V.O.: Query-relevant summarization using FAQs. In: Proceedings of the 38th Annual Meeting of ACL (2000)
18. Bosma, W.E.: Query-based summarization using rhetorical structure theory (2005)
19. Hachey, Murray, Reitter: The embra system at DUC 2005: Query-oriented multi-document summarization with a very large latent semantic space. In: Proceedings of the Document Understanding Conference (DUC) (2005)
20. Aickelin, U., Dasgupta, D., Gu, F.: Artificial immune systems. Search Methodologies, pp. 187–211. Springer US (2014)
21. Elsayed Mohamed, S., Ammar, R., Rajasekaran, S.: Artificial immune systems: models, applications, and challenges. In: Proceedings of the 27th Annual ACM Symposium on Applied Computing. ACM (2012)
22. Read, M., Andrews, P.S., Timmis, J.: An Introduction to Artificial Immune Systems. In: Handbook of Natural Computing. Springer (2012)
23. Dasgupta, D., Nino, F.: Immunological computation: theory and applications. CRC Press (2008)
24. Jerne, N.K.: Towards a network theory of the immune system. Annals of Immunology 125(C), 373–389 (1973)
25. Farmer, J.D., Packard, N.H., Perelson, S.A.: The Immune System, Adaptation and Machine Learning. Physica 22D, 187–204 (1986)
26. Jones, K.S.: Automatic summarising: factors and directions. In: Advances in Automatic Text Summarisation. MIT Press (1998)
27. Miller, G.A.: WordNet: A Lexical Database for English. Communication of the ACM 38(11) (November 1995)
28. Wu, Palmer: Verb Semantics and Lexical Selection. In: ACL 1994 Proceedings of the 32nd Annual Meeting on Association for Computational Linguistics, pp. 133–138 (1994)
29. Carbonell, Goldstein: The use of MMR, diversity-based reranking for reordering documents and producing summaries. In: Proceedings of the 21st Annual International ACM SIGIR Conference on Research and Development in Information Retrieval (1998)

A Novel Similarity Measure for Clustering Customer Transactions Using Ternary Sequence Vector

M.S.B. Phridviraj[1], Vangipuram Radhakrishna[2], K. Vinay Kumar[1], and C.V. GuruRao[3]

[1] Kakatiya Institute of Technology and Science, Warangal, India
[2] VNR Vignana Jyothi Institute of Engineering and Technology, India
[3] SR Engineering College, Warangal, India
{prudviraj.kits,vrkrishna2014,vinaykumar.kitswgl}@gmail.com, guru_cv_rao@hotmail.com

Abstract. Clustering Transactions in sequence, temporal and time series databases is achieving an important attention from the database researchers and software industry. Significant research is carried out towards defining and validating the suitability of new similarity measures for databases which can accurately and efficiently find the similarity between user transactions in the given database to predict the user behavior. The distribution of items present in the transactions contributes to a great extent in finding the degree of similarity between them. This forms the key idea for the proposed similarity measure. The main objective of this research is to design the similarity measure which essentially considers the distribution of the items in the item set over the entire transaction set, which is the major drawback in the Jaccard, Cosine, Euclidean similarity measures. We analyze the proposed measure for worst case, average case and best case situations. The proposed similarity measure may be used to both cluster and classify the user transactions and predict the user behaviors.

Keywords: Transaction Sequence vector, similarity measure, clustering.

1 Introduction

Clustering transactions in databases, is gaining an important attention from the database researchers and from the perspective of the software industry. The importance for clustering comes from the need for decision making such as classification, prediction. The input to clustering algorithm in databases is usually a set of user transactions with the output being set of clusters of user transactions. One of the important properties of clustering is, all the patterns within a cluster share similar properties in some sense and patterns in different clusters are dissimilar in corresponding sense.

The advantage of clustering in databases is that each user transaction has set of items defined only from item set, and do not change frequently. In other words,

R. Silhavy et al. (eds.), *Artificial Intelligence Perspectives and Applications*,
Advances in Intelligent Systems and Computing 347, DOI: 10.1007/978-3-319-18476-0_30

the item set is static. This eliminates the need for preprocessing the transaction dataset to reduce dimensionality.

In the recent years, clustering data streams has gained lot of research focus in academia and industry [1, 15, 16]. An approach for handling text data stream is discussed in [21]. A similarity measure for clustering and classification of the text which considers the distribution of words is discussed in [22] which helped us a lot in carrying out the work. A tree based approach for finding frequent patterns and clustering text stream data using the concept of ternary vector is discussed in [12–14].

The work in [2–4] discusses the clustering and classification of multiple data streams which is based on spectral component similarity and finding frequent items in data streams dynamically. In [7, 8], the authors discuss the approach of handling data streams using rough set theory and finding frequent items in data streams using sliding window approach. The method of dimensionality reduction in data streams is discussed in [9, 10]. An approach of mining data streams using decision trees is discussed in [11]. The research in [17, 18] discusses the method of clustering the click stream data and the method of clustering data streams using hierarchical approach. The approach of handling text stream data is handled in [19]. The method of clustering text stream data using maximum capturing mechanism is explained in [21] which helped a lot in carrying out our work. The research in [12–14, 22] forms the basis of designing the proposed similarity measure which discusses the properties of similarity measures and the need to consider the importance of feature distribution. An approach for clustering using commonality measure is discussed in [5, 6, 20].

In this paper, the objective is to design and analyze the similarity measure for clustering the user transactions which has the Gaussian property and considers the distribution of each item from the item set over the entire database of transactions. In case the transactions are arriving as a stream then we can first find the closed frequent item set and apply the similarity measure on the final set of transactions. Section 2 of this paper introduces basic terminology followed to arrive at the similarity measure and design of the proposed measure. Section 3 validates proposed measure. In section 4, we discuss clustering of transactions using sample case study. Section 5 concludes the paper.

2 Proposed Measure

2.1 Motivation

The idea for the present similarity measure comes from our previous work [5, 12–14] considering the feature distribution and commonality which also holds good between the pair of any two transactions.

2.2 Basic Terminology

In the subsequent sections, we adopt the following conventions for defining the similarity measure.

Let I = $\{I_1, I_2, I_3...I_m\}$ be the item set consisting of all items and T = $\{T_1, T_2, T_3..T_m\}$ be the set of all transactions from transaction set T. Now, if T_i and T_j are any two transactions then I_{ik} and I_{jk} denotes presence or absence of kth item in transaction T_i and T_j respectively. We can also store frequency of item in I_{ik} and I_{jk} . In case, frequency of kth item is considered then we call I_{ik} and I_{jk} as non binary. On the contrary, if we consider only the presence absence of kth item then we say it is binary.

We define I_{ik} for binary version as

$$I_{ik} = \begin{cases} 0 & ; k^{th} \; is \; not \; present \; in \; T_i \\ 1 & ; k^{th} \; is \; present \; in \; T_i \end{cases} \tag{1}$$

For non-binary case, I_{ik} denotes the frequency of occurrence of kth item in transaction T_i represented by C_{ik} or simply, C. Mathematically we define I_{ik} for non-binary case as

$$I_{ik} = \begin{cases} 0 & ; k^{th} \; is \; not \; present \; in \; T_i \\ C_{ik} & ; k^{th} \; is \; present \; in \; T_i \end{cases} \tag{2}$$

For the purpose of defining our similarity measure we define two functions $\Theta(I_{ik}, I_{jk})$ and $\Delta(I_{ik}, I_{jk})$. The function $\Delta(I_{ik}, I_{jk})$ is called the difference function and holds the positive difference between I_{ik} and I_{jk}. Similarly, $\Theta(I_{ik}, I_{jk})$ denotes the tri-state output function mapping I_{ik} and I_{jk} to any one of the output values from the set {0,1,U} and is formally denoted by $\Theta(I_{ik}, I_{jk}) :\rightarrow \{0, 1, U\}$.

The table.1 and table.2 shown below defines the functions $\Delta(I_{ik}, I_{jk})$ and $\Theta(I_{ik}, I_{jk})$ for binary and non-binary versions. In binary notion, we do not consider the frequency of kth item in transaction T_i where as for the non-binary case, we consider the frequency.

Table 1. Function Definitions Θ and Δ for transaction item set in binary form

I_{ik}	I_{jk}	$\Theta(I_{ik}, I_{jk})$	$\Delta(I_{ik}, I_{jk})$
0	0	U	0
0	1	0	1
1	0	0	1
1	1	1	0

In the table 2 above, C_{ik} and C_{jk} denote frequency of kth item in transactions T_i and T_j respectively. The output symbol U denotes k^{th} feature is absent in both transactions and hence can be discarded as its contribution in defining similarity will be zero.

Table 2. Function Definitions Θ and Δ for transaction item set in binary form

I_{ik}	I_{jk}	$\Theta(I_{ik}, I_{jk})$	$\Delta(I_{ik}, I_{jk})$
0	0	U	0
0	C_{ik}	0	C_{ik}
C_{ik}	0	0	C_{ik}
C_{ik}	C_{ik}	1	$C_{ik} - C_{jk}$

2.3 Transaction Vector (Γ_i)

We define the transaction vector (Γ_i) as a finite sequence set consisting 2-tuple elements of the form (C_{ik}, I_{ik}). The first element of the 2-tuple C_{ik}, denotes frequency of kth item for in T_i and the second element I_{ik} denotes presence or absence of kth item in T_i represented by 1 or 0 respectively. In the case of binary representation of transaction item set database, we consider the first element C_{ik} to be either 1 or 0 respectively.

Definition. Let T_i be any transaction from transaction set T, then the transaction vector is denoted by Γ_i and is formally represented as

$$\Gamma_i = \{(C_{11}, I_{11}), (C_{12}, I_{12})..(C_{1k}, I_{1k})\} \tag{3}$$

where the first element of 2-tuple, C_{ik} denotes the count of kth item in transaction, T_i and I_{ik} denotes presence or absence of kth item represented by 1 or 0 respectively.

Example: Let I=$\{I_1, I_2, I_3, ..., I_m\}$ be an item set consisting of m items and T=$\{T_1, T_2, ..., T_n\}$ be a transaction set. Now the Transaction Vector for T_1, T_2, ... T_n is given by Γ_1, $\Gamma_2 .... \Gamma_n$

$\Gamma_1 = \{(C_{11}, I_{11}), (C_{12}, I_{12})...(C_{1m}, I_{1m})\}$ and
$\Gamma_2 = \{(C_{21}, I_{21}), (C_{22}, I_{22})...(C_{2m}, I_{2m})\}$
......
$\Gamma_n = \{(C_{n1}, I_{n1}), (C_{n2}, I_{n2})...(C_{nm}, I_{nm})\}$

2.4 Sequence Vector($SV[\Gamma_i, \Gamma_j]$)

Definition 1: Let Γ_i and Γ_j be any two transaction vectors defined over item set consisting of m items. Then the sequence vector over Γ_i and Γ_j is union of all 2-tuple elements where every 2-tuple is a pair consisting, elements of the form $(\Delta(I_{ik}, I_{jk}), \Theta(I_{ik}, I_{jk}))$. Formally we represent the Sequence Vector as :

$$SV[\Gamma_i, \Gamma_j] = \{\Gamma < i, j >_1, \Gamma < i, j >_2, \Gamma < i, j >_3, ...\Gamma < i, j >_m\} \tag{4}$$

where

$$\Gamma < i, j >_k = (\Delta(I_{ik}, I_{jk}), \Theta(I_{ik}, I_{jk})) \tag{5}$$

$$\Delta(I_{ik}, I_{jk}) = (C_{ik} - C_{jk}) \tag{6}$$

and

$$\Theta(I_{ik}, I_{jk}) = (I_{ik}, I_{jk}) \tag{7}$$

where k varies from 1 to m.

Example: Let Γ_1 and Γ_2 be any two transaction vectors with items defined over the item set $I = \{I_1, I_2, I_3\}$ then the sequence vector over Γ_1 and Γ_2 is given by

$$SV[\Gamma_1, \Gamma_2] = \{T < 1,2 >_1, T < 1,2 >_2, T < 1,3 >_3, ...T < 1,2 >_m\}$$

where each

$$T < 1,2 >_1 = ((C_{11} - C_{21}), \Theta(I_{11}, I_{21}))$$
$$T < 1,2 >_2 = ((C_{12} - C_{22}), \Theta(I_{12}, I_{22}))$$
$$T < 1,2 >_3 = ((C_{13} - C_{23}), \Theta(I_{13}, I_{23}))$$

Definition 2: Formally, the Sequence Vector for any two transaction vectors Γ_i and Γ_j is given by

$$SV[\Gamma_i, \Gamma_j] = U_k\{\Gamma < i,j >_k\} = U_k\{\Delta_k^{i,j}, \Theta_k^{i,j}\} \tag{8}$$

where

$$\Delta_k^{i,j} = \Delta(I_{ik}, I_{jk}) \tag{9}$$

$$\Theta_k^{i,j} = \Theta(I_{ik}, I_{jk}) \tag{10}$$

$$\Delta(I_{ik}, I_{jk}) = |I_{ik}| - |I_{jk}| \tag{11}$$

with U_k denoting union of all 2-tuple elements and m is the no of items in the item set and k varying from 1 to m. The value of $\Theta(I_{ik}, I_{jk})$ is obtained from Table 1 or Table 2.

2.5 Similarity Measure

Having defined all the required terminology, we now define our similarity measure given by the equation 12.

$$TSIM = \frac{(1 + S(\alpha, \beta))}{2} \tag{12}$$

we define $S(\alpha,\beta)$ as given in equation 13

$$S(\alpha,\beta) = \frac{\sum_{k=1}^{k=m} \alpha(\Delta_k^{i,j}, \Theta_k^{i,j})}{\sum_{k=1}^{k=m} \beta(\Delta_k^{i,j}, \Theta_k^{i,j})} \tag{13}$$

where

$$\alpha(\Delta_k^{i,j}, \Theta_k^{i,j}) = \begin{cases} 0.5 * [1 + e^{-r^2}] & ; \Theta(I_{ik}, I_{jk}) = 1 \;\; and \;\; \Delta(I_{ik}, I_{jk}) = 0 \\ -e^{-r^2} & ; \Theta(I_{ik}, I_{jk}) = 0 \;\; and \;\; \Delta(I_{ik}, I_{jk}) = 1 \\ 0 & ; \Theta(I_{ik}, I_{jk}) = U \;\; and \;\; \Delta(I_{ik}, I_{jk}) = 0 \end{cases} \tag{14}$$

$$\sigma' = \frac{1}{\sigma_k} \tag{15}$$

and

$$\gamma = \frac{1 - \Delta(I_{ik}, I_{jk})}{\sigma'} \tag{16}$$

The above equation the value of γ in Equation 16 is used when transaction data set is in binary form, and σ_k = standard deviation of feature k in all transactions of training set.

$$\beta(\Delta_k^{i,j}, \Theta_k^{i,j}) = \begin{cases} 0 & \Theta(I_{ik}, I_{jk}) = U \\ 1 & \Theta(I_{ik}, I_{jk}) \neq U \end{cases} \tag{17}$$

The values of α and β are used to measure the contribution of each feature in finding similarity.

3 Validation of the Proposed Measure

3.1 Best Case Scenario

In the best case situation, all the items may be present in the pair of transactions considered. So, $T_1 = \{1, 1, ...m \; times\}$ and $T_2 = \{1, 1, ...m \; times\}$. The Transactions Vectors are given by $\Gamma_1 = \{(1,1), (1,1), ..., m \; items\}$ and $\Gamma_2 = \{(1,1), (1,1), ..., m \; items\}$. The Sequence Vector is denoted by $SV(\Gamma_1, \Gamma_2) = \{(0,1), (0,1), .., m \; times\}$. The value of $S(\alpha,\beta)$ is computed using Equation 14 to Equation 17 as given below

$$S(\alpha,\beta) = \frac{\alpha(\Delta_1^{1,2}, \Theta_1^{1,2}) + \alpha(\Delta_2^{1,2}, \Theta_2^{1,2}) + \alpha(\Delta_3^{1,2}, \Theta_3^{1,2}) + ... + \alpha(\Delta_m^{1,2}, \Theta_m^{1,2})}{\beta(\Delta_1^{1,2}, \Theta_1^{1,2}) + \beta(\Delta_2^{1,2}, \Theta_2^{1,2}) + \beta(\Delta_3^{1,2}, \Theta_3^{1,2}) + ... + \beta(\Delta_m^{1,2}, \Theta_m^{1,2})} \tag{18}$$

$$= \frac{0.5 * [(1 + e^{-\gamma_1^2}) + (1 + e^{-\gamma_2^2}) + (1 + e^{-\gamma_3^2}) + ... + (1 + e^{-\gamma_m^2})]}{(1 + 1 + 1 + 1..... + m \; times)} \tag{19}$$

for best case situation the values of γ_k for k = 1 to m, approaches zero. This makes the values of $e^{-\gamma_1^2}$, $e^{-\gamma_2^2}$, $e^{-\gamma_3^2}$, ..., $e^{-\gamma_m^2}$ becomes 1.

This means the above Equation 19 reduces to Equation 20

$$S(\alpha,\beta) = \frac{(0.5*m + 0.5*m)}{m} = \frac{m}{m} = 1 \tag{20}$$

In best case, the similarity measure is

$$TSIM = \frac{(S(\alpha,\beta)+1)}{2} = \frac{(1+1)}{2} = 1 \tag{21}$$

The value of TSIM = 1 in Equation 21 indicates that the two transactions or files, are most similar to each other.

3.2 Worst Case Scenario

In the worst case situation, all the items may be absent in the pair of transactions considered. So, $T_1 = \{0, 0, ...m\ times\}$ and $T_2 = \{0, 0, ...m\ times\}$. The Transactions Vectors are given by $\Gamma_1 = \{(0, U), (0, U), ..., m\ items\}$ and $\Gamma_2 = \{(0, U), (0, U), ..., m\ items\}$. The Sequence Vector is denoted by $SV(\Gamma_1, \Gamma_2) = \{(0, 1), (0, 1), .., m\ times\}$. The value of $S(\alpha, \beta)$ is computed using Equation 14 to Equation 17 as given below

$$S(\alpha,\beta) = \frac{\alpha(\Delta_1^{1,2},\Theta_1^{1,2}) + \alpha(\Delta_2^{1,2},\Theta_2^{1,2}) + ... + \alpha(\Delta_m^{1,2},\Theta_m^{1,2})}{\beta(\Delta_1^{1,2},\Theta_1^{1,2}) + \beta(\Delta_2^{1,2},\Theta_2^{1,2}) + ... + \beta(\Delta_m^{1,2},\Theta_m^{1,2})} = \frac{U}{U} \tag{22}$$

Note: However, this situation never occurs because there should be at least one item in each transaction.

$$S(\alpha,\beta) == -1(so\ return - 1) \tag{23}$$

$$TSIM = \frac{(S(\alpha,\beta)+1)}{2} = \frac{(-1+1)}{2} = 0 \tag{24}$$

The value of TSIM = 0 indicates that the two transactions or files are not similar.

3.3 Average Case Scenario

In average case situation, items may be present or absent in the pair of transactions considered. So, $T_1 = \{1, 0, 1, 0..m\ times\}$ and $T_2 = \{0, 1, 0, 1, 0, ...m\ times\}$. The Transactions Vectors are given by $\Gamma_1 = \{(1, 1), (0, 0), ..., m\ items\}$ and $\Gamma_2 = \{(0, 0), (1, 1), ..., m\ items\}$. The Sequence Vector is denoted by $SV(\Gamma_1, \Gamma_2) = \{(1, 0), (1, 0), .., m\ times\}$. The value of $S(\alpha, \beta)$ is computed using Equation 14 to Equation 17 as given below

$$S(\alpha,\beta) = \frac{(e^{-\gamma_1^2}) + (e^{-\gamma_2^2}) + (e^{-\gamma_3^2}) + ... + (e^{-\gamma_m^2})}{(1+1+1+1..... + m\ times)} = \frac{-me^{-\gamma^2}}{m} = -e^{-\gamma^2} \tag{25}$$

$$TSIM = \frac{(S(\alpha,\beta)+1)}{2} = \frac{(1-e^{-\gamma^2})}{2} \tag{26}$$

4 Case Study

Consider the transactions set given in Table 3. The Table.3 below shows the Binary representation of the transaction-item matrix for simplicity. Here 1 represents presence and 0 as absence. The Table 4 shows the computations for first row of similarity matrix of Table 5. Here $\sum \alpha$ and $\sum \beta$ indicates Numerator and Denominator of the function S(α,β) respectively.

Table 3. Transaction - Itemset matrix in Binary Form

	Bread	Butter	Jam	Coffee	Milk
T_1	1	1	1	0	0
T_2	0	0	1	1	1
T_3	0	1	1	1	1
T_4	1	1	1	0	1
T_5	0	0	1	1	0
T_6	1	1	0	0	1
T_7	1	1	0	1	0
T_8	0	1	0	1	0
T_9	0	1	1	0	1

Table 4. Sample Computations for First Row of the Similarity Matrix(Table 5)

	Sequence Vector	$\sum_{k=1}^{k=5}\alpha(\Delta_k^{i,j},\Theta_k^{i,j})$	$\sum_{k=1}^{k=5}\beta(\Delta_k^{i,j},\Theta_k^{i,j})$	TSIM
T_1-T_2	{(1,0), (1, 0), (0, 1), (1, 0), (1, 0)}	-3.1106	5	0.18894
T_1-T_3	{(1,0), (0, 1), (0, 1), (1, 0), (1, 0)}	-1.19895	5	0.380105
T_1-T_4	{(0,1), (0, 1), (0, 1), (0, U), (1, 0)}	1.925308	4	0.740663
T_1-T_5	{(1,0), (1, 0), (0, 1), (1, 0), (0, U)}	-2.1106	4	0.236175
T_1-T_6	{(0,1), (0, 1), (1, 0), (0, U), (1, 0)}	-0.21262	4	0.4734
T_1-T_7	{(0,1), (0, 1), (1, 0), (1, 0), (0, U)}	-0.21262	4	0.473423
T_1-T_8	{(1,0), (0, 1), (1, 0), (1, 0), (0, U)}	-2.02948	4	0.246315
T_1-T_9	{(1,0), (0, 1), (0, 1), (0, U), (1, 0)}	-0.19895	4	0.475131

4.1 Computations

The Table 4 above gives the computations of one row of the similarity matrix and table 5 shows the similarity matrix for each transaction pair. Here we compute only upper triangular values, because the similarity measure is symmetric.

Table 5. Similarity Matrix Showing Upper Triangular Values

	T_1	T_2	T_3	T_4	T_5	T_6	T_7	T_8	T_9
T_1	-	0.1889	0.3801	**0.7406**	0.2361	0.4645	0.4734	0.2463	0.4751
T_2	-	-	0.7058	0.3768	0.6280	0.1878	0.1878	0.2348	0.4710
T_3	-	-	-	0.5679	0.4710	0.3790	0.3790	0.4737	0.7099
T_4	-	-	-	-	0.1889	0.7082	0.3787	0.1911	0.7099
T_5	-	-	-	-	-	0	0.2348	0.3131	0.2361
T_6	-	-	-	-	-	-	0.4734	0.2389	0.4650
T_7	-	-	-	-	-	-	-	0.6975	0.1911
T_8	-	-	-	-	-	-	-	-	0.2389
T_9	-	-	-	-	-	-	-	-	-

Table 6. Similarity Matrix after Step 1

	T_2	T_3	T_5	T_6	T_7	T_8	T_9	
T_1	0.1889	0.3801	0.2361	0.4645	0.4734	0.2463	0.4751	
T_2	-	0.7058	0.6280	0.1878	0.1878	0.2348	0.4710	
T_3	-	-	0.4710	0.3790	0.3790	0.4737	**0.7099**	
T_4	-	-	0.1889	**0.7082**	0.3787	0.1911	**0.7099**	
T_5	-	-	-	0	0.2348	0.3131	0.2361	
T_6	-	-	-	-	0.4734	0.2389	0.4650	
T_7	-	-	-	-	-	0.6975	0.1911	
T_8	-	-	-	-	-	-	0.2389	
T_9	-	-	-	-	-	-	-	-

Table 7. Similarity Matrix after Step 2

	T_2	T_5	T_7	T_8	
T_1	0.1889	0.2361	0.4734	0.2463	
T_2	-	0.6280	0.1878	0.2348	
T_3	-	0.4710	0.3790	0.4737	
T_4	-	0.1889	0.3787	0.1911	
T_5	-	-	0.2348	0.3131	
T_6	-	-	0.4734	0.2389	
T_7	-	-	-	**0.6975**	
T_8	-	-	-	-	
T_9	-	-	-	-	-

Table 8. Similarity Matrix after Step 3

	T_2	T_5
T_1	0.1889	0.2361
T_2	-	**0.6280**
T_3	-	0.4710
T_4	-	0.1889
T_5	-	-
T_6	-	-
T_7	-	-
T_8	-	-
T_9	-	-

4.2 Clustering

We carry out the clustering process using the procedure outlined in our previous work [5, 12, 20]. The main difference between clustering process carried out in this work and the one in [12] is the similarity measure designed in this paper for clustering transactions and it can be also be used to cluster data streams dynamically.

Here we assume user defined threshold for similarity measure to be 0.6. So, any transaction pair with similarity value greater than or equal to 0.6 can be grouped into one cluster and those less than the threshold will be placed as separate clusters.

Step 1: Choose the element value from the similarity matrix of Table 5 with highest value. Mark it as VISITED. Here T_1, T_4 has maximum similarity value. This is shown in Table 5 above marked as BOLD. Now place T_1, T_4 in one cluster. Eliminate Column 1,4 of similarity matrix, so as to stop considering column 1,4 in future. At end of step 1 we get cluster containing transactions (T_1, T_4). i.e Cluster-1 $= \{T_1, T_4\}$ and Table 5 is reduced to Table 6.

Step 2: Choose the next maximum element from the similarity matrix. Here (T_3, T_9), (T_4, T_9), have similarity value of 0.7099. This is shown in Table 6 above marked as BOLD. So place all these transactions in one cluster to get cluster containing transactions (T_3, T_4, T_9). Now add T_1 from cluster 1 to this cluster because of Transitivity. i.e Cluster-2$= \{T_1, T_3, T_4, T_9\}$. Next choose 0.7082 grouping T_4, T_6 into Cluster-3 $= \{T_4, T_6\}$. From Transitivity property we can combine Cluster-1 and Cluster-2 to get a single Cluster A$= \{T_1, T_3, T_4, T_6, T_9\}$. Eliminate Columns 1,3,4,6,9 of similarity matrix, to discard considering the same in the future. And Table 6 is reduced to Table 7. This forms input for step 3.

Step 3: Choose the next maximum element from the similarity matrix. Here (T_7, T_8) has next maximum similarity value of 0.6975. This is shown in Table 7 above marked as BOLD. So place (T_7, T_8) together to form a new cluster, denoted as Cluster B = $\{T_7, T_8\}$. Eliminate Columns 7,8 to discard considering the same in the future. This forms input for step 4.

Step 4: At the end of Step 3 we are left with only T_5 as shown in Table 8. So place T_2, T_5 in new cluster Cluster C = $\{T_2, T_5\}$. The final set of clusters formed are :

Cluster A: $\{T_1, T_3, T_4, T_6, T_9\}$; Cluster B: $\{T_7, T_8\}$; Cluster C: $\{T_2, T_5\}$

5 Conclusion

The objective of this research is to propose a new similarity measure which considers the distribution of the items of the transaction over the entire transaction dataset and can be used for clustering of transactions and users based on the customer transactions. This helps in predicting the user behaviors in advance. The similarity measure is analyzed for worst case, average case and best case situations. In future we may extend the research to handle the data streams and evaluate the suitability of proposed similarity measure to perform the classification.

References

1. Wang, C.D., Huang, D.: A Support Vector Based Algorithm for Clustering Data Streams. IEEE Transactions on Knowledge and Data Engineering 25(6), 1410–1424 (2013)
2. Aggarwal, C.C., Han, J., Wang, J., Philip, S.: On Demand Classification of Data Streams. In: ACM SIGKDD International Conference on Knowledge Discovery and Data Mining, pp. 503–508. ACM (2004)
3. Ling, C., Jun, Z.L., Li, T.: Clustering Algorithm For Multiple Data Streams Based on Spectral Component Similarity. Information Sciences 18(3), 35–47 (2012)
4. Jin, C., Qian, W., Sha, C., Yu, J.X., Aoying, Z.: Dynamically maintaining frequent items over a data stream. In: International Conference on Information and Knowledge Management, pp. 287–294 (2003)
5. Srinivas, C., Radhakrishna, V., Guru Rao, C.V.: Clustering Software Components for Program Restructuring and Component Reuse Using Hybrid XNOR Similarity Function. Procedia Technology Journal 12, 246–254 (2014)
6. Srinivas, C., Radhakrishna, V., Guru Rao, C.V.: Clustering and Classification of Software Component for Efficient Component Retrieval and Building Component Reuse Libraries. Procedia Computer Science Journal 31, 1044–1050 (2014)
7. Zhou, H., Bai, X., Shan, J.: A Rough Set based Clustering Algorithm for Multi-Stream. Procedia Engineering 15, 1854–1858 (2011)
8. Lam, H.T., Calders, T.: Mining Top-K Frequent Items in a Data Streamwith Flexible Sliding Windows. In: ACM SIGKDD International Conference on Knowledge Discovery and Data Mining, pp. 283–292 (2010)

9. Yan, J., Zhang, B., et al.: A Scalable Supervised Algorithm for Dimensionality Reduction on Streaming Data. Information Sciences 17(6), 2042–2065 (2006)
10. Yan, J., Zhang, B., Liu, N., Yan, S., Cheng, Q.: Effective and Efficient Dimensionality Reduction for Large-Scale and Streaming Data Preprocessing. IEEE Transactions on Knowledge and Data Engineering 18, 320–333 (2006)
11. Rutkowski, L., Pietruczuk, L., Duda, P., Jaworski, M.: Decision trees for Mining Data Streams Based on McDiarmid's Bound. IEEE Transactions on Knowledge and Data Engineering 25, 1272–1279 (2013)
12. PhridviRaj, M.S.B., Srinivas, C., Guru Rao, C.V.: Clustering Text Data Streams – A Tree based Approach with Ternary Function and Ternary Feature Vector. Procedia Computer Science 31, 976–984 (2014)
13. PhridviRaj, M.S.B., Guru Rao, C.V.: Data Mining – Past, Present and Future, A Typical Survey on Data Streams. Procedia Technology 12, 255–263 (2014)
14. PhridviRaj, M.S.B., Guru Rao, C.V.: Mining Top-K Rank Frequent Patterns in Data Streams-A Tree Based Approach with Ternary Function and Ternary Feature Vector. In: ACM International Conference on Innovative Computing and Cloud Computing, pp. 271–277 (2013)
15. Gaber, M.M.: Advances in Data stream mining. Wiley Interdisciplinary Reviews: Data Mining and Knowledge Discovery 2(1), 79–85 (2012)
16. Jiang, N., Grunewald, L.: Research Issues in Data Stream Association Rule Mining. SIGMOD Record 35(1) (2006)
17. Antonellis, P., Makris, C., Tsirakis, N.: Algorithms for Clustering Click Stream Data. Information Processing Letters 109(8), 381–385 (2009)
18. Rodrigues, P.P., Gama, J., Pedro, J.P.: Hierarchical Clustering of Time Series Data Streams. IEEE Transactions on Knowledge and Data Engineering 20(5) (2008), ISSN: 1041–4347
19. Zhong, S.: Efficient Streaming Text Clustering. Neural Networks 18(5), 790–798 (2005)
20. Radhakrishna, V., Srinivas, C., Guru Rao, C.V.: Document Clustering Using Hybrid XOR Similarity Function for Efficient Software Component Reuse. Procedia Computer Science Journal 17, 121–128 (2013)
21. Liu, Y.B., et al.: Clustering Text data streams. Journal of Computer Science and Technology 23(1), 112–128 (2008)
22. Lin, Y.S., Jiang, J.Y., Lee, S.J.: A Similarity Measure for Text Clustering and Classification. IEEE Transactions on Knowledge and Data Engineering 26(7), 320–333 (2014)

On the Value of Parameters of Use Case Points Method

Tomas Urbanek, Zdenka Prokopova, and Radek Silhavy

Faculty of Applied Informatics
Tomas Bata University in Zlin
Nad Stranemi 4511
Czech Republic
turbanek@fai.utb.cz

Abstract. Accurate effort estimates plays crucial role in software development process. These estimates are used for planning, controlling and managing resources. This paper deals with the statistical value of Use Case Points method parameters, while analytical programming for effort estimation is used. The main question of this paper is : Are there any parameters in Use Case Points method, which can be omitted from the calculation and the results will be better? The experimental results show that this method improving accuracy of Use Case Points method if and only if UUCW parameter is present in the calculation.

Keywords: analytical programming, differential evolution, use case points, effort estimation.

1 Introduction

Effort estimation is the activity of predicting the amount of effort that is required to complete a software development project [1]. Accurate and consistent prediction of effort estimation is a crucial point in project management for effective planning, monitoring and controlling of software projects. Better management decisions could be made with more accurate effort estimates. It is also very important to predict these estimates in the early stages of software development [2]. Atkinson et al. [3] claims that regression analysis does not provide enough accuracy. Therefore, the use of artificial intelligence may be a promising way to improve accuracy of effort estimations.

Attarzadeh et al. [4] declare that effort estimation in software engineering is divided into two categories.

- Algorithmic methods
- Non-algorithmic methods

Algorithmic methods are based on a mathematical formula and relies on historical datasets. The most famous methods are COCOMO [5], FP [3] and UCP method [2]. To the second category belong methods like expert judgement and analogy based methods. The most famous method of this group is Delphi [6].

R. Silhavy et al. (eds.), *Artificial Intelligence Perspectives and Applications*,
Advances in Intelligent Systems and Computing 347, DOI: 10.1007/978-3-319-18476-0_31

The quality of software could be reduced, when we underestimate the effort. This could cause dysfunctional software and raise the cost for testing and maintenance [7]. With more accurate effort estimations the management of software projects could be less challenging [2]. Therefore, we need more accurate predictions to effectively manage software projects [8].

This study offers some important insights into the statistical value of Use Case Points parameter. No previous study has investigated the importance of parameters of the Use Case Points method, while analytical programming method is used. What is not yet clear is the impact of UUCW parameter of this method. Therefore, this study makes a major contribution to research on the statistical value of Use Case Points method parameters, while analytical programming method is used.

1.1 Related Work

Despite of a lot of effort of scientists, there is no optimal and effective method for every software project. Very promising way is a research of Kocuganeli et al. [9], this paper shows, that ensemble of effort estimation methods could provide better results then a single estimator. Because we need an effort estimate as soon as possible, there is a strong pressure to predict these estimates after requirements are built. There is a work of Silhavy et al. [10] which proposed a method for automatic complexity estimation based on requirements. The work of Kaushik et al. [8] and Attarzadeh et al. [4] uses neural networks and COCOMO [5] method for prediction. COCOMO method is widely used for testing and calibrating in cooperation with artificial intelligence or with fuzzy logic [11]. Neural networks in these cases search parameters of the regression function. Unlike presented method, which search for the regression function itself. Differential evolution and analytical programming are used for this task. Because it is very difficulty to obtain a reliable dataset in case of Use Case Points method, this paper shows results on a dataset from Poznan University of Technology [12] and from this paper [13]. Presented approach is an evolution of my previous work [14], in this research is used differential evolution instead of self-organizing migration algorithms.

1.2 Use Case Points Method

This method was presented in 1993 by Gustav Karner. This method is based on the similar principle as function point method. The project manager have to estimate project parameters to four tables. These tables are following

- Unadjusted Use Case Weight (UUCW) can be seen in Table 1
- Unadjusted Actor Weight (UAW)can be seen in Table 2
- Technical Complexity Factor (TCF) can be seen in Table 3
- Environmental Complexity Factor (ECF) can be seen in Table 4

Table 1. UCP table for estimation unadjusted use case weight

Use Case Classification	No. of Transactions	Weight
Simple	1 to 3 transactions	5
Average	4 to 7 transactions	10
Complex	8 or more transactions	15

Table 2. UCP table for actor classification

Actor Classification	Weight
Simple	1
Average	2
Complex	3

1.3 Differential Evolution

Differential evolution is a optimization algorithm introduced by Storn and Price in 1995 [15]. This optimization technique is evolutionary algorithm based on population, mutation and recombination. Differential evolution is simple to implement and have only four parameters to set. These parameters are Generations, NP, F and Cr. Parameter Generations determines the number of generations, NP is population size, parameter F is weighting factor and parameter CR is crossover probability.[16]

1.4 Analytical Programming

Analytical programming (AP) is a tool for symbolic regression. The core of analytical programming is a set of functions and operands. These mathematical objects are used for synthesis a new function. Every function in the set of analytical programming core has various numbers of parameters. Functions are sorted by these parameters into general function sets (GFS). For example GFS_{1par} contains functions that have only 1 parameter like $sin()$, $cos()$ and other functions. AP must be used with any evolutionary algorithm that consists of a population of individuals for its run [17] [18]. In this paper is used differential evolution (DE) as the evolutionary algorithm for analytical programming [15]. The function of AP is following:

A new individual is generated by evolutionary algorithms. Then this individual (the list of integer numbers) is passed to the function of analytical programming. These integer numbers serves as an index into the general function set, where the functions are defined. Then the algorithm of analytical programming creates a new function from these indexes. After that this new function is evaluated by cost function. The evolutionary algorithm decides either this new equation is suited or not for the next evolution.

Table 3. UCP table for technical factor specification

Factor	Description	Weight
T1	Distributed system	2.0
T2	Response time/performance objectives	1.0
T3	End-user efficiency	1.0
T4	Internal processing complexity	1.0
T5	Code re-usability	1.0
T6	Easy to install	0.5
T7	Easy to use	0.5
T8	Portability to other platforms	2.0
T9	System maintenance	1.0
T10	Concurrent/parallel processing	1.0
T11	Security features	1.0
T12	Access for third parties	1.0
T13	End user training	1.0

Table 4. UCP table for environmental factor specification

Factor	Description	Weight
E1	Familiarity with development process used	1.5
E2	Application experience	0.5
E3	Object-oriented experience of team	1.0
E4	Lead analyst capability	0.5
E5	Motivation of the team	1.0
E6	Stability of requirements	2.0
E7	Part-time staff	-1.0
E8	Difficult programming language	-1.0

2 Problem Definition

Dataset with values of Use Case Points method was obtained from Poznan University of Technology [12] and from this paper [13]. The Table 5 shows Use Case Points method data from 24 projects. Only data of Use Case Points method with transitions have been used in this paper in case of Poznan University of Technology dataset. There are 4 values for each software project UUCW, UAW, TCF and ECF.

Gustav Karner in his work [2] derived nominal value for calculation of staff/hours from Use Case Points method. This value was set to 20. Thus, effort estimate in staff/hours is calculated as

$$estimate = UCP * 20$$

Table 5 shows calculated differences. The equation for calculation of MRE is Equation 1.

Table 5. Data used for effort estimation

ID	Act. Effort [h]	$UCP * 20$	MRE [%]
1	3037	2971	2
2	1917	1094	43
3	1173	1531	31
4	742	2103	183
5	614	1257	105
6	492	883	79
7	277	446	61
8	3593	6117	70
9	1681	1599	5
10	1344	1472	10
11	1220	1776	46
12	720	1011	40
13	514	627	22
14	397	1884	375
15	3684	6410	74
16	1980	2711	37
17	3950	6901	75
18	1925	2125	10
19	2175	2692	24
20	2226	2862	29
21	2640	3901	48
22	2568	3216	25
23	3042	5444	79
24	1696	2127	25
MMRE			62

$$MRE = \frac{|ActualEffort - (UCP * 20)|}{ActualEffort}, \tag{1}$$

where MRE is calculated error for each project in Table 5. Results from MRE calculation and Equation 2 were used for calculation of MMRE.

$$MMRE = \frac{1}{n}\sum_{i=1}^{n} MRE, \tag{2}$$

where $MMRE$ is mean magnitude of relative error through all project in Table 5.

$$MMRE = 62\%$$

The question is, there is a parameter in this dataset which can be omitted from calculation and the accuracy of analytical programming method for effort estimation will show better results.

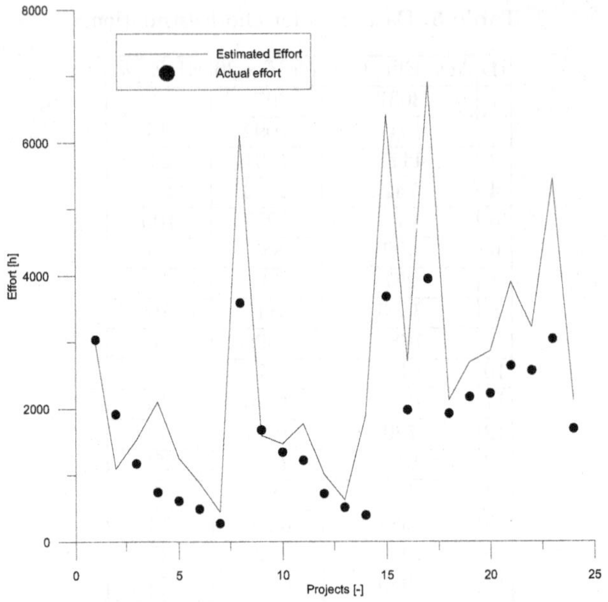

Fig. 1. Difference between estimated and real effort

3 Method

From dataset was constructed matrix A. This matrix has size MxN, where $M = 5$ and $N = 24$. Every row of this matrix A contains a calculation of Use Case Points method and actual effort.

The columns of the matrix A from beginning to end were UUCW, UAW, TCF, ECF and actual effort. Whole dataset could not be optimized by the evolutionary algorithm, because no data was remained for testing purposes. Because of this problem, the matrix A was divided into two matrices. Matrix B is training dataset and matrix C is testing dataset.

The Matrix B contains 12 rows of data for training purposes and the matrix C contains 12 rows for testing purposes. The matrix B was processed by analytical programming with the differential evolution algorithm. The Result of this process was a new equation. This new equation describes relationships between variables in training dataset, moreover in testing dataset.

Table 6 shows the set-up of differential evolution. The set-up of differential evolution is subject of further research.

Table 7 shows the set-up of analytical programming. Number of used function was set to 30, because we want to find function, which can perform better and also function complexity is not a problem. There is no need to generate short and easily memorable equations, but equations, that will be more accurate and predict effort estimation better. There was chosen linear functions like plus(), multiply() because there was a possibility that final estimation will be linear,

Table 6. Set-up of differential evolution

Parameter	Value
NP	20
Generations	60
F	0.7
Cr	0.7

Table 7. Set-up of analytical programming

Parameter	Value
Function number	30
Functions	Plus, Subtract, Divide, Multiply, Tan, Sin, Cos, Exp

Table 8. Parameter groups

One parameter	Two parameters	Three parameters	Four parameters
UUCW	UUCW,UAW	UUCW,UAW,TCF	UUCW,UAW,TCF,ECF
UAW	UUCW,TCF	UUCW,TCF,ECF	
TCF	UUCW,ECF	UUCW,UAW,ECF	
ECF	UAW,TCF	UAW,TCF,ECF	
	UAW,ECF		
	TCF,ECF		

on the other hand, there was being chosen also functions non-linear because the data from Poznan university contains strong non-linear behaviour.

3.1 Parameter Groups

Parameters of Use Case Points method were divided into parameter groups. There were 15 parameter groups and these can be seen in the Table 9. Each of these groups were calculated by an analytical programming method.

3.2 Cost Function

The new equation that is generated by the method of analytical programming contains these parameters UUCW, UAW, TCF and ECF. There is no force applied to analytical programming, that equations generated by this method have to contain all of these parameters. Cost function that is used for this task is following:

$$CF = \sum_{i=1}^{n} \left| B_{n,5} - f(B_{n,1}, B_{n,2}, \ldots, B_{n,4}) \right| , \tag{3}$$

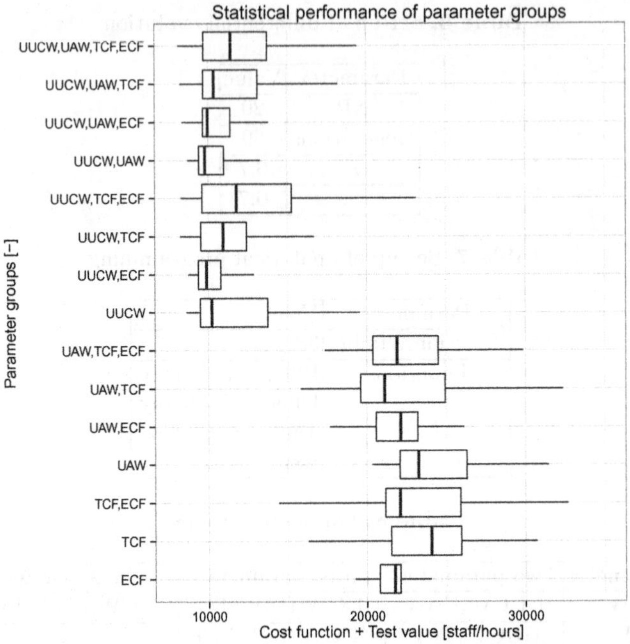

Fig. 2. Statistical performance of each parameter group

where n is equal to the number of projects in training dataset, $B_{n,5}$ is actual effort, $B_{n,1}$ is UUCW, $B_{n,2}$ is UAW, $B_{n,3}$ is TCF, $B_{n,4}$ is ECF.

4 Results

It was calculated 100 equations for each parameter group. Each calculation was generated in approximately 22 seconds. Simple statistical analysis was used to evaluate these calculations.

The Figure 2 shows the statistical performance of each parameter group. What is interesting in this data is that the performance of these calculations can be divided into two groups. The parameter groups which contain UUCW parameter, perform better than parameter groups without UUCW parameter. Further analysis showed that the best results give parameter groups UUCW,ECF and UUCW,UAW. On the other hand the worst result gives parameter group TCF,ECF.

The Figure 3 shows generated data for 15000 equations after removing extreme values. These extreme values have not been calculated in training data, because these values have been removed by natural selection of the differential evolution algorithm. On this figure can be seen a lot of calculations, which have estimation

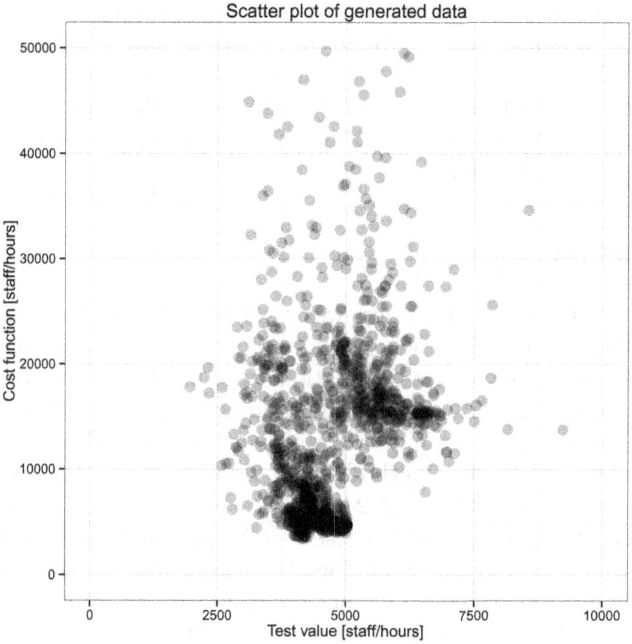

Fig. 3. Scatter plot of generated data

Table 9. Simple statistical analysis of parameter groups

Parameter group	N	Minimum	Quartile 25	Median	Quartile 75	Maximum
UUCW,UAW	100	8704	9439.25	9916.5	13362.00	1000000
UUCW,ECF	100	8750	9474.00	10072.0	12906.00	1000000
UUCW,UAW,ECF	100	8355	9731.00	10115.5	12296.25	1000000
UUCW,UAW,TCF	100	7642	9632.50	10440.5	14325.25	1000000
UUCW,TCF	100	7620	9371.50	10946.0	13262.25	1000000
UUCW,UAW,TCF,ECF	100	7722	9621.50	11630.5	15158.00	1000000
UUCW,TCF,ECF	100	7678	9604.25	12068.5	15525.00	1000000
UUCW	100	8611	9647.50	12551.5	27409.25	1000000
ECF	100	14693	20932.00	21801.5	22644.50	1000000
UAW,ECF	100	14142	21071.75	22919.0	42457.50	1000000
TCF,ECF	100	14433	21632.00	23308.5	27956.75	1000000
UAW,TCF,ECF	100	17327	21009.25	23345.0	55521.25	1000000
TCF	100	14142	21953.00	24351.5	26871.25	1000000
UAW,TCF	100	15804	20498.00	24886.0	48705.00	1000000
UAW	100	20348	22621.00	25537.0	40015.50	1000000

improving parameters. The cluster of calculations near point [5000, 5000] were the calculation of parameter groups which contained parameter UUCW. And the cluster of calculation near point [5000,16000] were calculated without UUCW parameter.

The Table 9 contains simple statistical analysis for each parameter group. As can be seen the minimum can be found in UUCW,TCF group. Nevertheless the median with minimal value can be found in UUCW,UAW parameter group. The same maximum value in each parameter group means that at least one equation of 100 calculations of each parameter group contains pathological equation and value 1000000 is penalization.

5 Conclusion

The main goal of the current study was to determine the statistical value of Use Case Points method parameters, while the analytical programming method is used. This study shown that calculation without UUCW parameter results in significant statistical differences. This finding suggests that in general the UUCW parameter is the most important parameter in Use Case Points method. It was also shown that this method could improve estimation with error of 7620 staff/hours in this dataset of 24 software projects. This research extends our knowledge about significance of the UUCW parameter in Use Case Points method. Further research might explore the significance of these parameters, whether some of the projects will be omitted from the calculations.This research has thrown up many questions in need of further investigation.

Acknowledgement. This study was supported by the internal grant of TBU in Zlin No. IGA/FAI/2014/019 funded from the resources of specific university research.

References

1. Keung, J.W.: Theoretical Maximum Prediction Accuracy for Analogy-Based Software Cost Estimation. In: 2008 15th Asia-Pacific Software Engineering Conference, pp. 495–502 (2008)
2. Karner, G.: Resource estimation for objectory projects. Objective Systems SF AB (1993)
3. Atkinson, K., Shepperd, M.: Using Function Points to Find Cost Analogies. In: 5th European Software Cost Modelling Meeting, Ivrea, Italy, pp. 1–5 (1994)
4. Attarzadeh, I., Ow, S.H.: Software development cost and time forecasting using a high performance artificial neural network model. In: Chen, R. (ed.) ICICIS 2011 Part I. CCIS, vol. 134, pp. 18–26. Springer, Heidelberg (2011)
5. Boehm, B.W.: Software Engineering Economics. IEEE Transactions on Software Engineering SE-10, 4–21 (1984)
6. Rowe, G., Wright, G.: The Delphi technique as a forecasting tool: issues and analysis. International Journal of Forecasting 15, 353–375 (1999)
7. Jiang, Z., Naudé, P., Jiang, B.: The effects of software size on development effort and software quality. Journal of Computer and Information Science, 492–496 (2007)
8. Kaushik, A., Soni, K., Soni, R.: An adaptive learning approach to software cost estimation. In: 2012 National Conference on Computing and Communication Systems, pp. 1–6 (November 2012)

9. Kocaguneli, E., Menzies, T., Keung, J.W.: On the value of ensemble effort estimation. IEEE Transactions on Software Engineering 38(6), 1403–1416 (2011)
10. Silhavy, R., Silhavy, P., Prokopova, Z.: Automatic complexity estimation based on requirements. In: Latest Trends on Systems, Santorini, Greece, vol. II, p. 4 (2014)
11. Reddy, C., Raju, K.: Improving the accuracy of effort estimation through fuzzy set combination of size and cost drivers. WSEAS Transactions on Computers 8(6), 926–936 (2009)
12. Ochodek, M., Nawrocki, J., Kwarciak, K.: Simplifying effort estimation based on Use Case Points. Information and Software Technology 53, 200–213 (2011)
13. Subriadi, A.P., Ningrum, P.A.: Critical review of the effort rate value in use case point method for estimating software development effort. Journal of Theroretical and Applied Information Technology 59(3), 735–744 (2014)
14. Urbanek, T., Prokopova, Z., Silhavy, R., Sehnalek, S.: Using Analytical Programming and UCP Method for Effort Estimation. In: Modern Trends and Techniques in Computer Science. Springer International Publishing (2014)
15. Storn, R., Price, K.: Differential evolution-a simple and efficient adaptive scheme for global optimization over continuous spaces. Technical Report TR-95-012 (1995)
16. Storn, R.: On the usage of differential evolution for function optimization. In: Fuzzy Information Processing Society, NAFIPS (1996)
17. Zelinka, I., Davendra, D., Senkerik, R., Jasek, R., Oplatkova, Z.: Analytical programming-a novel approach for evolutionary synthesis of symbolic structures. InTech, Rijeka (2011)
18. Zelinka, I., Oplatkova, Z., Nolle, L.: Analytic programming-symbolic regression by means of arbitrary evolutionary algorithms. Int. J. of Simulation, Systems, ... 6 (2005)

Intelligent Integrated Decision Support Systems for Territory Management

Boris V. Sokolov[1,3], Vyacheslav A. Zelentsov[1], Olga Brovkina[2], Alexsander N. Pavlov[1], Victor F. Mochalov[1], and Semyon A. Potryasaev[1]

[1] Russian Academy of Science,
Saint Petersburg Institute of Informatics and Automation (SPIIRAS), Russia
sokol@iias.spb.su
[2] Global Change Research Centre Academy of Science of the Czech Republic
[3] University ITMO, St. Petersburg, Russia

Abstract. This study proposes a scientific and methodical approach to problem solving automation, and intellectualization of multi-criteria decision-making processes for territory management. The substantiation of the composition and structure of the Integrated Intelligent Decision Support System (IIDSS) in management areas is based on a methodology developed by the authors and technologies of proactive monitoring and management of the structural dynamics of complex objects.

Keywords: intelligent geoinformation technology, multi-criteria selection, decision support systems.

1 Introduction

At present, the central role in ensuring the necessary quality control of Complex Objects (CO) of natural and artificial origin belongs to the Integrated Intelligent Decision Support Systems (IIDSS) and the kernel which is Special Mathematical Software (SMS) for decision making [1,2,6,11,12,13,16,23,25-27,36,39].

An IIDSS is required for informational, methodical and instrumental support for the decision-training and decision-making processes in all stages of management. The purpose of the introduction of an IIDSS is to improve the efficiency and quality of functional and management activities through the use of advanced intelligent information technologies, formation of complex analytical information and knowledge, to allow us to develop and make informed decisions in a dynamically changing environment.

To achieve this goal, the following tasks have to be resolved: 1) creation of a single feature space and indicators characterizing the state of CO management based on a centralized data repository, information and knowledge with accumulation, storage, access, and manipulation; 2) integration of existing local databases as part of this centralized information repository; 3) collection, accumulation and application of expert knowledge in distributed bases to form conclusions and recommendations; 4)

R. Silhavy et al. (eds.), *Artificial Intelligence Perspectives and Applications*,
Advances in Intelligent Systems and Computing 347, DOI: 10.1007/978-3-319-18476-0_32

continuous monitoring (complex analysis) of the current situation; 5) forecasting the developing situation on the basis of complex proactive modelling; 6) increasing the efficiency and quality of management decisions based on the use of analytical and forecasting tools; 7) process automation for preparation of analytical reports; 8) data visualization with the use of cognitive graphics (including the use of geographic information systems, etc); 9) expert instrumental and informational support and analytical activities.

Currently, there are a large number of options for creating and using an IIDSS, and they are widely used in various subject areas [3-6,8-13,16,20-24]. IIDSS's oriented for territory management and IIDSS's based on geographic information technologies have an important place [13,24,27,32,39]. This is due to the fact that spatial data has a key role in territory management. Fundamental spatial data are the basis for all other types of spatial data. These include coordinate systems data, digital maps, orthophotos, digital elevation models, and others.

Software packages with automated solutions were recently developed [3,5,6,7,9,10,13,14,29-32,35-40]: 1) identification of oil-polluted water and soil with detection of qualitative and quantitative characteristics; 2) identification of illegal garbage and detection of their morphological composition; 3) determination of the degree of clogging of clean fallow weed; 4) identification of degradation of vegetation; 5) identification of fire-damaged forest and deforestation; 6) classification of the forest's species composition; 7) determination of a lake's state (degree of overgrowth); 8) determination of an agricultural field's status.

The general structure of typical Automated Aerospace Monitoring (AAM) technology of complex spatial objects for territory management is shown in Figure 1.

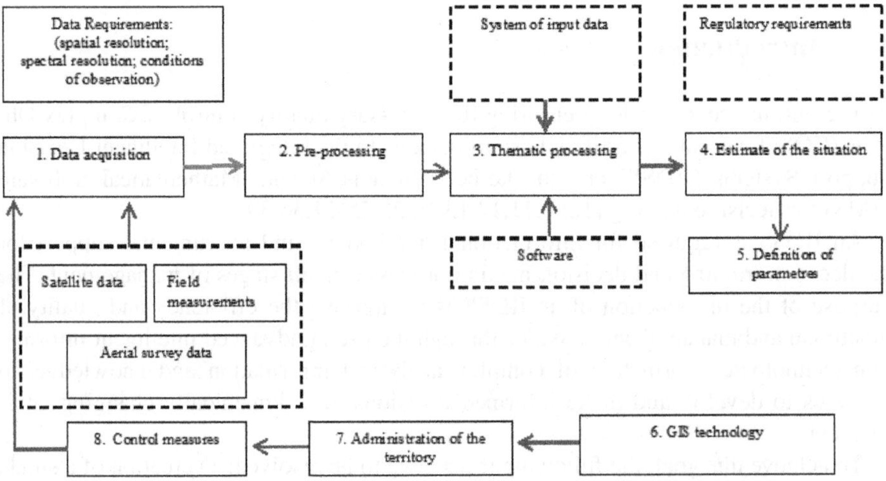

Fig. 1. Technology of Automated Aerospace Monitoring

There are several weaknesses in existing technologies of AAM: the need to manually choose the means of acquisition of baseline data; assessing the current state of the control object; the development of proposals to respond to dynamically changing parameters; characterization of the control object; the need to attract highly qualified professionals to the operations of data processing; the exploitation of an automated Aerospace Monitoring System (AMS) requires a significant time and financial investment and assumes a certain subjectivity in making current management decisions; the predicted situational modeling of the functioning of the system and the state of the control object is not performed; the criterial basis for assessing the quality of the operation of the AMS is not developed which constrains hardware and software improvements.

Therefore, the issues related to the creation of an IIDSS have to fend off the above disadvantages of existing AMS's. In this paper we start by discussing the scientific and methodical nature of the proposed approach to solving the problems of automation and intellectualization of multi-criteria decision-making processes for territory management. Further, the composition and structure of IIDSS territory management is suggested based on the approach of a realization of multi-criteria selection.

2 Intelligent Integrated Decision Support System for Territory Management

During the design phase of an IIDSS it is important to quantitatively and qualitatively analyze the effectiveness of the relevant control technology. At the same time, management will focus on the concept of proactive monitoring and management of the structural dynamics of complex objects (CO). The proactive monitoring and management of the structural dynamics of the CO is the processes of assessment, analysis and control of the current multi-structural macro-states of those objects, as well as the formation and implementation of control actions to ensure the transition from the current CO to a synthesized multi-structural macro-state [10,12-14,22]. Proactive management of a CO, in contrast to the reactive control of a CO that is traditionally used in practice, is based upon a rapid response and subsequent prevention of incidents by the creation of innovative predictive and proactive capabilities in the formulation and implementation of control actions in the existing monitoring and management system. Proactive management of a CO also relies on the methodology and technology of complex modelling, which involves a multi-domain description of the study, the combined use of methods, algorithms and techniques of multi-criteria evaluation, and analysis and selection of preferential solutions.

With regard to management of the previously mentioned environmental systems, the following observations about the components of the corresponding vector performance of proactive monitoring and management can be realized. The structure of this vector includes [24,39]: 1) a generalized qualitative indicator of the natural territory system with a special indicator for forests bonitet?; 2) a generalized quantitative indicator of the lake; 3) a quantitative assessment of the quality of the agricultural field.

Economic estimates can also be considered, such as the cadastral value of the territory, economic benefits from forest management, agricultural field management, water resources management, and the costs of liquidation of ecological disturbances.

Indicators for assessments for cost of resources are important and are special indicators of the AMS. Some of these indicators include the cost of developing hardware and software for the system; the value of the primary flow of informational materials and the results of ground measurements; the cost of training staff; and others. The private indicators for quality management areas include time-cycle operations and the maximum frequency of operation of the system; time required to bring the system into operation; time required for the structural and algorithmic adaptation of operations, and so on.

The most important step when estimating the effectiveness of territory management is the construction stage of the generalized indicator for quality control. At present, a wide variety of methods for solving multi-criteria evaluation of the efficiency of a CO have been developed [1,3-4,8-19,29-35,37-39]. These tasks are usually performed as a form of convolution for private assessments, integrated to a scalar characteristic of quality. We refer to these methods: constructing an additive convolution multiplicative convolution functions of distance, the minimum function, and so on.

One of the main drawbacks of the convolution of vector indices is ignoring the possibility of mutual compensation of assessments by functions with different criteria. Key features and relevant issues related to the solution of problems of multi-criteria evaluation, tend to have a computational and conceptual character. Additional information from the decision maker (the expert) is needed for a correct solution to the construction of a convolution.

Depending on the specific problem of estimating the efficiency of a CO, the private parameters are usually a non-linear influence on each other and on the whole integral indicator. On the other hand, the calculation of the resulting figure allows us to determine that the efficiency of the layer is complicated by inaccurate and illegible interval values of private indicators (in the general case, private indicators are defined by the linguistic variables).

For the solutions to these problems of territory management and assessment of the effectiveness of territory management, we propose a fuzzy-possibilistic approach; an approach using fuzzy logic methods and a theory of experiment design [14,34,35].

From the perspective of a *fuzzy-possibilistic approach,* the private indicators efficiency of the layer can be represented in the form of fuzzy events. Then, the task of estimating the efficiency of the CO is to construct the integral index, which is an operation on the fuzzy events with private indicators and their impact on the assessment of the CO's effectiveness.

To take into account the non-linear nature of the impact of these private indicators of the management areas f_i on the effectiveness of the generalized index of the CO, as well as a fuzzy-possibilistic approach of private indicators, we propose the construction of a generalized index using fuzzy-possibilistic convolution. Fuzzy-possibilistic convolution is based on fuzzy measures and fuzzy integrals, and allows us the flexibility to take into account the non-linear nature of the influence of private indicators. The method of multi-criteria evaluation of the efficiency of a CO consists of the following steps [14]:

Step 1: Construction of the evaluation function for different subsets of indicators of the CO by normalizing private fuzzy indicators.

Step 2: Conduction of an expert survey to find the coefficients of the importance of individual indicators.

Step 3: Construction of fuzzy measures (Sugeno λ-measures), which characterize the importance of the various subsets of the CO.

Step 4: Calculation of the integral indicator of the efficiency of the CO based on fuzzy convolution of private indicators using a fuzzy integral evaluation function from subsets of indicators from fuzzy measures.

If the *coefficients of efficiency* of a CO is given by *linguistic variables* (LV) the scale of linguistic assessments should be mentioned with the importance of the criteria and with the method of information convolution. To implement this approach, we propose a combined method for solving the problem of multi-criteria evaluation of the efficiency of a CO using the fuzzy method, and a theory of experiment planning for the flexible convolution of indicators.

The efficiency of a CO is an estimated set of indicators, or linguistic variables. A set of several indicators with the corresponding values of the terms of the decision-maker's view, characterizes the resulting indices expressing a general view of the CO's performance. LV-knowledge, in general terms, can be represented by production rules.

To construct the integral indicator it is necessary to translate the values of each parameter to a single scale [-1, 1]. The extreme values of the linguistic variable's ordinal scale are marked as -1 and +1, the point of "0" corresponds to the linguistic definition of the middle of the scale, according to the physical meaning of this indicator.

In accordance to the theory of experiment planning, to build the convolution matrix of indicators it is necessary to build an expert survey of extreme values of particular indicators that reflect the expert opinion in the form of production models (e.g. If indicator 1 is set to "high", indicator 2 is set to "low", ... indicator n has a value of "low", then the integral indicator is rated as "below average").

To determine the coefficients of the importance of the convolution of indicators, taking into account the impact of both individual performance and the impact of sets of two, three, and so on, indicators, an orthogonal expert survey plan is formed. The polynomial coefficients are calculated according to the rules adopted in the theory of experimental planning. Averaged scalar multiplications of the respective columns of the orthogonal matrix and vector values of the resulting indicator are calculated.

3 Rationale for the Composition and Structure of IIDSS Territory Management

In addition to hardware and software from Fig. 1, it is proposed that components are included in the system that reflect the state of the control object (the territory) and that realize the performance of additional operations modeling and that can control the operation of the system itself. These components should be prioritized:

- A predictive model of situation development on the whole test object and its components.
- A situational model of the whole test object and its components (in a controlled implementation of control actions aimed at maintaining a control object within the specified parameters).
- A model of the integrated decision support system that provides a mapping process monitoring system operation.

A natural complex with components of forests, lakes, and fields, could be an example of an IIDSS. For a natural complex there is a predictive model which represents the situation over time. The situation is described by parameters (metadata), part of which is defined in the automatic mode. In particular, the following characteristics are determined by the forest: area, species composition, age, timber stock, general biological condition, and the degree of fire hazard (or fire safety). In addition, fire, felling, and windfall are determined.

Table 1. Parameters for territory monitoring

Forest	Lake	Field
- area, perimeter - species composition - age - wood volume - biological state - fire rating - fire - forest cut - windfall	- area, perimeter - transparency - indented coastline - water protection zone - plums, sewage, pollution	- area, perimeter - degree of overgrowth - intensity of fertilizer application - flooding, erosion processes - career, dump burns

It is known that the quality of remote sensing data with regard to energy calibration, spatial correction and geo-referencing, is determined by the following main characteristics: the spatial resolution, the number and boundaries of spectral channels, the energy resolution (number of gradation of signal levels), and the width and area of the frame. Major thematic processing techniques aimed at detection and monitoring the parameters of objects are divided, in general, for the spatial, textural, and spectral brightness techniques. They can include both common software image processing systems such as ENVI Erdas Imagine, and specialized software modules that are configured to determine one or more parameters.

In practice, an IIDSS can be implemented as a set of application modules with standardized interfaces with intelligent software interaction. An analysis has shown that the construction of the modular system under consideration is carried out on a Service-Oriented Architecture basis (SOA) [5-7, 13, 20-22]. For this purpose, application modules are executed as web services with the ability to exchange data via a standard protocol (SOAP). A general structure created by the IIDSS and oriented to solve problems of territory management is presented (Fig. 2). A central role of this system belongs to the "Coordination" module. The objectives of the module are: a choice of a given set and the implementation of a specific computer program for calculating partial indicators (values of monitored parameters), and a generalized indicator of territory management quality. Computer programs vary in completeness of estimates (estimated number of private values), by error of the generalized indicator of the territory, and by efficiency of estimation. Selection of an appropriate computer program is according to the dynamics of the generalized indicator of territory management quality, the time of year (growing season), and meteorological conditions, is carried out in an automatic mode using production rules based on expert knowledge. The application modules are predictive models and situational development software customized to the test object (forest, field, lake).

There are a variety of forecast and situational models that differ in accuracy, speed, and the requirements of the original data. Traditionally, the expert decides to use a particular software implementation of these models depending on the composition of the original data, the purpose of the main computer program, and time and financial constraints. To automate the calculation of the values of particular indicators of the territory, a so-called "intelligent interface" is used. It is the software that performs the selection of a particular algorithm for calculating indicators on the basis of formal expertise.

In terms of the proposed service-oriented architecture "Coordination" module, is the synergy of the concepts of the Enterprise Service Bus (ESB) and the management application. An ESB is a link component providing a centralized and unified event-driven data exchange between different application modules using the SOAP protocol. A technical specification of each web-service is stored in a WSDL-file (Web Services Description Language). Thus, we now have a catalogue of web-services, ready to be invoked and able to exchange SOAP messages.

A management application defines the logic of the system due to the selection and implementation of a computer program. The standardized language for describing the processes is used to create and correct the computer programs by experts in the subject area (eg, ecology).

There are several standards describing the logic of web services execution.

XLANG, an XML-based language developed by Microsoft, was initially designed to describe successive, parallel and multi-variant workflows for BizTalk Server. The main task for XLANG is defining the business processes and the organization of message exchange amongst Web-services. XLANG has tools for processing extreme situations and supports long-term transactions.

The WSFL (Web Services Flow Language) from Microsoft allows the description of both public and private processes. It determines the data exchange, the succession of execution (a flow model), and the expression of each step in the flow in the form of specific operations (the global model).

The United Nations Organization center for co-operation in the area of commerce and electronic business (UN/CEFACT) developed the Electronic Business Extensible Markup Language – ebXML, which is also XML-based.

Hewlett-Packard has offered a standard for modeling sequences of interacting Web-services – Web Services Conversion Language (WSCL), released in March 2002. BPML (Business Process Modeling Language) is a language designed to describe business processes, which again, is based on XML. Its specification was developed by the Business Process Management Initiative. It enables task planning and contains functions for long-term storage of data. Data exchange among the participants is realized in the XML format, using roles and definitions of the partners, similarly to BPEL constructions. BPML supports a recursive composition, intended to form integrated processes from components, and provides both long-term and short-term transactions. In May 2003, Microsoft, IBM, Siebel, BEA Systems, and SAP, collaborated to develop a version of the specification for the BPEL4WS language. This specification, called BPEL, enables the modeling of the behavior of Web-services in the interaction of the business processes. The uniting of Web-services with Business Process Execution Language for Web Services increases the efficiency of the integration of applications.

The main result of the IIDSS we have considered, is a forecast of values of the generalized indicator for the quality management area. During program execution a set of geospatial data is generated, which is of interest to the end user. An important place in the IIDSS is assigned for visualization of generated data - information and knowledge about the monitored territory. Analysis shows that the analytic data visualization on the graphs and diagrams has undergone a minor change during the past decade [5-7, 20-24]. However, significant changes have been made recently in the field of spatial data visualization[39]. The majority of geographical information systems were difficult desktop applications that were available only to professionals. The rapid development of web-based technology, widespread propagation, wireless internet access, the release of new high-performance mobile devices, and new modes of interaction with the user interface, have all influenced a modern view of geographical informational systems. [24, 39].

Modern web-cartography offers a large variety of methods for delivering spatial data to end-users. The Open Geospatial Consortium (OGC) is an international industry consortium of 479 companies, government agencies, and universities, all participating in a congruent process to develop publicly available interface standards. OGC Standards support interoperable solutions that "geo-enable" the Web, wireless and location-based services, and mainstream I.T. Those standards empower technology developers to make complex spatial information and services accessible and useful to all kinds of applications.

Thus, the proposed technology allows us to implement a decision support system with intelligent control at various levels and to ensure prompt delivery of results to the end user.

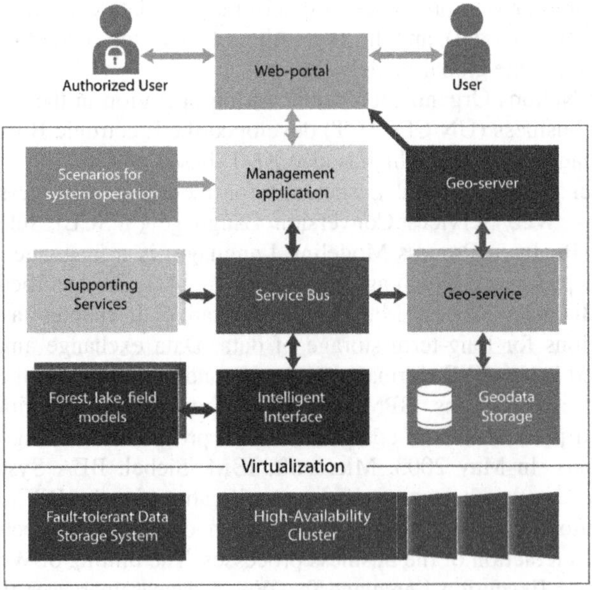

Fig. 2. General structure of the developed Intelligent Integrated Decision Support System for territory management

4 Conclusions

The general structure of an IIDSS was presented for a wide range of tasks for territory management. The proposed architecture of an IIDSS allows: 1) to execute all developed software modules with incompatible runtime requirements under various virtual machines within a single hardware server to overcome the problems of heterogeneity on one hand, and the convenience of deploying an IIDSS on the other; 2) to provide an interaction between modules through network communication, taking into account the prospects for further development of the IIDSS towards geographically distributed architectures. We propose to create a software "wrapper" that consolidates a private system for input-output of each module as a standardized data exchange interface. There are advantages of using the intelligent technology for support: 1) use a combination of quantitative and qualitative (fuzzy) information about the management effectiveness of the territory, which will significantly improve the quality of decisions and conclusions; 2) formalization of expert information provided to the expert in a natural language by introducing linguistic variables, which allow an adequate display of the approximate verbal description of objects and phenomena, even in cases where there is no deterministic description; 3) the proposed method in a fuzzy environment with a synthesized theory of experiments design, and a fuzzy-logic linguistic description of statements, allows us to carry out the formalization of a constructive level of expert experience (an expert group) in the form of predictive models in a multidimensional space of linguistic performance. The proposed method minimizes the number of calls to the experts and takes into account the complex non-linear nature of the influence of efficiency indicators on a CO indicator.

Acknowledgments. The research is partially supported by the Russian Foundation for Basic Research (grants 13-07-00279, 13-08-00702, 13-08-01250, 13-06-00877, 13-07-12120-офи-м, 15-29-01294-офи-м, 15-07-08391, 15-08-08459); grant 074-U01 supported by Government of Russian Federation; Program "5-100-2020" supported by Government of Russian Federation, Department of nanotechnologies and information technologies of the RAS (project 2.11); by the Ministry of Education, Youth and Sports of CR within the National Sustainability Program I (NPU I) (grant number LO1415); by ESTLATRUS projects 1.2./ELRI-121/2011/13 «Baltic ICT Platform» and 2.1/ELRI-184/2011/14 «Integrated Intelligent Platform for Monitoring the Cross-Border Natural-Technological Systems».

References

1. Balashov, E.P.: Evolutsionnyj sintez system.-M. Radio i svyaz, 328 p. (1985) (in Russian)
2. Beer, S.: Brain of the Firm (2003)
3. Vasilev, S.N.: Ot klassicheskih zadach regulirovaniya k intellektualnomu upravleniyu. Teoriya i Sistemy Upravleniya (1), 5–22, (2), 5–21 (2001) (in Russian)
4. Vasilev, S.N., Zherlov, A.K., Fedosov, E.A., Fedunov, B.E.: Intellektualnoe upravlenie dinamicheskimi sistemami. M. Fizmatlit (2000) (in Russian)

5. Want, R., Pering, T., Tennenhouse, D.: Comparing autonomic and proactive computing. IBM Systems Journal 42(1), 129–135 (2003)
6. Gorodetskiy, V., Karsaev, O., Samoilov, V., Serebryakov, S.: Interaction of Agents and Data Mining in Ubiquitous Environment. In: International Workshop on Agents and Data Mining Inter-actions (ADMI 2008), Sydney, Australia, December 9-12 (2008)
7. Chappell, D.: Enterprise Service Bus: Theory in Practice Paperback, July 2 (2004)
8. Kalinin, V.N., Sokolov, B.V.: Mnogomodelnoe opisanie protsessov upravleniya kosmicheskimi sredstvami. Teoriya i Sistemy Upravleniya (1), 149–156 (1995) (in Russian)
9. George, J.: Klir Architecture of systems problem solving. State University of New York at Binghamton (1985)
10. Majdanovich, O.V., Ohtilev, M.Y., Kussul, N.N., Sokolov, B.V., Tsyvirko, E.G., Yusupov, R.M.: Mezhdistsiplinarnyj podchd k upravleniyu i analizu effektivnosti informatsionnych technologij i system. Priborostroenie 53(11), 7–16 (2010) (in Russian)
11. Morozov, V.P., Dymarskij, Y.S.: Elementy teorii upravleniya GAP: Matematicheskoe obespechenie-Mashinostroenie, 245 p. (1984) (in Russian)
12. Okhtilev, M.Y., Sokolov, B.V.: Problems of the development and using of automation monitoring systems of complex technical objects. SPIIRAS Proceedings (1), 167–180 (2002), http://proceedings.spiiras.nw.ru/ojs/index.php/sp/article/view/1085
13. Okhtilev, M.Y., Sokolov, B.V., Yusupov, R.M.: Intellekualnye technologii monitoringa i upravleniya stukturnoj dinamikoj slozhnych technicheskih obektov. M. Nauka, 410 p. (2006) (in Russian)
14. Pavlov, A.N.: Integrated modelling of the structural and functional reconfiguration of complex objects. SPIIRAS Proceedings (28), 143–168 (2013), http://proceedings.spiiras.nw.ru/ojs/index.php/sp/article/view/1735
15. Zaychik, E.M., Ikonnikova, A.V., Petrova, I.A., Potrjasaev, S.A.: Algorithm for multicriteria synthesis of control technologies managing information systems of virtual enterprise. Institution of Russian Academy of Sciences Institute for Systems Analysis RAS 35, 8–15 (2008)
16. Reznikov, B.A.: Sistemnyj analiz i metody sistemotechniki. M.MO, 522 p. (1990) (in Russian)
17. Sokolov, B.V., Yusupov, R.M.: The Role and Place of Neocybernetics in Modern Structure of System Knowledge. Mechatronics, Automation, Control (6), 11–21 (2009)
18. Sokolov, B.V., Yusupov, R.M.: Kvalimetriya modelej i polimodelnych kompleksov: kontseptualnye osnovy i puti rasvitiya. Mechatronics, Automation, Control (12), 2–10 (2004) (in Russian)
19. Sokolov, B.V., Yusupov, R.M.: Kontseptualnye osnovy otsenivaniya i analiza kachestva modelej. Teoriya I Sistemy Upravleniya (6), 5–16 (2004) (in Russian)
20. Chernyak, L.: SOA-shag za gorizont. Otkrytye Sistemy (9), 34–40 (2003) (in Russian)
21. Chernyak, L.: Adaptiruemost i adaptivnost. Otkrytye Sistemy (9), 30–35 (2004) (in Russian)
22. Chernyak, L.: Ot adaptivnoj infrastruktury k adaptivnomu predpriyatiyu. Otkrytye Sistemy (10), 32–39 (2003) (in Russian)
23. Yusupov, R.M., Sokolov, B.V.: Kompleksnoe modelirovanie riskov pri vybore upravlencheskih reshenij v slozhnych organizatsionno-tehnicheskih sistemach. Problemy Upravleniya i Informatiki (1-2), 39–59 (2006) (in Russian)
24. Merkuryev, Y., Merkuryeva, G., Sokolov, B., Zelentsov, V. (eds.): Information Technologies and Tools for Space-Ground Monitoring of Natural and Technological Objects. Riga Technical University, 110 p. (2014)

25. Dmitry, I., Boris, S., Dilou, R.E.A.: Integrated dynamic scheduling of material flows and distributed information services in collaborative cyberphysical supply networks. International Journal of Systems Science: Operations & Logistics 1(1), 18–26 (2014)
26. Merkuryev, Y.A., Sokolov, B.V., Zelentsov, V.A., Yusupov, R.M.: Multiple models of information fusion processes: Quality definition and estimation. Journal of Computational Science 5(3), 380–386 (2014)
27. Merkuryeva, G.V., Merkuryev, Y.A., Lectauers, A., Sokolov, B.V., Potryasaev, S.A., Zelentsov, V.A.: Advanced timer flood monitoring, modeling and forecasting. Journal of Computational Science (October 2014)
28. OASIS Standard: Web Services Business Process Execution Language (2007)
29. Potryasaev, S.A., Sokolov, B.V., Ivanov, D.A.: A Dynamic Model and an Algorithm for Supply Chain Scheduling Problem Solving. In: Proceedings of the 16th International Conference on Harbor, Maritime & Multimodal Logistics Modelling and Simulation, Bordeaux, France, September 10-12, pp. 85–91 (2014)
30. Potryasaev, S.A., Sokolov, B.V., Ivanov, D.A.: Analysis of dynamic scheduling robustness with the help of attainable sets. In: 2014 International Conference on Computer Technologies in Physical and Engineering Applications, ICCTPEA 2014 – Proceedings 6893327, pp. 143–144 (2014)
31. Potryasaev, S.A., Sokolov, B.V., Ivanov, D.A.: Control theory application to spacecraft scheduling problem. In: 2014 International Conference on Computer Technologies in Physical and Engineering Applications, ICCTPEA 2014 – Proceedings 6893328, pp. 145–146 (2014)
32. Romanovs, A., Sokolov, B.V., Lektauers, A., Potryasaev, S., Shkodyrev, V.: Crowdsourcing interactive technology for natural-technical objects integrated monitoring. In: Ronzhin, A., Potapova, R., Delic, V. (eds.) SPECOM 2014. LNCS (LNAI), vol. 8773, pp. 176–183. Springer, Heidelberg (2014)
33. St. Laurent, S., Johnson, J., Dumbill., E.: Programming Web Services with XML-RPC, 1st edn. O'Reily (June 2001)
34. Sokolov, B.V., Ivanov, D.A., Pavlov, A.N.: Optimal distribution (re)planning in a centralized multi-stage supply network in the presence of the ripple effect. European Journal of Operational Research 237(2), 758–770 (2014)
35. Sokolov, B.V., Ivanov, D.A., Pavlov, A.N.: Reconfiguration model for production-inventory-transportation planning in a supply network. In: Proceedings of the 16th Intnternational Conference on Harbor, Maritime & Multimodal Logistics Modelling and Simulation, Bordeaux, France, September 10-12, pp. 34–39 (2014)
36. Sokolov, B.V., Yusupov, R.M.: Influence of Computer Science and Information Technologies on Progress in Theory and Control Systems for Complex Plants. Keynote Papers of the 13th IFAC Symposium on Information Control Problems in Manufacturing, Moscow, Russia, June 3-5, pp. 54–69 (2009)
37. Sokolov, B., Ivanov, D.: Adaptive Supply Chain Management, 269 p. Springer, London (2010)
38. Sokolov, B., Ivanov, D., Kaeschel, J.: A multi-structural framework for adaptive supply chain planning and operations with structure dynamics considerations. European Journal of Operational Research 200(2), 409–420 (2010)
39. Sokolov, B.V., Zelentsov, V.A., Brovkina, O.V., Mochalov, V.F., Potryasaev, S.A.: Complex Objects Remote Sensing Forest Monitoring and Modeling. In: Silhavy, R., Senkerik, R., Oplatkova, Z.K., Silhavy, P., Prokopova, Z. (eds.) Modern Trends and Techniques in Computer Science. AISC, vol. 285, pp. 445–453. Springer, Heidelberg (2014)
40. Vasiliev, Y.: SOA and WS-BPEL: Composing Service-Oriented Solution with PHP and ActiveBPEL. Packt Publishing (2007)

Contextual Soft Classification Approaches for Crops Identification Using Multi-sensory Remote Sensing Data: Machine Learning Perspective for Satellite Images

Anand N. Khobragade[1] and Mukesh M. Raghuwanshi[2]

[1] Maharashtra Remote Sensing Applications Centre, Nagpur, India
anand.khobragade@mrsac.maharashtra.gov.in
[2] Computer Technology Department, YC College of Engineering, Nagpur, India
m_raghuwanshi@rediffmail.com

Abstract. Agriculture department plays vital role on forecasting crop production and acreage estimation in the State. Commodity market estimates crop production on the basis of crop mass arrival in market and field prediction from authorized sources like Crop Advisory Boards. However, it is obvious that estimates from such board and government are often remains unmatched due to non-qualitative and unreliable approaches. The timely and accurate acreage estimation of crop is the pre-requisite for the purpose of better management upon crop production estimation. The conventional methods of gathering information on crop acreage are cumbersome, costly, and protracted, especially when the extent of work is whole county. The crop acreage statistics proves more crucial in event of natural calamity for taking strategic decisions like compensations to farmers based on losses they come up with. In a nutshell, non-availibity of accurate and finely estimated forecast necessitates the formation of coherent policy on fixing up agricultural commodity prices. Finally, soft classification approaches proved to be an alternative to error prone crop statistics by virtue of machine learning algorithms that applied on remote sensing images, a third eye technology which never lies. This paper conferred about gamut of machine learning algorithms for satellite data applications and envisages future trends that would be a magnet for researchers in upcoming years.

Keywords: sub pixel, mixed pixel, machine learning, soft classification, contextual, multi-source, FERM, kappa, PCM, FCM, ANN, NDVI, GA, SAR.

1 Introduction

1.1 Motivation

Remotely sensed data is an ideal source for mapping land cover and land uses at a variety of spatial and temporal scales (Raugh-garden et al. 1991, Foody 1995, 1996a). Due to its multi-spectral nature and repetitive coverage, remote sensing data is the most suitable for rapid assessment of crops [B-42]. Department of Space (SAC, DOS)

R. Silhavy et al. (eds.), *Artificial Intelligence Perspectives and Applications,*
Advances in Intelligent Systems and Computing 347, DOI: 10.1007/978-3-319-18476-0_33

executes FASAL project in participation with Ministry of Agriculture and Cooperation, GOI (Government of India) and State Remote Sensing Applications Centers and State Agriculture Department for forecasting crop production of major crops in the country viz. rice, cotton, sugarcane, mustard, groundnut, etc. has been completed for year 2007-08 [B-43]. They used multi-date single sensor remote sensing data for estimating specific crop, for example, RADARSAT data used for rice whereas RESOURCESAT data used for cotton and sugarcane. Moreover, methodology/model proposed by SAC uses manual mapping procedures for crop acreage estimation and hence prone to errors at ease. Other issues include use of pure statistical Maximum Likelihood Classifier and NDVI algorithm for estimating crop production. Briefly, hard classification approaches were adapted under this model. Multi-date single sensor data used for identification of single crop, whereas no estimation provided due to non-availability of optical remote sensing data in monsoon for kharip crops. Lack of applicability of any advance sub-pixel classification as well as learning based classifier, appeal to contemplate for improvement in model. By now, no attempt is made towards incorporating contextual information into this knowledge engine. It's really a matter of investigation that how model behave with the fusion of contextual classifier, Genetic Algorithms, and Active Learning Algorithms in this soft classification approach.

As a final point, outcomes of this research could be validated using Land Suitability Model of National Bureau Soil Survey & Land Use Planning, (NBSS & LUP, GOI). One can crosscheck the feasibility of remote sensing results with ground condition of the area under study. The study may encourage soil scientist, agro-economist, agriculturist, and decision makers in entire agriculture domain towards effective use of Remote Sensing and Geographical Information System (GIS) technologies in their day-to-day planning.

1.2 Technical Confront

As per usual assumption, each pixel represents a homogeneous area on the ground. However, it is found to be heterogeneous in nature in real world. In fact, spatial variation of real landscapes may cause numerous 'mixed pixels' in remotely sensed images due to the spatial resolution of the sensor under consideration and spatio-temporal deviation of earth's features. This research review is an attempt to address the problem of mixed pixel for want of optimally identifying specific crop. However, sensitivity of commercially available digital image processing softwares towards different topography, varied data dimensionality, algorithms adequacy, etc. are still dubious. More manpower and timely product delivery were two prominent challenges in any project execution. Particularly this reason strikes a chord with me and appeal to think why not we should develop an error free automated machine learning approach that will map entire classes in landuse/landcover classification accurately, specifically crop acreage mapping.

1.3 Correlation of Mixed Pixel with Classification Accuracy

The relationship between the 'grain' of a natural scene and the resolution of a remote sensing sensor is important in the recognition of boundaries and transitions between homogeneous fields. Mixed pixels at boundaries are recognized as an important factor influencing classification accuracy [37]. As in the following figure 1(a), the object i.e., a forest area is giving the impression of agriculture pixel, but is actually a mixed pixel. In figure 1(b), as resolution becomes finer, these features can be distinguished, and mixed pixels are seen at the corners. With still finer resolution as in figure 1(c), it is possible to see a transition between the two land cover types, covered by a band of pixels.

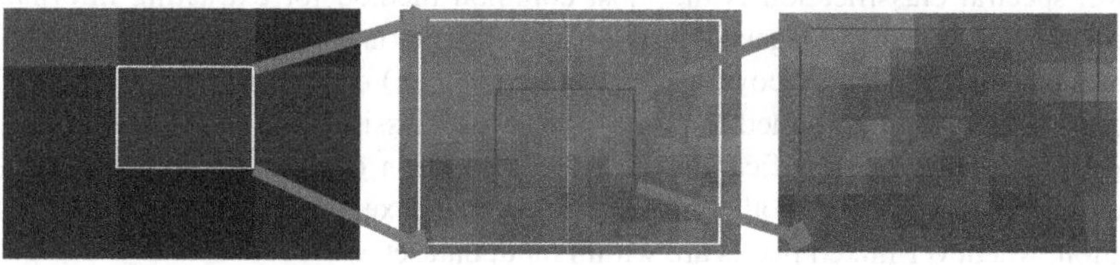

(a) (b) (c)

Fig. 1. Schematic representation of mixed pixel (a) AWiFs-56m (b) LISS IV – 5.8m (c) LISS-IV + CARTOSAT-I Merged Product (2.5m)

Hence, these mixed pixels are treated as noise or uncertainty in class allocation of a pixel. Conventional hard classification may thus produce inaccurate classification results. The application of sub-pixel or soft classification methods such as those based on spectral mixture analysis, fuzzy set theory and artificial neural network may be adopted for classification of images acquired in complex and uncertain environment. The outputs from these methods are in the form of membership values for each pixel.

1.4 Importance of Classifiers

Employing a difference metric or a classifier function performs the classification of pixels in a satellite image. Most of the conventional methods in literature are based on a predetermined classifier like Euclidian distance from the class mean, k-nearest neighbour algorithm, minimum distance classifier, gaussian maximum likelihood, etc. However, due to the several reasons including noise in the images, non-unique distribution of the data into classes, existence of texture type of features, dependence of the brightness values to the angle of the projection, etc., the performance of those conventional algorithms are around 65 % which is not sufficient for many applications. A broader approach to this problem is finding the optimal discriminants from a larger range of functions. An implementable application of this approach can be using threshold decomposition and Boolean discriminant functions to partition the feature space into subspaces each corresponding to a class. In general, basically unsupervised and supervised learning techniques were applied to understand land cover classification of the study area using multispectral images. The methods were implemented and overall accuracy, kappa statistics were calculated for each classifier: k-means, Gaussian Maximum Likelihood and Boolean discriminant function to get better performance of the used algorithms and compare results among the classifiers [B-55].

A framework for combining classifiers was proposed, which use distinct pattern representations. An experimental comparison of various classifier combination schemes demonstrates that the combination rule developed under the most restrictive assumptions the sum rule outperforms other classifier combinations schemes [A-28].

2 Conventional Methodology

Multi-spectral classification is the most common method for extracting information from remote sensing images. Based on the multivariate statistical parameters (i.e. mean, standard deviation, covariance, correlation, etc.) of each training site, the classification algorithm is trained to perform land use classification of the rest of the image. However, such classification results in allocation of one and only one class to each pixel, which ignores other classes within that pixel. This leads to loss of information, whenever mixed pixels are within input dataset.

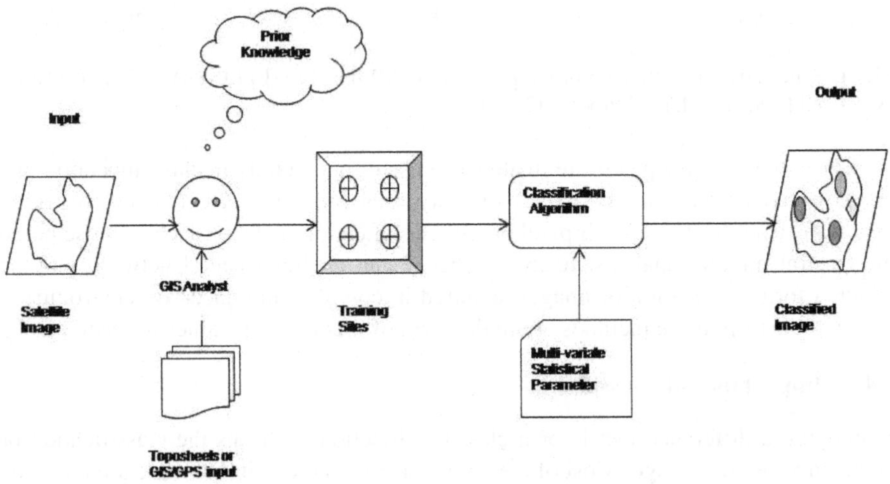

Fig. 2. Traditional Supervised Classification

3 Proposed Methodology

The proposed research work encourages use of multi-sensory multi-source satellite images for desired applications. For instance, high dimensional hyperspectral data set (e.g. ASTER), microwave data sets (RADARSAT/RISAT), and multi-spectral remote sensing data of ResourceSat-I (IRS P6) satellite, of same area could be used for classification problem. In addition, different density function based, statistical learning based and contextual based algorithms may have been investigated for extracting material of interest at sub-pixel level. These algorithms usually found useful in supervised classification mode; it means reference data will be generated from input high resolution multi spectral image. The reference data will be used for estimating statistical/learning parameters of targeted algorithms. Even, various vegetation indices (e.g. NDVI/SaVI/ TaVI) may have immense impact on remote sensing image classification. At last, the output so generated from this multi-source approaches in the form of land use and land cover map will be evaluated using well accepted means.

The accuracy of the output generated in this methodology will be evaluated in different ways. In first step accuracy of the output will be checked using higher resolution multi-spectral data of same temporal resolution and geometry from Resource Sat (P6) satellite. In the second step the accuracy will be checked using ground reference data collected from Mobile GIS/GPS/Geodetic Single Frequency GPS. The output will also be analyzed using statistical approach through fuzzy (error) confusion matrix as well as through multivariate analysis known as Kappa (K) coefficient. The methodology flow chart is given in figure 3.

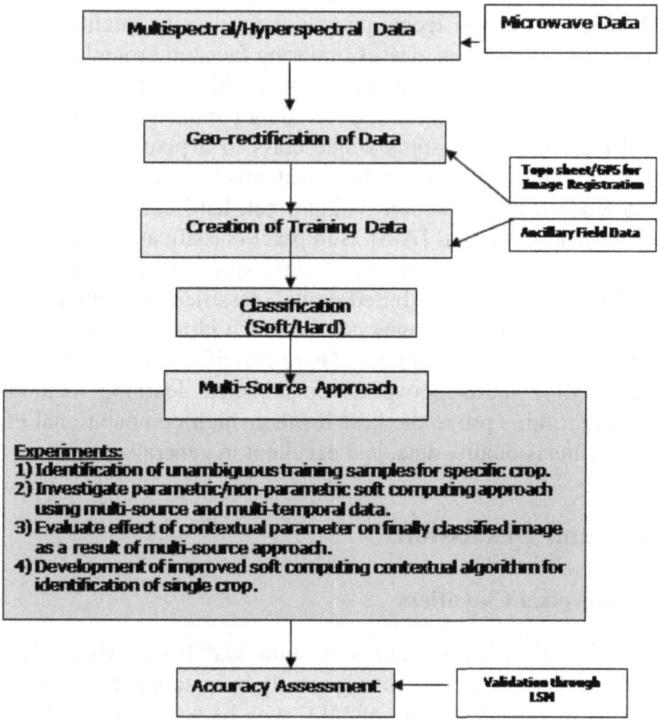

Fig. 3. Process flow diagram for methodology to be adopted

3.1 Why to Adopt Sub-pixel Based Soft Classification Approaches?

Generally, Remote Sensing data have mixed information at pixel level, which is a function of spatial resolution of the sensor. The satellite sensor system normally records the reflected or emitted radiant flux from heterogeneous mixtures of biophysical material such as soil, water, and vegetation. Also, land cover classes usually mixed into one another without sharp, hard boundaries. Thus, real world information on ground is very imprecise and heterogeneous. Perhaps, each pixel may contain heterogeneous in-formation and hence encounter the problem of mixed pixel. For example, first forecast on sugarcane acreage is done in month of December, during which surrounding places are covered with thick vegetation. The result of sugarcane mapping suffers with problem of mixed pixel and hence affects the estimation of sugarcane acreage. Traditional image classification techniques assumes that pixels are pure, however, in reality, different amounts of mixing of land cover within a pixel can be found due to the continuation of variation in landscape and intrinsic mixed nature of most classes [A-12]. Therefore, pixels with mixture of classes (i.e. mixed pixels) may be problematic. In a nutshell, mixed pixels at boundaries are recognized as an important factor influencing the classification accuracy.

Usually, hard classification techniques used to classify satellite images into discrete, homogeneous information classes, ignoring facts on ground on mixed pixel and hence results in inaccurate classification. The multi spectral classification may be performed using a variety of algorithms, including parametric and non-parametric [A-5]. Traditional classification assigns single class to a pixel may have ‘m’ possible membership grade values depending how that pixel is close to the class probably. That’s why, ac-counting information within pixel, hard classifiers is inadequate and hence soft classification needed [A-8]. Sub pixel classification based soft classification accounts for the mixed proportion of class values within a pixel. It is obvious from studies [A-36] that sub-pixel based digital classification output from coarse spatial resolution remote sensing images can be much closer to ground information as compared to hard classification output. These classifiers are based on statistical theory, fuzzy set theory, neural networks, or statistical learning methods. Statistical learning based algorithms prove the best for their high computational efficiency, robustness in high dimensionality data, and excellent in generalization.

4 Results and Discussions

4.1 About Sub-pixel Classifiers

Well-known statistical classifiers are maximum likelihood (MLC), Linear Mixture Model (LMM) etc. The MLC [A-7] has generally been adopted to provide hard classification output. However, the output of MLC may be softened, which may depict the partial and multiple class membership of each pixel, as reported by a number of studies [A-13]; [A-11]; [A-17]; [A-18]; [A-23]; [A-7]; [A-20]. Other statistical sub-pixel classifiers linear mixture model (LMM), which has been used in a number of studies for various applications [A-13]; [B-46]. It states that number of classes to be extracted from the data should be less than or equal to dimensionality of the data plus one.

Other soft classification method like Artificial Neural Network algorithm acts as supervised as well as unsupervised, which is known as Multi-Layer Perceptron (MLP) based on Back Propagation Algorithm (BPNN) [A-8], Radial Basis Function (RBF) [A-4], Probabilistic Neural Network, Fuzzy ARTMAP Neural Network etc. Neural Network Classifier can be trained well to extract single Landcover/Landuse class (i.e. Water spread in event of flood disaster and sugarcane mapping in crop acreage estimation) as compared to conventional classifications. Perhaps, Neural Network suffers from lacunae of dependency on a range of parameters related to the design of neural network architecture. Moreover, neural network are very slow in learning phase of the classification, which is the serious drawback when dealing with large-scale images.

Artificial Neural Network is a learning algorithm but does not al-ways generalize well. So it is important to have a learning algorithm for sub-pixel classification of remote sensing data which will be computational efficient, robust in high dimensions and good in generalization performance. Statistical based, fuzzy based and artificial neural network based algorithms have been incorporated in different commercially available softwares; like neural network and fuzzy c-mean (FCM) in PCI Geomatica, Linear Mixture Model (LMM) in ERDAS, ENVI etc.

Fuzzy c- Means (FCM) is basically a clustering technique where each pixel belongs to a cluster to some degree that is specified by membership grade (range). The sum of the membership for each pixel must be equal to unity. ([A-18];[A-20];[A-12]; [A-3];[A-8]. This can be achieved by minimizing the generalized least square error objective function. FCM is an iterative method that is employed to partition pixels of input data/image into different class membership values having probabilistic constrained. Further studies found that fuzzy set based PCM algorithm is more suitable than FCM for extracting single class [A-34].

In research on soft classification, while selecting Fuzzy c-Means (FCM) as a base soft classifier entropy parameter has been added. Soft classified outputs from entropy based FCM classifiers for single sensor ResourceSat datasets have been evaluated using sub-pixel confusion uncertainty matrix (SCM). It has been observed that output from FCM classifier has higher classification accuracy with higher uncertainty but entropy based classifier with optimum value of regularizing parameter generates classified output with minimum uncertainty [A-35].

In Probabilistic c-Means (PCM), the membership for representative feature points to be as high as possible, while unrepresentative points should have low membership in all clusters. PCM [A-37] relaxes probabilistic constraints as used in FCM. This leads to membership function to be as high as possible such that pixels with high degree of relevance with respects to the cluster may have high value of membership function, whereas pixel with less relevance may have low value of membership function. Besides, class based rationing technique is used to eliminate shadow effects in classification outputs. In return, this output is feed as in-put to PCM classifier. Input data was IRS-P6 LISS III, whereas IRS P6 LISS IV sensor image is used as testing data. The classified image is evaluated using Fuzzy Error Matrix (FERM). Recently, the concept of fuzzy error matrix (FERM) has been put forth to assess the accuracy of soft classification [A-24]. The result shows that effect of pixel having varying illuminations due to shadow within class was minimized. i.e. shadow pixels are not mixing with other classes under investigations. The experimentation is carried out in SMIC

(Sub-pixel Multi-spectral Image Classifier) [B-56] indigenous software developed by Dr. Anil Kumar at IIRS Dehradun.

Mahesh & Mathur have studied the sub-pixel classification of hyperspectral DAIS and LandSat-7 ETM+ data, where they apply four sub-pixel algorithms viz. Maximum Likelihood, Decision Tree, Artificial Neural Network and Support Vector Machine [A-38]. Dr. Anil Kumar has studied sub-pixel classification algorithms like LMM, ANN, & SVM and its behavior on effect of dimensionality of feature space and effect of training sample size using ASTER data. All above algorithms were examined w.r.t. number of bands used to achieve better classification accuracy. The finding of his research illustrate that upcoming statistical learning classifier SVM can produce better accuracies as compared to other algorithms, even if number of bands increases to 150 [A-29]. Research further illustrate that SVM kernel type sigmoid perform better than other SVM kernel types like Polynomial, Radial Basic, and Linear kernel at fixed training data size of 300 pixel/class. It is most appropriate in situations where the training sample data are difficult to collect, as it works well even with small number of training samples. Perhaps, 75 to 100 pixels per class are recommended for accuracy assessment of classification. Classification performed on various training sample size like 100, 200, 300, 400 and 500 pixels/class demonstrate interesting results [A-29]. Mahesh & Mather concluded in their research that no reduction in classification accuracies observed as the number of bands was increased even if training samples is small i.e. 100 pixels/class. In a nutshell, it is observed from various studies that classification accuracy depends on parameters like nature/number of training samples, number of bands used, number of classes to be identified, properties of classification algorithms, and optimum parameters of classification algorithm.

Other emerging classifiers are non-parametric learning algorithm known as Support Vector Machine (SVM). Support Vector Machines are used for the estimation of probability density functions [A-32]. SVM has been used for estimating the parameters of a Bayes classifier for remote sensing multi-spectral data [B-57]

The three stages in supervised digital classification of remote sensing data are training, classification, and testing. The commonly adopted approaches assume that boundaries between classes are crisp and hard classification is applied. In the real world, however, as spatial resolution decreases significantly, the proportion of mixed pixels increases. This leads to vagueness or fuzziness in the data, and in such situations researchers have applied the fuzzy approach at the classification stage. Some researchers have tried fuzzy approaches at the training, classification, and testing stages (full fuzzy concept) using statistical and artificial neural network methods. In this paper a full fuzzy concept has been presented, at a sub-pixel level, using statistical learning methods. This combined approach using SVM and FCM is evaluated with respect to a fuzzy weighted matrix and found that a SVM function using a Euclidean norm yields the best accuracy [A-29].

The research has been carried out to study the effect of feature dimensionality using statistical learning classifier - support vector machine (SVM with sigmoid kernel) while using different single and composite operators in fuzzy-based error matrices generation. In this work, mixed pixels have been used at allocation and testing stages and sub-pixel classification outputs have been evaluated using fuzzy-based error matrixes applying single and composite operators for generating matrix. As subpixel accuracy assessment were not available in commercial software, so in-house SMIC

(Sub-pixel Multispectral Image Classifier) package has been used. Data used for this research work was from HySI sensor at 506 m spatial resolution from Indian Mini Satellite-1 (IMS-1) satellite launched on April 28, 2008 by Indian Space Re-search Organization using Polar Satellite Launch Vehicle (PSLV) C9, acquired on 18th May 2008 for classification output and IRS-P6, AWIFS data for testing at sub-pixel reference data. The finding of this research illustrate that the uncertainty estimation at accuracy assessment stage can be carried while using single and composite operators and overall maximum accuracy was achieved while using 40 (13 to 52 bands) band data of HySI (IMS-1) [A-36].

4.2 Value Addition by Means of Contextual Algorithms

Modeling spatial context (e.g., autocorrelation) is a key challenge in classification problems that arise in geospatial domains. Markov random fields (MRF) is a popular model for incorporating spatial class dependencies (spatial context) between neighboring pixels in an image, and temporal class dependencies between different images of the same scene. The spatial autoregression (SAR) model, which is an extension of the classical regression model for incorporating spatial dependence, is popular for prediction and classification of spatial data in regional economics, natural resources, and ecological studies. It is argued that the SAR model makes more restrictive assumptions about the distribution of feature values and class boundaries than MRF. The relationship between SAR and MRF is analogous to the relationship between regression and Bayesian classifiers. [A-27]

Earlier for the hard classification techniques contextual information was used to improve classification accuracy. While modeling the spatial contextual information for hard classifiers using Markov Random Field it has been found that Metropolis algorithm is easier to program and it performs better in comparison to the Gibbs sampler. In the present study it has been found that in case of soft contextual classification Metropolis algorithm fails to sample from a random field efficiently and from the analysis it was found that Metropolis algorithm is not suitable for soft con-textual classification due to the high dimensionality of the soft outputs [A-34].

Information concerning the spatial variation in crop yield has become necessary for site specific crop management. Traditional satellite imagery has long been used to monitor crop growing conditions and to estimate crop yields over large geographic areas. However, this type of imagery has limited use for assessing within-field yield variability because of its coarse spatial resolution. Therefore, high-resolution airborne multispectral and hyperspectral imagery has been used for this purpose [B-54]. Hyperspectral analysis of vegetation involves obtaining spectral reflectance measurements in hundreds of bands in the electro-magnetic spectrum. Hyperspectral remote sensing provides valuable information about vegetation type, leaf area index, biomass, chlorophyll, and leaf nutrient concentration which are used to understand ecosystem functions, vegetation growth, and nutrient cycling [B-53].

Hyperspectral AVIRIS data from Blythe were acquired in June 1997 to study the agricultural spectra from different crops and for identification of crops in other areas with similar environmental factors and similar spectral properties. In this research, the spectral and radiometric characteristics of AVIRIS data for agriculture crops and the use of AVIRIS images in identifying agricultural crops were evaluated thoroughly.

The results of this study showed that there is a significant correlation between the data that were collected by AVIRIS image scene in 1997 and spectral data collected by the FieldSpec spectrometer. This correlation allowed us to build a spectral library to be used in ENVI-IDL software. This leads to identification of different crops and in particular the visible part of the spectra. AVIRIS data are in agreement with FieldSpec data. Using IDL algorithms showed that there is an excellent agreement between the predicted and the actual crop type (i.e. the correlation is between 85-90% match) and hence useful for crops identification or crops yield [A-31].

A national level project on kharif rice identification and acreage estimation is being carried out successfully for several states in India. Though the main growing season is predominantly winter but the uncertainty of getting cloud free data during the season has resulted in the use of microwave data. A feasibility study was taken up for early forecasting of the rabi rice area using microwave data. Hierarchical decision rule classification technique was used for the identification of the different land cover classes. The increase or decrease in the SAR backscatters due to progress in the crop phenology or due to delay sowing respectively forms the basis for identifying the rice areas. This study emphasizes the synergistic use of SAR and optical data for delineating the rabi rice areas which is of immense use in giving an early forecast [A-33]

Solberg & Jain proposed a method for statistical classification of multisource (i.e Landsat TM images and ERS-1-SAR images) data of different dates. The performance of the model is evaluated for land-use classification and founds significant improvements in the classification error rates compared to the conventional single-source classifiers [A-22]. When RADAR data used with NASA's Soil Moisture Active and Passive (SMAP) mission for soil-moisture estimates, it obtained better soil-moisture accuracy at a high resolution [B-47].

Another proposed multisource classification model based on Markov random fields (MRF), fused optical images, synthetic aperture radar (SAR) images, and GIS ground cover data. For change detection in agricultural areas, the MRF model detects 75% of the actual class changes [A-25].

Another study on multisource approach with RadarSAT data combined with Landsat TM stated that such fusion improves the discrimination of land cover classes on an average by 10% over using Landsat TM data alone. [A-30]

A "mutual" MRF approach is proposed that aims at improving both the accuracy and the reliability of the classification process by means of a better exploitation of the temporal information using both multi-temporal and multi-sensor approach. The presented method to automatically estimate the MRF parameters yielded significant results that make it an attractive alternative to the usual trial-and-error search procedure. [A-32]

Being surveyed literature so far, it is obvious that few attempt in multisource approach had been done using maximum two sensor data, but no research found to be blend of multi-spectral, hyper-spectral, and microwave data altogether. This will be a novel approach to endeavor fusing of all three types of remote sensing data for identification of specific crop. On the other hand, addition of contextual sensitivity will be a value addition to the system. Third objective of this proposed method is to validate outcome of this research with Land Suitability Model so as to prove its feasibility with real world scenario.

5 Conclusion and Future Scope

In nutshell, statistical learning algorithms reveal to be the most robust for classification problems. Out of all, SVM model uncovers almost all the dimensions for feature extraction and classification of satellite data as well. It is strongly demonstrated and accepted globally amongst almost researchers to be the best suitable algorithm for land use classification using remote sensing images. The only exception that sets bounds would be the limitation of testing data size; else undoubtedly it is paramount algorithm.

Innovative use of optimization algorithms in amalgamation with recent trends in machine learning algorithms would definitely set milestones in enhancing performance of the contextual statistical learning algorithms. For example, Genetic Algorithms (GA) exhibits improvement when used with SVM for optimizing its training part. Only 2% improvement in case of Cotton crop, but seems to have no effect in case of Gram crop. The possible reason behind may be the limited number of training sets taken for experimentation. However, it needs to drill down into root cause for not giving the results at par [A-39]. Even, use of Differential Evolution Algorithms (DE), Active Learning Algorithms, Comprehensive Learning Particle Swarm Optimization (CLPSO) [A-40], may impact upon optimizing kernel parameters setting for established machine learning algorithms. The last but not the least futuristic scope towards expanding horizon of this research would be the generation of crop acreage and crop yield model based upon the state of the art machine learning algorithms.

References

A: Relevant Bibliography

1. Aplin, P., Atkinson, P.M.: Sub-pixel land cover map-ping for per-field classification. International Journal Remote Sensing 22(14), 2853–2858 (2001)
2. Atkinson, P.M., Cutler, M.E.J., Lewis, H.: Mapping sub-pixel proportional land cover with AVHRR imagery. International Journal Remote Sensing 18(4), 917–935 (1997)
3. Bastin, L.: Comparison of fuzzy c-means classification, linear mixture modeling and MLC probabilities as tools for unmixing coarse pixels. International Journal Remote Sensing 18(17), 3629–3648 (1997)
4. Bruzzone, L., Cossu, R., Vernazza, G.: Combining parametric and non-parametric algorithms for a partially unsupervised classification of multi-temporal remote sensing images. Information Fusion 3, 289–297 (2002)
5. Burke, E.J., Simmonds, L.P.: Effects of sub-pixel heterogeneity on the retrieval of soil moisture from passive microwave radiometry. International Journal Remote Sensing 24(10), 2085–2104 (2003)
6. Cracknell, A.P.: Synergy in remote sensing - what's in a pixel? International Journal of Remote Sensing 19(11), 2025–2047 (1998)
7. Eastman, J.R., Laney, R.M.: Bayesian soft classification for sub-pixel analysis: a critical evaluation. Photogrammetric Engineering and Remote Sensing 68, 1149–1154 (2002)

8. Foody, G.M., Arora, M.K.: An evaluation of some factors affecting the accu-racy of classification by an artificial neural network. International Journal of Remote Sensing 18, 799–810 (1997)
9. Foody, G.M.: Sharpening fuzzy classification output to refine the representation of sub-pixel land cover distribution. International Journal Remote Sensing 19(13), 2593–2599 (1998)
10. Kumar, G.J.: Automated interpretation of sub-pixel vegetation from IRS LISS-II images. International Journal Remote Sensing 25(6), 1207–1222 (2004)
11. Guerschman, J.P., Paruelo, J.M., Di Bella, C., Giallorenzi, M.C., Pacin, F.: Land cover classification in the Argentine Pampas using multi-temporal machines for land cover classification. International Journal of Remote Sensing 23, 725–749 (2003)
12. Ju, J., Kolczyk, E.D., Gopal, S.: Guassian mixture discriminant analysis and sub-pixel land cover characterization in remote sensing. Remote Sensing of Environment 84, 550–560 (2003)
13. Lu, D., Mausel, P., Batistella, M., Moran, E.: Comparison of land cover classifi-cation methods in the Brazilian Amazon basin. Photogrammetric Engineering and Remote Sensing 70, 723–732 (2004)
14. Mertens, K.C., Verbeke, L.P.C., Ducheyne, E.I., De Wulf, R.R.: Using genetic algorithms in sub-pixel mapping. International Journal Remote Sensing 24(21), 4241–4247 (2003)
15. Pergola, N., Tramutoli, V.: SANA: sub-pixel automatic navigation of AVHRR im-agery. International Journal Remote Sensing 21(12), 2519–2524 (2000)
16. Sa, A.C.L., Pereira, J.M.C., Vasconcelos, M.J.P., Silva1, J.M.N., Ribeiro, N., Awasse, A.: Assessing the feasibility of sub-pixel burned area mapping in miombo wood-lands of northern Mozambique using MODIS imagery. International Journal Remote Sensing 24(8), 1783–1796 (2003)
17. Seera, P., Pons, X., Sauri, D.: Post-classification change detection with data from different sensors: some accuracy considerations. International Journal of Remote Sensing 24, 3311–3340 (2003)
18. Shalan, M.A., Arora, M.K., Ghosh, S.K.: An evaluation of fuzzy classification from IRS IC LISS III data. International Journal of Remote Sensing 23, 3179–3186 (2003)
19. Shokr, M.E., Moucha, R.: Co-location of pixels in satellite remote sensing images with demonstrations using sea ice data. International Journal Remote Sensing 19(5), 855–869 (1998)
20. Zhang, J., Foody, G.M.: Fully-fuzzy supervised classification of sub-urban land cover from remotely sensed imagery: statistical and artificial neural network approach. International Journal of Remote Sensing 22, 615–628 (2001)
21. Zhang, X., Van Genderen, J.L., Kroonenberg, S.B.: A method to evaluate the capability of Landsat-5 TM band 6 data for sub-pixel coal re-detection. International Journal Remote Sensing 18(15), 3279–3288 (1997)
22. Solberg, A.H.S., Jain, A.K., Taxt, T.: Multisource classification of remotely sensed data: fusion of Landsat TM and SAR images. IEEE Transactions on Geoscience and Remote Sensing 32(4) (2002) abstract
23. Bruzzone, L., Prieto, D.F., Serpico, S.B.: A neural-statistical approach to multi-temporal and multisource re-mote-sensing image classification. IEEE Transactions on Geo-science and Remote Sensing 37(3 pt.1) (1999)
24. Binaghi, E., Madella, P., Grazia Montesano, M., Rampini, A.: Fuzzy contextual classification of multisource re-mote sensingimages, IEEE transactions on Geoscience and Remote Sensing vol. IEEE Transactions on Geoscience and Remote Sensing 35(2), rc53 (1997)

25. Solberg, T., et al.: A Markov random field model for classification of multisource satellite imagery. IEEE Transactions on Geoscience and Remote Sensing (1996)
26. Duin, R.P.W., Tax, D.M.J.: statistical pattern recognition in remote sensing. IEEE Transactions on Pattern Analysis and Machine Intelligence 41(9), 226–239 (2008)
27. Shekhar, S., Member, S., Schrater, P.R., Vatsavai, R.R., Wu, W., Chawla, S.: Spatial Contextual Classification and Prediction Models for Mining Geospatial Data. IEEE Transactions on Multimedia, rc22 (2002)
28. Kittler, J., Hatef, M., Duin, R.P.W., Matas, J.: On Combining Classifiers, Spatial Contextual Classification and Prediction Models for Mining Geospatial Data. Pattern Analysis and Machine Intelligence 20(3), 226–239 (1998)
29. Kumar, A., Ghosh, S.K., Dadhwal, V.K.: Full fuzzy land cover mapping using remote sensing data based on fuzzyc-means and density estimation. Canadian Journal of Remote Sensing 33(2), 81–87 (2007)
30. Supervised classification of multisource spectral and texture data for agricultural crop mapping in Buenos Aires Province, Argentina. Canadian Journal of Remote Sensing 27(6), 679–685 (2001)
31. Hanna, S.H.S., Rethwisch, M.D.: Characteristics of AVIRIS bands measurements in agricultural crops at Blythe area, California: III. Studies on Teff Grass, The International Society for Optical Engineering (2003)
32. Melgani, F., Serpico, S.B.: A Markov Random Field Approach to Spa-tio-Temporal Contextual Image Classification. IEEE Transactions on Geoscience and Remote Sensing 41(11), 2478–2487 (2003)
33. Haldar, D., Patnaik, C.: Synergistic use of Multi-temporal Radarsat SAR and AWiFS data for Rabi rice identification. J. Indian Soc. Remote Sensing 153–160 (2009)
34. Dutta, A., Kumar, A., Sarkar, S.: Some Issues in Con-textual Fuzzy c-Means Classification of Remotely Sensed Data for Land Cover Mapping. IJRS 38, 109–118 (2010)
35. Kumar, A., Dadhwal, V.K.: Entropy-based Fuzzy Classification Parameter Optimization using uncertainty variation across Spatial Resolution. IJRS, 179–192 (2010)
36. Kumar, A., Saha, A., Dadhwal, V.K.: Some Issues Related with Sub-pixel Classifica-tion using HYSI Data from IMS-1 Satellite. IJRS, 203–210 (2010)
37. Kumar, A., Ghosh, S.K., Dadhwal, V.K.: ALCM: Automatic Land Cover Mapping. Journal of Indian Society of Remote Sensing, 239–245 (2010)
38. Mahesh, Mather: Some issues in the classification of DAIS hyperspectral data. IJRS 27(14), 2895–2916 (2006)
39. Athawale, P., Khobragade, A.N., Tehre, V.A.: Study of Classification Techniques on Multispectral Remote Sensing Data for Agricultural Application. IOSR Journal of Electrical and Electronics Engineering (IOSR-JEEE), 76–78 e-ISSN: 2278-1676, p-ISSN: 2320-3331 B

B: Parallel References

40. Bannari, A., Morin, D.: Bonn F: A Review of vegetation indices. A Remote Sensing Review 12, 95–120 (1995)
41. Shirish Raven, A.: Geomatics to estimate crop acreage estimation to assess water cess in the irrigation command areas. A Remote Sensing Review, 60–87 (2001)
42. `http://www.isro.org/scripts/rsa_fasal.aspx`
43. Evaluation of the application potential of IRS-P6 (Recourse Sat-1), Action Plan Document, NRSA (2003)
44. Jenson, J.R.: Introductory Digital Image Processing – A Remote Sensing Perspective, 2nd edn (1996)

45. Sanjeevi, S., Barnsley, M.J.: Spectral unmixing of Compact Airborne Spectro-graphic Imager (CASI) data for quantifying sub-pixel proportions of biophysical parame-ters in a coastal dune system. Photonirvachak Journal of Indian Society of Remote Sensing 28, 187–203 (2000)
46. An Algorithm for Merging SMAP Radiometer and Radar Data for High-Resolution Soil-Moisture Retrieval IEEE Transactions on Geoscience and Remote Sensing 49(5), abstract (2011)
47. `http://www.grss-ieee.org/`
48. Fukunaga, K.: Introduction to statistical pattern recognition, 2nd edn. Academic Press Professional, Inc., San Diego (1990)
49. Kuncheva, L.I.: Combining Pattern Classifiers: Methods and Algorithms. Wiley-Interscience (2004)
50. Duin, R.P.W., Tax, D.M.J.: Statistical pattern recognition. In: Chen, C.H., Wang, P.S.P. (eds.) Handbook of Pattern Recognition and Computer Vision. World Scientific Publishing, pp. 3–24. World Scientific Publishing, Singapore
51. `http://citeseerx.ist.psu.edu/`
52. Im, J., Jensen, J.R.: Hyper spectral Remote Sensing of Vegetation, By State University of New York, College of Environmental Science and Forestry University of South Carolina (2008)
53. Yang, C.: Airborne Hyperspectral Imagery for Mapping Crop Yield Variability, USDA, Agriculture Research Service, Garza Subtropical Research Centre (2009)
54. Gürol, S.: a, *, H. Öktem a, T. Özalpa, B. Karasözen a, Statistical Learning And Optimization Methods For Improving The Efficiency. In: Landscape Image Clustering And Classification Problems, Commission VI, WG VI/4
55. Kumar, A., Ghosh, S.K., Dhadhwal, V.K.: Sub-Pixel Land Cover Mapping: SMIC System. In: ISRPS, Goa (2006)
56. Refaat, M.M., Farag, A.A.: Mean Field Theory for Density Estimation Using Support Vector Machines. Computer Vision and Image Processing Laboratory, University of Louisville, Louisville, KY, 40292 (2004)

Using the Deterministic Chaos in Variable Mode of Operation of Block Ciphers

Petr Zacek, Roman Jasek, and David Malanik

Faculty of Applied Informatics, Tomas Bata University in Zlin, Zlin, Czech Republic
{zacek,jasek,dmalanik}@fai.utb.cz

Abstract. This paper describes creating the new variable mode of operation for block cipher, and it also discusses possibilities of using the deterministic chaos in this mode of operation. It discusses the use of deterministic chaos to make the mode of operation variable and pseudo-random. Main idea of this article is to design the mode of operation dependent on all parts of encryption process as plaintext, cipher text and previous used key, and to create the function controlled by deterministic chaos, where the chaotic system has role of pseudo-random number generator, which manages the rules how these parts are used. The logistic map as the simple deterministic chaos system is used for showing the example of the application.

Keywords: Deterministic Chaos, Block Cipher, Mode of Operation, Logistic Map, Variability.

1 Introduction

Most of the block ciphers used these days are fixed to their structure, and all of them use one of the five modes of operation. All of these modes of operation are also fixed to their structure. We can learn all information about used block cipher from its crypt-analysis. It may not lead to vulnerabilities, but it would be interesting to avoid this phenomenon and make the mode of operation variable and unpredictable without knowledge of input values. There are many possible ways to make the block ciphers mode of operation variable. The deterministic chaos can be used as one of them.

The situation about using chaotic systems as pseudo-random number generators is described in material [5]. In other pages we will call it with acronym CPRNG.

All modes of operations, which are used for encryption these days, work with only one part of the element of block ciphers as the "changing value" for the next used block of data. Every mode of operation has fixed structure. For example, the CBC – Cipher Block Chaining uses the previous cipher text as the "changing value" for the next plaintext before its encryption. OFB – Output Feedback mode uses the encrypted previous key (IV – Initialization vector) as the data for encryption for the next block, and the output of encryption is exclusive-ORed with the next block of plaintext. As we can see, all of these modes use the plaintext or the cipher text or the previous used key (IV) as "changing value". Our designed variable mode uses all of these elements.

R. Silhavy et al. (eds.), *Artificial Intelligence Perspectives and Applications*,
Advances in Intelligent Systems and Computing 347, DOI: 10.1007/978-3-319-18476-0_34

[1]. The most of mode of operation developed before are mentioned at NIST pages, section Modes development [3], and as we can see, none of those modes of operation seem to be variable or be driven by chaotic system, there.

2 Definition of the Variable Block Ciphers Mode of Operation Using the Deterministic Chaos

As it was mentioned, new variable block ciphers mode of operation uses all of the three main elements as a "changing value" for the encryption of the next block of plaintext. The mode of operation variably uses previous plaintext, the last cipher text and previous key.

In the beginning, the IV is used for the encryption of the first block. All the next blocks of plaintext are encrypted with the modified previous key, which is made of the previous plaintext, the last cipher text and the previous key. These elements are variable chosen by generated value from CPRNG.

Important part of definition is to define the CPRNG. Thus the CPRNG is used for generating pseudo-random numbers that are later used in the function for determining the next key for encryption of the next block of plaintext. The whole process is shown on the following diagrams. The first diagram on figure 1 presents the mode of operation for encryption; figure 2 shows the mode of operation for decryption.

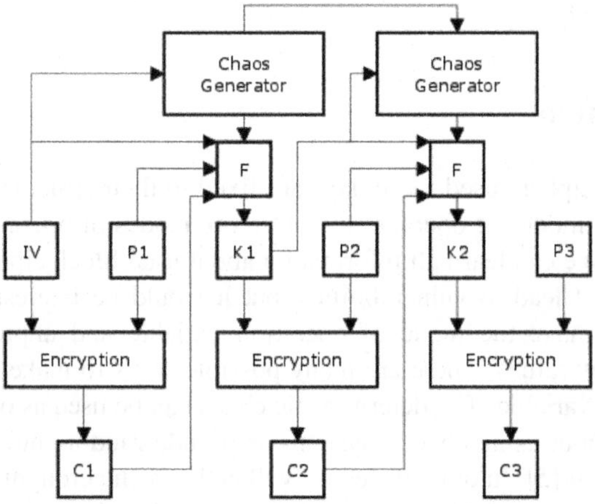

Fig. 1. Mode of operation for encryption (F - Function for generating keys; IV - Initialization vector; P - Plaintext; C - Cipher text, K – Previous key)

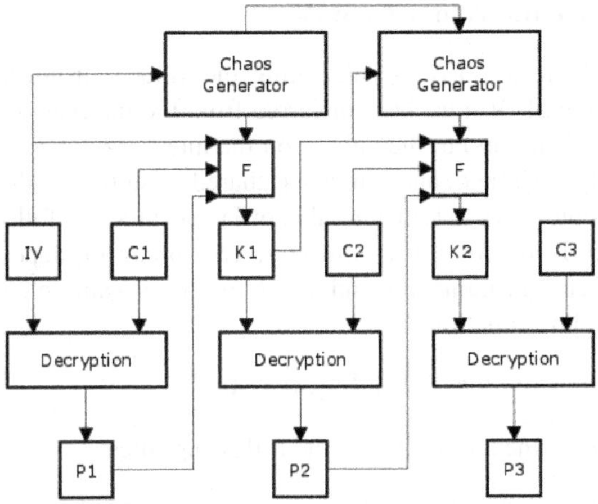

Fig. 2. Mode of operation for decryption (F - Function for generating keys; IV - Initialization vector; P - Plaintext; C - Cipher text, K – Previous key)

3 Designing of the CPRNG

The main premise is that the CPRNG should be designed to generate values for pseudo-random variable behavior of the mode of operation. The deterministic chaos behaves deterministic and strongly depends on the initial value. When using the deterministic chaos, the changes in variability of mode of operation are greater even with only small changes in IV. For our purpose, the deterministic chaos is utilized as pseudo-random number generator of values.

One of the types of deterministic chaos is called logistic map and defined by following equation (1).

$$x_n = r \cdot x_{n-1} \cdot (1 - x_{n-1}) \tag{1}$$

Where r is control parameter, x_n is actual value, and x_{n-1} is previous generated value. If the r is above 3.57, the system behaves chaotically for almost values[1]. We choose values on the interval <3.9, 4.0) to avoid values near the 3.82842712... In this CPRNG, the r value is generated from IV. Optimal interval of the values x should be (0, 1). Thus, the first value $x \rightarrow x_0$ is calculated by following equation. [2]

$$x_0 = \begin{cases} (r - 3.9) \cdot 10, & \text{for } r > 3.9 \\ 10^{-15}, & \text{for } r = 3.9 \end{cases} \tag{2}$$

The value x_n is used as a parameter in determining function.

[1] For the r = 3.82842712... the system oscillates among three values [2].

3.1 Calculation of the Value *r* from IV

Another main problem to solve is calculating the value *r* from value of the IV. It means, that the value of IV must be transferred from the interval <0, 2^*b* – 1> (where the *b* is size of the IV in bits) to the value r on the interval <3.9, 4.0). The value of IV should be randomly generated, and it must be shared or stored for decryption.

One of the possible ways to transfer value of IV to value *r* is following. Let us consider the precision of the control parameter *r*. This precision depends on number of decimal places after 9 in number *r* and we define it as value *p*[2]. On this basis, the value *r* is calculated by equation 3.

$$r = 3.9 + \sum_{i=1}^{p} v_i \cdot 10^{-(i+1)} \tag{3}$$

The values *v* are defined according to the following rules:

1. The values *v* are set of numbers $<v_1, v_2, \dots, v_{n-1}, v_n>$ where *n* is on the interval <1, *p*>.
2. v_i is defined as natural number on the interval <0, 9>.
3. The values of *v* are stored as digits of IV in binary. If the length of IV in binary exceeded the number of precision, the IV must be split into blocks and the blocks have length same as precision *p*. Then the values v_i are calculated as follows.

   (a) We split value of IV in binary to l_{IV}/p blocks, where the l_{IV} is length of IV in binary and the *p* is precision. The remainder we define as *d*. We get set of blocks represented as binary numbers of length *p*. Every block is the set of the binary digits and we define it as values $b_j = <b_{j,1}, b_{j,2}, \dots, b_{j,n-1}, b_{j,n}>$ where b_j is natural number on the interval <0, 9> and *j* is natural number on the interval <1, l_{IV}/p>. The values $v = <v_1, v_2, \dots, v_{n-1}, v_n>$ are calculated by following equation (4).

$$v_i = (v_i + \sum_{j=1}^{l_{IV}/p} b_{j,i}) \ mod\ 10 \tag{4}$$

   (b) For the remainder it is similar, except that the first $p - l_r$ numbers are zeros, where l_r is length of the remainder. The remainder is also represented as the set of the binary digits and we define it as $rem = <rem_1, rem_2, \dots, rem_{n-1}, rem_n>$ where *n* is natural number on the interval <1, *p*>. Then we modify the values *v* as follows.

$$v_i = (v_i + rem_i) \ mod\ 10 \tag{5}$$

Example of Calculating the Value *r*

We define the precision *p* = 10 and the IV as the binary number of the length 21. The IV is chosen as the value on the interval <0, 2^21 – 1> and it is written as binary number. Now, the number of IV is randomly chosen – 101100101111110010001. Generated IV has length of 21; thus we can split the IV into two blocks b_1, b_2 of length 10, and into remaining block of length 1 as the remainder *rem*. Since the length

[2] The *r* = 3.989891 has precision *p* = 5.

of IV is not dividable by 10, we must add 9 zeros before the remainder. Result of splitting IV into these blocks and remainder is shown below.

$$b = [b_1 = 1011001011, b_2 = 1111001000]$$

$$b_1 = 1011001011$$
$$b_1 = [b_{1,1} = 1;\ b_{1,2} = 0;\ b_{1,3} = 1;\ \ldots;\ b_{1,9} = 1;\ b_{1,10} = 1]$$

$$b_2 = 1111001000$$
$$b_2 = [b_{2,1} = 1;\ b_{2,2} = 1;\ b_{2,3} = 1;\ \ldots;\ b_{2,9} = 0;\ b_{2,10} = 0]$$

$$rem = 0000000001$$
$$rem = [rem_1 = 0;\ rem_2 = 1;\ rem_3 = 1;\ \ldots;\ rem_9 = 0;\ rem_{10} = 0]$$

Now we can calculate values v from *b1*, *b2* and *rem,* and the result from equations (4) and (5) is shown below.

$$v = 2122002012$$
$$v = [v_1 = 2;\ v_2 = 1;\ v_3 = 2;\ \ldots;\ v_9 = 1;\ v_{10} = 2]$$

The value r is calculated from v as result from equation (3) → $r = 3.92122002012$. There is not good ratio between length of IV and precision p in this example.

3.2 Features of the CPRNG

Ways to change and adjust the CPRNG are plenty. CPRNG has a role of pseudo-random behavior how the function generates the next keys. On this value from CPRNG depends on ways how the previous plaintext, the last cipher text and the previous key are used to calculate the next key for encryption. We have these options:

- We can change the type of deterministic chaos used in the CPRNG.
- With changing the value p (precision), we can change number of possible CPRNGs, which we can create from IV. The bigger precision means more ways how to start generating values of CPRNG.
- We can change the way of the derivation values r and x from IV.

Precision *p* of Value *r*

Ideally, the precision should be ten times smaller than the length of IV; hence we can get all variations of value r. Based on this, the optimal precision p should be around 10-15 because we usually use blocks and keys (IV) with the length at least 128 bits. As can be seen, the precision p may also be variable. Another way is using the length of IV as we want. For example, we can simply use IV with unlimited length (e.g. external file as IV), the first encryption will use part of IV with the same length as the blocks used for encryption algorithm. If we have precision p at least ten times smaller, we can get this number of values r (number of different CPRNGs). Then the number of possibly made CPRNGs is calculated from following equation.

$$Number\ of\ generators = 10^p - 1 \tag{6}$$

4 Function for Generating Next Keys

As it was mentioned, function for generating next keys depends on value x_n from the CPRNG, and x_n presents the main parameter. The function also uses the previous key, previous plaintext, and the last cipher text and the way in which they will be used depends on this value x_n. The ways, how we can use those three elements, are plenty. We can change it as is required.

For our example, we use the last digit of value x_n from the CPRNG and we will use one equation of function based on the last digit defined as d, which will work with three mentioned elements. These features make mode of operation variable. The result of function is key for encryption (decryption) n - 1 block of plaintext according to value x_n.

4.1 Example of Function for Generating Next Keys

As it was mentioned, we use value x_n from the CPRNG as a main parameter. We use the last digit of it as value d. For our example, we define equations according to this value d. All equations will operate with previous plaintext, the last cipher text and the previous key as a field of byte values. For all bytes of previous plaintext, the last cipher text, and previous key we define this list of equation, which will be used depending on the value d. We define value o as byte of previous plaintext, value c as byte of the last cipher text, value k_p as byte of previous key and value k_n as byte of next key. The values o, c, k_p, and k_n are natural numbers on the interval <0, 255> (type byte). The values k_n are calculated from these following equations depending on value d.

$$k_n = \begin{cases} (o + c + k_p) \bmod 256, & \text{for } d = 1 \\ (o + c - k_p) \bmod 256, & \text{for } d = 2 \\ (o - c - k_p) \bmod 256, & \text{for } d = 3 \\ (o + (c+1) \cdot (k_p + 1)) \bmod 256, & \text{for } d = 4 \\ ((o+1) \cdot (c+1) + k_p) \bmod 256, & \text{for } d = 5 \\ ((o+1) \cdot (c+1) \cdot (k_p + 1)) \bmod 256, & \text{for } d = 6 \\ (o - (c+1) \cdot (k_p + 1)) \bmod 256, & \text{for } d = 7 \\ ((o+1) \cdot (c+1) - k_p) \bmod 256, & \text{for } d = 8 \\ (o + c + 2 \cdot (k_p + 1)) \bmod 256, & \text{for } d = 9 \end{cases} \tag{7}$$ [3]

[3] d can not be zero, because the last digit of x_n can not be zero, too.

Equations of the function during the decryption are the same because we use the same keys for encryption as keys for decryption. There is the addition a one before multiplication because the values could be zero and without this addition there is bigger probability to obtain zero from equation and to obtain zero in the next key.

4.2 Features of the Function

Ways, how to make this function based on the value of CPRNG, are plenty. We can change form of equations, or their number, or way of using the value x from CPRNG. For example, we can use last two digits from value x.

This function and its equations make the mode of operation variable. Function changes the next keys also. We want to change the next keys as much as possible. Ideally, we want to change about half of bits of the next key against the last used key. Because the function uses the last cipher text, the important role is in encryption and changes from plaintext to cipher text. Form of function and its equations are important to avoid from generating weak keys. For example, if we use XOR function, we will not want to have key with all zeros.

5 Conclusion

It this article was described the designing the variable mode of operation of block ciphers controlled by using the deterministic chaos. It was also showed how to use the deterministic chaos to control variability in mode of operation of block ciphers. As a type of deterministic chaos, there was used logistic map.

The variable mode of operation was designed to use all elements of block ciphers – plaintext, cipher text and key for calculating the next keys. Two main parts were designed – CPRNG and the function for calculation of the next keys. The function makes the mode of operation variable and the CPRNG makes the behavior of the function dependent on key and unpredictable without knowledge of key. We can say that the CPRNG has role of pseudo-random number generator, and function has the role of variability.

There are many ways to adjust the CPRNG and the function for generating the next keys. There was also shown an example how to get values which are required for starting the CPRNG from IV. Principle of using values from CPRNG and the way how to the make function were also presented to demonstrate that the use of deterministic chaos in variable mode of operation is possible. The exact way, how to ideally construct function, could be part of future research. Also the ways of using another deterministic chaos types could be part of future research.

Number of ways how to modify designed variable mode of operations above is unlimited, but not all of them are right. We must consider to avoid making weak keys and the other factors.

If the mode of operation is variable and strongly dependable on small changes of input setting of CPRNG, the attacker would have no knowledge about the structure without knowing the keys (IV). This is the main advantage why to make this kind of mode of operation.

Acknowledgement. This work was supported by Internal Grant Agency of Tomas Bata University under the project No. IGA/FAI/2015/47; further it was supported by financial support of research project NPU I No. MSMT-7778/2014 by the Ministry of Education of the Czech Republic; also by the European Regional Development Fund under the Project CEBIA-Tech No. CZ.1.05/2.1.00/03.0089.

References

1. Bellovin, S.M.: Modes of Operation. Columbia University, Columbia, USA (2009), `https://www.cs.columbia.edu/~smb/classes/s09/105.pdf` (cit. January 30, 2015)
2. Weisstein, E.W.: Logistic Map. From MathWorld–A Wolfram Web Resource, `http://mathworld.wolfram.com/LogisticMap.html`
3. Modes Development. In: National Institute of Standards and Technology: Computer Security Resource Center (2001) (October 16, 2014), `http://csrc.nist.gov/groups/ST/toolkit/BCM/modes_development.html` (cit. February 03, 2015)
4. Sprott, J.C.: Chaos and Time-Series Analysis. Oxford University Press (2003)
5. Senkerik, R., Pluhacek, M., Zelinka, I., Davendra, D., Oplatkova, Z.: A Brief Survey on the Chaotic Systems as the Pseudo Random Number Generators. In: Sanayei, A., Rössler, O.E., Zelinka, I. (eds.) ISCS 2014: Interdisciplinary Symposium on Complex Systems. Emergence, Complexity and Computation, vol. 14, pp. 205–214. Springer International Publishing (2015), doi:10.1007/978-3-319-10759-2_22

An Initial Study on the New Adaptive Approach for Multi-chaotic Differential Evolution

Roman Senkerik, Michal Pluhacek, and Zuzana Kominkova Oplatkova

Tomas Bata University in Zlin, Faculty of Applied Informatics,
Nam T.G. Masaryka 5555, 760 01 Zlin, Czech Republic
{senkerik,pluhacek,oplatkova}@fai.utb.cz

Abstract. This paper aims on the initial investigations on the novel adaptive multi-chaos-driven evolutionary algorithm Differential Evolution (DE). This paper is focused on the embedding and adaptive alternating of set of two discrete dissipative chaotic systems in the form of chaotic pseudo random number generators for the DE. In this paper the novel adaptive concept of DE/rand/1/bin strategy driven alternately by two chaotic maps (systems) is introduced. Repeated simulations were performed and analyzed on the well known test function in higher dimension setting.

Keywords: Differential Evolution, Deterministic chaos.

1 Introduction

This research is focused on the interconnection of the two different softcomputing fields, which are theory of chaos and evolutionary computation. This paper is aimed at initial investigating of the novel adaptive control scheme for multi-chaos driven Differential Evolution algorithm (DE) [1]. Although a number of DE variants have been recently developed, the focus of this paper is the further development of ChaosDE concept, which is based on the embedding of chaotic systems in the form of Chaos Pseudo Random Number Generators (CPRNG) into the DE.

A chaotic approach generally uses the discrete chaotic map as a pseudo random number generator [2]. This causes the meta-heuristic to map unique regions, since the discrete chaotic map iterates to new regions. The difficult task is then to select a very good chaotic map at the place of the pseudo random number generator.

The focus of our research is the embedding of direct chaotic dynamics in the form of CPRNG for DE. The initial concept of embedding chaotic dynamics into the evolutionary algorithms is given in [3]. Later, the initial study [4] was focused on the simple embedding of chaotic systems in the form of chaos pseudo random number generator (CPRNG) for DE and Self Organizing Migration Algorithm (SOMA) [5] in the task of optimal PID tuning. Also the PSO (Particle Swarm Optimization) algorithm with elements of chaos was introduced as CPSO [6]. The concept of ChaosDE proved itself to be a powerful heuristic also in combinatorial problems domain [7].

R. Silhavy et al. (eds.), *Artificial Intelligence Perspectives and Applications*,
Advances in Intelligent Systems and Computing 347, DOI: 10.1007/978-3-319-18476-0_35

At the same time the chaos embedded PSO with inertia weigh strategy was closely investigated [8], followed by the introduction of a PSO strategy driven alternately by two chaotic systems [9].

Based on the promising experimental results with PSO algorithm driven adaptively by two different chaotic systems, the idea was to extend this adaptive alternating approach also on DE the algorithm.

Firstly, the motivation for this research is proposed. The next sections are focused on the description of evolutionary algorithm DE, the concept of multi-chaos driven DE and the experiment description. Results and conclusion follow afterwards.

2 Related Work and Motivation

This research is an extension and continuation of the previous successful initial experiments with both multi-chaos driven PSO and DE algorithms [10], [11].

In this paper the novel adaptive control concept for DE/rand/1/bin strategy driven alternately by two chaotic maps (systems) is experimentally investigated. From the aforementioned previous research it follows, that very promising experimental results were obtained through the usage of different chaotic dynamics as the CPRNGs. And at the same time it was obvious that different chaotic systems have different effects on the performance of the algorithm. The idea was then to connect these two different influences to the performance of DE into the one multi-chaotic concept. The previous research was aimed at the determining of the switching time (certain number of generations/iterations) between different chaotic systems. Such a "manual" approach proved to be successful, but also many open issues have arisen (i.e. when to start the switching, how many times, etc.). The novelty of the proposed adaptive approach is that, the chaotic pseudorandom number generators are switched over automatically without prior knowledge of the optimization problem and without any manual setting of the "switching point".

3 Differential Evolution

DE is a population-based optimization method that works on real-number-coded individuals [1]. DE is quite robust, fast, and effective, with global optimization ability. It does not require the objective function to be differentiable, and it works well even with noisy and time-dependent objective functions. Due to a limited space and the aims of this paper, the detailed description of well known differential evolution algorithm basic principles is insignificant and hence omitted. Please refer to [1], [12] for the detailed description of the used DERand1Bin strategy (both for ChaosDE and Canonical DE) as well as for the complete description of all other strategies.

4 The Concept of Adaptive Multi-chaotic DE

The general idea of ChaosDE and CPRNG is to replace the default pseudorandom number generator (PRNG) with the discrete chaotic map. As the discrete chaotic map is a set of equations with a static start position, we created a random start position of the map, in order to have different start position for different experiments (runs of EA's). This random position is initialized with the default PRNG, as a one-off randomizer. Once the start position of the chaotic map has been obtained, the map generates the next sequence using its current position.

In this research, direct output iterations of the chaotic maps were used for the generation of real numbers in the process of crossover based on the user defined CR value and for the generation of the integer values used for selection of individuals.

Previous successful initial experiments with multi-chaos driven PSO and DE algorithms [10], [11] have manifested that very promising experimental results were obtained through the utilization of Delayed Logistic, Lozi, Burgers and Tinkerbelt maps. The last two mentioned chaotic maps have unique properties with connection to DE: strong progress towards global extreme, but weak overall statistical results, like average cost function (CF) value and std. dev., and tendency to premature stagnation. While through the utilization of the Lozi and Delayed Logistic map the continuously stable and very satisfactory performance of ChaosDE was achieved. The above described influences around switching point (500 or 1500 generations) are visible from the Fig.1, which depicts the illustrative example of time evolution of average CF values for all 50 runs of simple single ChaosDE (no alternation) and two combinations of Multi-Chaotic DE and canonical DE/rand/1/bin strategies.

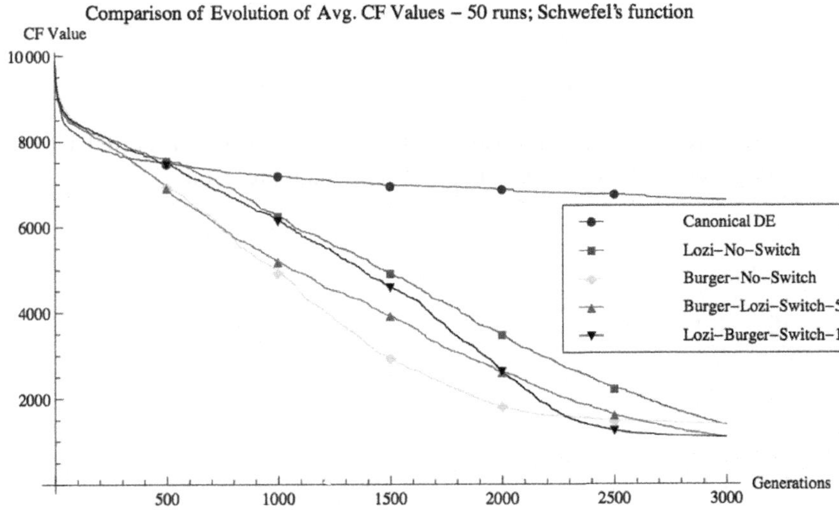

Fig. 1. Comparison of the time evolution of avg. CF values for the all 50 runs of Canonical DE, Simple ChaosDE and two versions of Multi-ChaosDE; Schwefel's function, $D = 30$

To maximize the benefit from the influences of different chaotic dynamics a new adaptive control approach was developed. It does not require any prior knowledge of the optimization problem and any manual setting of the one ore more "switching points". The exact transition point is determined by following simple rule: If the change of global best value between two subsequent generations is less than 0.001 over more than 1% of total number of generations, the chaotic systems used as the CPRNGs are alternated.

5 Chaotic Maps

This section contains the description of discrete dissipative chaotic maps used as the chaotic pseudo random generators for DE. Following chaotic maps were used: Burgers (1), and Lozi map (2).

The Burgers mapping is a discretization of a pair of coupled differential equations which were used by Burgers [13] to illustrate the relevance of the concept of bifurcation to the study of hydrodynamics flows. The map equations are given in (4) with control parameters $a = 0.75$ and $b = 1.75$ as suggested in [14].

$$\begin{aligned} X_{n+1} &= aX_n - Y_n^2 \\ Y_{n+1} &= bY_n + X_nY_n \end{aligned} \tag{1}$$

The Lozi map is a discrete two-dimensional chaotic map. The map equations are given in (2). The parameters used in this work are: $a = 1.7$ and $b = 0.5$ as suggested in [14]. For these values, the system exhibits typical chaotic behavior and with this parameter setting it is used in the most research papers and other literature sources.

$$\begin{aligned} X_{n+1} &= 1 - a|X_n| + bY_n \\ Y_{n+1} &= X_n \end{aligned} \tag{2}$$

The illustrative histograms of the distribution of real numbers transferred into the range <0 - 1> generated by means of studied chaotic maps are in Fig. 3.

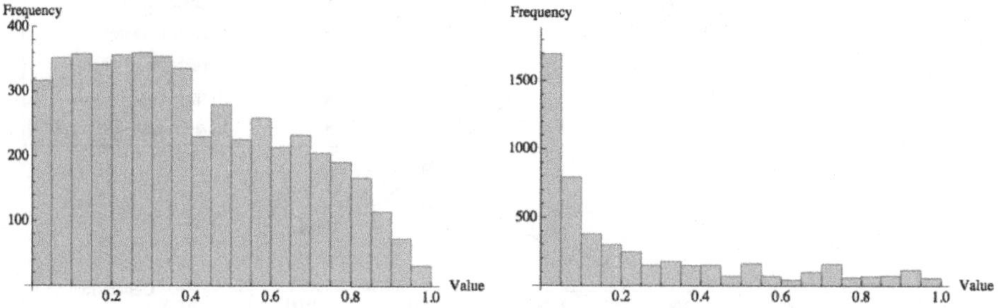

Fig. 2. Histogram of the distribution of real numbers transferred into the range <0 - 1> generated by means of the chaotic Lozi map (left) and Burgers map (right) – 5000 samples

6 Results

Simple Schwefel's test function (3) was utilized within this initial experimental research for the purpose of performance investigation of Multi-Chaotic DE.

$$f(x)=\sum_{i=1}^{Dim}-x_i \sin\left(\sqrt{|x_i|}\right) \tag{3}$$

Function minimum:
Position for E_n: $(x_1, x_2 \ldots x_n) = (420.969, 420.969, \ldots, 420.969)$
Value for E_n: $y = -418.983 \cdot Dim$
Function interval: <-512, 512>.

Experiments were performed in the combined environments of Wolfram Mathematica and C language; Canonical DE in comparisons therefore used the built-in C language pseudo random number generator Mersenne Twister C representing traditional pseudorandom number generators in comparisons. All experiments used different initialization, i.e. different initial population was generated in each run.

Within this research, one type of experiment was performed. It utilizes the maximum number of generations fixed at 3000 generations, Population size of 75 and dimension *dim* = 30. This allowed the possibility to analyze the progress of all studied DE variants within a limited number of generations and cost function evaluations.

The parameter settings (see Table 1) for Multi-ChaosDE was obtained based on numerous experiments and simulations. In general, ChaosDE requires lower values of *CR* and *F* for almost any CPRNG and benchmark test function (See Fig 3). The similar settings (except different values of *CR* and *F*) was used also for the canonical DE.

Table 1. Parameter set up for Multi-Chaos DE and Canonical DE

DE Parameter	Value
Popsize	75
F (for ChaosDE)	0.4
CR (for ChaosDE)	0.4
F (for Canonical DE)	0.5
CR (for Canonical DE)	0.9
Dim	30
Max. Generations	3000
Max Cost Function Evaluations (CFE)	225000

Statistical results of the selected experiments are shown in comprehensive Table 2 which represents the simple statistics for CF values, e.g. average, median, maximum values, standard deviations and minimum values representing the best individual solution for all 50 repeated runs of DE.

Table 3 compares the progress of two versions of adaptive Multi-ChaosDE and Canonical DE. This table contains the average CF values for the generation No. 750, 1500, 2250 and 3000 from all 50 runs. The bold values within the both Tables 2 and 3 depict the best obtained results. Following versions of adaptive Multi-ChaosDE were studied:

- *Burgers-Lozi-Switch-Adaptive*: Start with Burgers map CPRNG, and then switch to the Lozi map CPRNG when the adaptive rule will be fulfilled.
- *Lozi-Burgers-Switch-Adaptive*: Start with Lozi map CPRNG, then switch to the Burgers map CPRNG when the adaptive rule will be fulfilled.

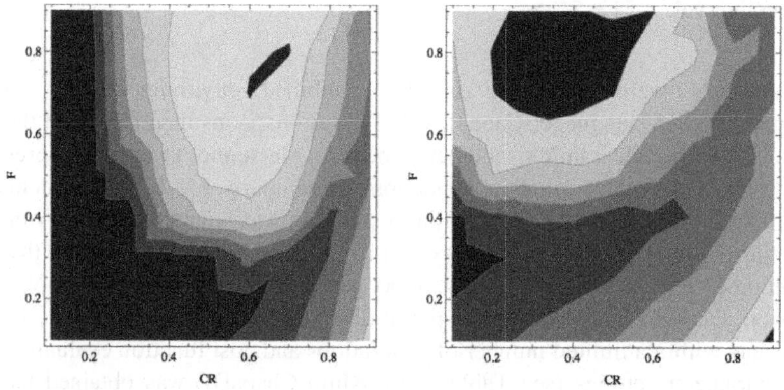

Fig. 3. Contours plots for *F* and *CR* combinations of average results for 50 repeated runs of ChaosDE, from left to right: Lozi map CPRNG, Burgers map CPRNG, Schwefel's test function, dim = 30

Table 2. Simple results statistics for the Canonical DE and adaptive Multi-ChaosDE Schwefel's function, *dim* = 30, max. Generations = 1500

DE Version	Avg CF	Median CF	Max CF	Min CF	StdDev
Canonical DE	-6844.78	-6986.41	-5189.84	-9895.53	1174.671
Burger-Lozi-Switch-Adaptive	-12067.9	-12120	-11344.9	**-12569.5**	306.334
Lozi-Burger-Switch-Adaptive	**-12566.7**	**-12569.5**	**-12445.9**	**-12569.5**	**17.4708**

Table 3. Comparison of progress towards the minimum for the Schwefel's function, *dim* = 30, max. Generations = 1500

DE Version	Generation No. 750	Generation No. 1500	Generation No. 2250	Generation No. 3000
Canonical DE	-4909.08	-5332.98	-5947.35	-6844.78
Burger-Lozi-Switch-Adaptive	**-7582.95**	**-10698.2**	**-11918.4**	-12067.9
Lozi-Burger-Switch-Adaptive	-7115.44	-8619.48	-11327.3	**-12566.7**

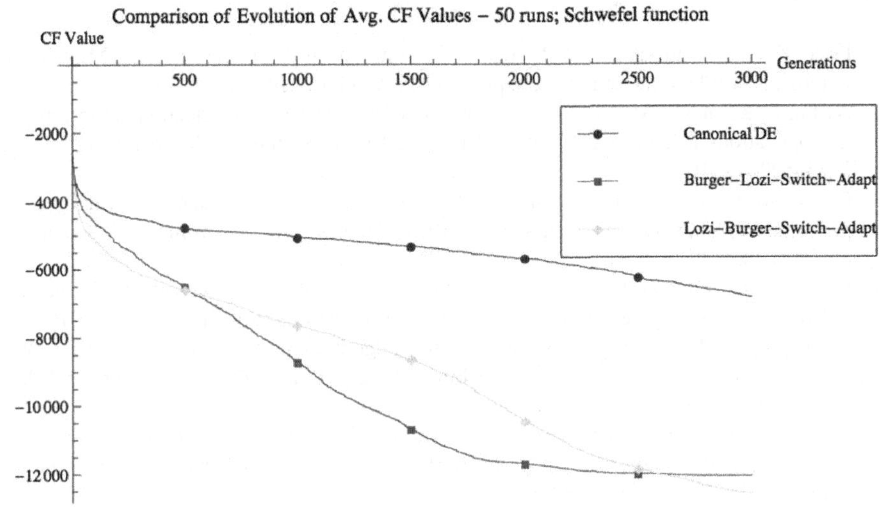

Fig. 4. Comparison of the time evolution of avg. CF values for the all 50 runs of Canonical DE, and two versions of adaptive Multi-ChaosDE. Schwefel's function, *Dim* = 30.

7 Conclusion

The primary focus of this research is to investigate the unconventional hybridization of natural chaotic dynamics with evolutionary algorithm as the multi-chaotic pseudo random number generator. In this paper the novel adaptive concept for DE/rand/1/bin strategy driven alternately by two discrete chaotic maps (systems) is introduced. These two different influences to the performance of DE were connected here into the one adaptive multi-chaotic concept with automatic switching without prior knowledge of the optimization problem and without any manual setting of the "switching point". The obtained results were compared with the original canonical DE. The high sensitivity of the DE to the internal dynamics of the chaotic PRNG is fully manifested here.

Acknowledgements. This work was supported by Grant Agency of the Czech Republic - GACR P103/15/06700S, further by financial support of research project NPU I No. MSMT-7778/2014 by the Ministry of Education of the Czech Republic; also by the European Regional Development Fund under the Project CEBIA-Tech No. CZ.1.05/2.1.00/03.0089; and by Internal Grant Agency of Tomas Bata University under the project No. IGA/FAI/2015/057.

References

1. Price, K.V.: An Introduction to Differential Evolution. In: Corne, D., Dorigo, M., Glover, F. (eds.) New Ideas in Optimization, pp. 79–108. McGraw-Hill Ltd. (1999)
2. Aydin, I., Karakose, M., Akin, E.: Chaotic-based hybrid negative selection algorithm and its applications in fault and anomaly detection. Expert Systems with Applications 37(7), 5285–5294 (2010)
3. Caponetto, R., Fortuna, L., Fazzino, S., Xibilia, M.G.: Chaotic sequences to improve the performance of evolutionary algorithms. IEEE Transactions on Evolutionary Computation 7(3), 289–304 (2003)
4. Davendra, D., Zelinka, I., Senkerik, R.: Chaos driven evolutionary algorithms for the task of PID control. Computers & Mathematics with Applications 60(4), 1088–1104 (2010)
5. Zelinka, I.: SOMA — Self-Organizing Migrating Algorithm. In: Onwubolu, G.C., Babu, B.V. (eds.) New Optimization Techniques in Engineering. STUDFUZZ, vol. 141, pp. 167–217. Springer, Heidelberg (2004)
6. Coelho, L.d.S., Mariani, V.C.: A novel chaotic particle swarm optimization approach using Hénon map and implicit filtering local search for economic load dispatch. Chaos, Solitons & Fractals 39(2), 510–518 (2009)
7. Davendra, D., Bialic-Davendra, M., Senkerik, R.: Scheduling the Lot-Streaming Flowshop scheduling problem with setup time with the chaos-induced Enhanced Differential Evolution. In: 2013 IEEE Symposium on Differential Evolution (SDE), April 16-19, pp. 119–126 (2013)
8. Pluhacek, M., Senkerik, R., Davendra, D., Kominkova Oplatkova, Z., Zelinka, I.: On the behavior and performance of chaos driven PSO algorithm with inertia weight. Computers & Mathematics with Applications 66(2), 122–134 (2013)
9. Pluhacek, M., Senkerik, R., Zelinka, I., Davendra, D.: Chaos PSO algorithm driven alternately by two different chaotic maps - An initial study. In: 2013 IEEE Congress on Evolutionary Computation (CEC), June 20-23, pp. 2444–2449 (2013)
10. Senkerik, R., Pluhacek, M., Davendra, D., Zelinka, I., Oplatkova, Z.K.: Performance Testing of Multi-Chaotic Differential Evolution Concept on Shifted Benchmark Functions. In: Polycarpou, M., de Carvalho, A.C.P.L.F., Pan, J.-S., Woźniak, M., Quintian, H., Corchado, E. (eds.) HAIS 2014. LNCS, vol. 8480, pp. 306–317. Springer, Heidelberg (2014)
11. Pluhacek, M., Senkerik, R., Zelinka, I., Davendra, D.: New Adaptive Approach for Chaos PSO Algorithm Driven Alternately by Two Different Chaotic Maps – An Initial Study. In: Zelinka, I., Chen, G., Rössler, O.E., Snasel, V., Abraham, A. (eds.) Nostradamus 2013: Prediction, Model. & Analysis. AISC, vol. 210, pp. 77–87. Springer, Heidelberg (2013)
12. Price, K.V., Storn, R.M., Lampinen, J.A.: Differential Evolution - A Practical Approach to Global Optimization. Natural Computing Series. Springer, Heidelberg (2005)
13. ELabbasy, E., Agiza, H., EL-Metwally, H., Elsadany, A.: Bifurcation Analysis, Chaos and Control in the Burgers Mapping. International Journal of Nonlinear Science 4(3), 171–185 (2007)
14. Sprott, J.C.: Chaos and Time-Series Analysis. Oxford University Press (2003)

Author Index

Printed in the United States
By Bookmasters